Gas Tungsten Arc Welding Handbook

Sixth Edition

GTAW

William H. Minnick
Mark A. Prosser

Welding/Fabrication Instructor
Blackhawk Technical College
Janesville, Wisconsin

Publisher
The Goodheart-Willcox Company, Inc.
Tinley Park, IL
www.g-w.com

Copyright © 2013
by
The Goodheart-Willcox Company, Inc.

Previous editions copyright 2006, 2000, 1996, 1992, 1985

Manufactured in the United States of America.

Library of Congress Catalog Card Number 2011043757

ISBN 978-1-60525-793-8

1 2 3 4 5 6 7 8 9 – 13 – 17 16 15 14 13 12

The Goodheart-Willcox Company, Inc. Brand Disclaimer: Brand names, company names, and illustrations for products and services included in this text are provided for educational purposes only and do not represent or imply endorsement or recommendation by the author or the publisher.

The Goodheart-Willcox Company, Inc. Safety Notice: The reader is expressly advised to carefully read, understand, and apply all safety precautions and warnings described in this book or that might also be indicated in undertaking the activities and exercises described herein to minimize risk of personal injury or injury to others. Common sense and good judgment should also be exercised and applied to help avoid all potential hazards. The reader should always refer to the appropriate manufacturer's technical information, directions, and recommendations; then proceed with care to follow specific equipment operating instructions. The reader should understand these notices and cautions are not exhaustive.

The publisher makes no warranty or representation whatsoever, either expressed or implied, including but not limited to equipment, procedures, and applications described or referred to herein, their quality, performance, merchantability, or fitness for a particular purpose. The publisher assumes no responsibility for any changes, errors, or omissions in this book. The publisher specifically disclaims any liability whatsoever, including any direct, indirect, incidental, consequential, special, or exemplary damages resulting, in whole or in part, from the reader's use or reliance upon the information, instructions, procedures, warnings, cautions, applications, or other matter contained in this book. The publisher assumes no responsibility for the activities of the reader.

The Goodheart-Willcox Company, Inc. Internet Disclaimer: The Internet resources and listings in this Goodheart-Willcox Publisher product are provided solely as a convenience to you. These resources and listings were reviewed at the time of publication to provide you with accurate, safe, and appropriate information. Goodheart-Willcox Publisher has no control over the referenced websites and, due to the dynamic nature of the Internet, is not responsible or liable for the content, products, or performance of links to other websites or resources. Goodheart-Willcox Publisher makes no representation, either expressed or implied, regarding the content of these websites, and such references do not constitute an endorsement or recommendation of the information or content presented. It is your responsibility to take all protective measures to guard against inappropriate content, viruses, or other destructive elements.

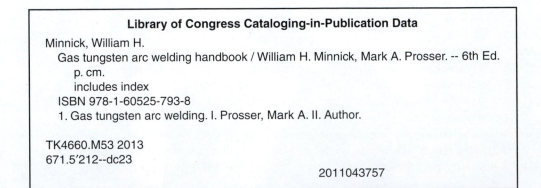

Library of Congress Cataloging-in-Publication Data

Minnick, William H.
Gas tungsten arc welding handbook / William H. Minnick, Mark A. Prosser. -- 6th Ed.
p. cm.
includes index
ISBN 978-1-60525-793-8
1. Gas tungsten arc welding. I. Prosser, Mark A. II. Author.

TK4660.M53 2013
671.5′212--dc23

2011043757

Introduction

The **Gas Tungsten Arc Welding Handbook** is designed to impart the knowledge and skills required to begin a career in welding. The text includes information about safety, equipment, processes, techniques, materials, and quality control in order to teach you to make successful welds.

This text presents basic skills and proper procedures in easy-to-understand language. Welding terms used in the text comply with the AWS *Standard Welding Terms and Definitions* publication. Tables and charts within chapters offer useful data, including dimensions, information about materials, and problems and corrective measures. Hundreds of illustrations help guide your learning, including many images of welds in the process of being made.

The chapters lead you through GTAW principles and practices in a logical sequence. The following areas are included in the text:

- welding hazards and safety procedures
- power sources and auxiliary equipment
- weld types and joint design
- shielding gases and filler metals
- tooling
- setup and maintenance of equipment
- techniques and procedures for welding various types of metals
- pipe welding
- semiautomatic and automatic welding systems
- inspection, testing, and repair of welds
- qualification and certification
- employment
- estimating cost

A Reference Section, located at the back of the text, contains many useful tables and charts. You will also find an extensive Glossary of Welding Terms, which contains the definitions of the key terms presented in each chapter.

Every weld you will make represents your skill as a welder. Take pride in your work and make each weld as if your career depends on it. The **Gas Tungsten Arc Welding Handbook** will provide you with the information you need to make that weld.

About the Author

Mark Prosser grew up in Michigan and learned how to weld in his father's collision/restoration business. As a teenager, Mark attended a local vocational school and competed in SkillsUSA welding competitions, becoming one of the six finalists for the world competition in the early 1990s.

Mark earned his associate's degree in welding technology from Ferris State University. He then transferred to Central Michigan University, where he obtained his bachelor's degree in Industrial Education. Mark has also earned a Master's degree in Adult Education and is a CWI/CWE through the American Welding Society.

Mark has taught professionally for nine years in high schools, private colleges, and public technical colleges. He has worked in the automotive industry, done governmental contract work, and welded structural steel and high-pressure piping for the chemical industry. Mark is also the author of *Full-Bore Sheet Metal*, a sheet metal fabrication book for custom car and bike builders.

Mark is currently a welding/fabrication instructor at Blackhawk Technical College in Janesville, Wisconsin. In addition to writing and teaching, he runs a small custom paint business and welds every day. Mark continually strives to inspire young people to become excited about the welding trade.

Acknowledgments

The authors gratefully acknowledge the assistance of the following companies who contributed suggestions, ideas, photographs, and information to this textbook.

Aquasol Corporation
Aronson Machine Co.
Ceramic Nozzles, Inc.
CFM/VR-TESCO LLC
CK Worldwide, Inc.
Dow Chemical Co.
International Magnesium Assoc.
Jetline Engineering
Junction Tool Supplies Pty. Ltd.
Lincoln Electric Company
Magnaflux Corp.
Miller Electric Mfg. Co.
Qualitest International LC
Techalloy Maryland, Inc.
The Harris Products Group
Tinius Olsen
VJ Technologies
Walhonde Tools, Inc.
Weldcraft, Inc.
Welding Material Sales, Inc.

Brief Contents

Table of Contents

Chapter

Gas Tungsten Arc Welding History and Safety

Objectives

After completing this chapter, you will be able to:

- ☐ Discuss the history of GTAW welding.
- ☐ Recall the different types of light produced by welding and the dangers associated with each type.
- ☐ Identify the different types of burns.
- ☐ Identify electrical hazards in the weld shop.
- ☐ Identify the respiratory hazards associated with welding.
- ☐ Recall the proper work clothing and personal protective equipment (PPE) to be worn while welding.
- ☐ Recognize fire hazards and use fire extinguishers appropriately.
- ☐ Explain the importance of material safety data sheets (MSDS).
- ☐ Determine ventilation requirements for welding different materials.
- ☐ Recall the dangers associated with high-pressure gas cylinders and regulators.
- ☐ Identify proper storage and set up procedures for high pressure gas cylinders.

Key Terms

cracking
Dewar flasks
filter lens
first degree burns
flash burn
flowmeters
forced ventilation
fusion
gas tungsten arc welding (GTAW)
infrared (IR) light
material safety data sheet (MSDS)
micro bulk systems
natural ventilation
nonconsumable
Occupational Safety and Health Administration (OSHA)
second degree burns
third degree burns
tungsten
ultraviolet (UV) light
visible light
weld zone

Introduction

Gas tungsten arc welding (GTAW) is a welding process that fuses metals by melting them using an arc from a *nonconsumable* tungsten electrode (one that does not melt). The heat necessary for *fusion* (mixing or combining of molten metals) is provided by an arcing electric current between the tungsten electrode and the base metal. See **Figure 1-1**.

This type of welding is usually done with a single electrode. However, it may be done with several electrodes. The *tungsten* electrode and the *weld zone* (area being welded) are shielded from the atmosphere by an inert gas, such as argon or helium. An inert gas is a gas that does not react chemically with other materials. Additional metal, called filler metal, may be added to build up the weld.

The hazards associated with GTAW include first, second, and third degree burns, which can be caused by hot metal, a hot torch, or the different types of light produced from the welding arc. GTAW hazards and safety guidelines for promoting a safe working environment are discussed in this chapter.

History

Gas tungsten arc welding began in 1939 and was made practical in 1941. An engineer named Russell Meredith, who worked for the Northrop Aircraft Company, needed to find an improved way

of welding magnesium and aluminum in a protective inert cover gas. Meredith used a tungsten electrode, direct current arc for heat, and helium gas to protect the molten metal from the atmosphere. *Heliarc* was the trade name given to the process because of the use of helium as the cover gas. When the patents for the process were sold to the Linde Company in 1942, *Heliarc* became a brand name.

Later, the American Welding Society (AWS) called the Heliarc process *tungsten inert gas welding (TIG)*. The letters TIG were used to designate the process. The name of the process was later changed to gas tungsten arc welding, and the letters GTAW came into popular use. Today, both of the names and letters are used. However, the American Welding Society recommends the use of gas tungsten arc welding, or GTAW. This book will follow current AWS practice and refer to the process as GTAW.

GTAW was first done with rotating direct-current welding machines. Rotating direct-current welding machines were electrical generators that produced direct current by converting mechanical energy to electrical energy. These machines had many moving parts. They were large and heavy because of the massive copper windings and a large rotating mass to generate the current.

In the 1950s, rectifier welding machines were introduced. These machines created direct current from an alternating current supply using a current switching component called a *rectifier*. These machines had fewer moving parts and functioned in a different way than the rotating machines. They converted an existing alternating current instead of creating a direct current. Rectifier welding machines were smaller and more manageable than rotating machines.

The newest type of machines, called *inverter machines*, are essentially the opposite of rectifier machines. Inverter machines convert direct current into alternating current. They are easily identifiable by their small footprint (size). However, they have capabilities that far exceed those of rectifier machines, which can be twice as large. Also, they have no moving parts. These machines are far superior to previous welding machines in many ways, which we will discuss later in the book. The disadvantage of these machines is their cost, which can be two to three times as much as a traditional rectifier machine. An inverter machine is shown in **Figure 1-2**.

With the successful introduction of GTAW, aluminum and magnesium became popular materials for manufacturing. They are lightweight, strong, have good corrosion resistance, and with the introduction of GTAW, could be easily welded. Besides being the only process that could be used to acceptably weld reactive

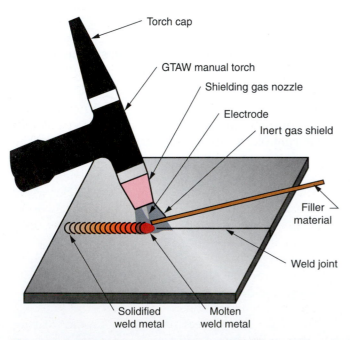

Figure 1-1. The basic components and function of the GTAW process.

Figure 1-2. A square-wave inverter-type machine. (Miller Electric Mfg. Co.)

metals (metals that undergo a chemical reaction when combined with elements such as oxygen, hydrogen, or nitrogen), GTAW also grew in popularity because of its versatility, cleanliness, and the fact it produced no smoke, hot flying sparks, or slag.

Until the gas metal arc welding (GMAW) process was developed in the late 1940s, GTAW was the only acceptable means of welding reactive metals such as titanium, aluminum, magnesium, and stainless steels. Although GTAW was economical for welding sheet metals, it was expensive for joining metals thicker than 1/4″ (6.4 mm) because of slow metal deposition rates. A skilled welder could complete a weld much faster by using GMAW rather GTAW. In addition, GTAW required more training of personnel, a higher skill level, and better hand-eye coordination than previous welding processes. GMAW has replaced GTAW in many heavy manufacturing roles. However, GTAW is currently very popular in many different industries, including the process piping industry, food industry, and custom automotive industry because it produces very clean and precise welds.

GTAW Hazards

GTAW does not produce large amounts of smoke or red hot sparks. This can create a dangerous misconception that can lead to injury if welders are not properly trained in the hazards of GTAW. The hazards associated with GTAW can range from flash burns to eye injuries. The risks from those hazards can be minimized by using proper procedures and protective equipment. Safety guidelines specific to GTAW welding are covered in the ANSI Z49.1 *Safety in Welding, Cutting, and Allied Processes* publication.

Arc Flash

There are three types of light radiation emitted from a welding arc: infrared, visible, and ultraviolet. Ultraviolet and infrared light are dangerous and can cause burns if proper protective actions are not taken. The least dangerous type of light is the light that we see, which is called *visible light*. It is produced from the welding arc and comes in different colors and amounts. Too much visible light can cause temporary blindness, similar to looking directly into a bright spotlight at night. However, visible light is generally not hazardous to the eyes or skin.

Infrared (IR) light has a wavelength that is longer than the wavelengths of visible light. Infrared light is emitted from any hot object such as a heat lamp. Infrared light is often used to heat large objects that cannot fit in an oven or furnace and in the automotive industry to dry and cure paint. Infrared light can cause burns, but its effects can be felt as the damage is occurring, so protective measures can be taken to prevent serious injuries. The concern with infrared light from the welding arc is that you are probably being exposed to ultraviolet light at the same time, which is a larger problem.

Ultraviolet (UV) light is light that has a wavelength shorter than visible light. This is the type of light emitted by black lights and tanning bed lamps. The wavelength and intensity of the UV light emitted by black lights is relatively harmless, but the wavelength and intensity of the UV light emitted by a welding arc is extremely hazardous. Ultraviolet radiation is especially dangerous because, although it cannot be seen or felt, it can severely burn your skin and eyes.

Flash burn is a burn that results from exposure to the UV light generated by a welding arc. Flash burn of the eyes is especially dangerous. Without proper protection or barriers, flash burn occurs quite easily. It can result from wearing an improper welding lens shade, wearing a welding hood with cracks, or working next to someone who is welding. Flash burn

Figure 1-3. Arc rays can damage your eyes. Always wear the proper lens shade to protect against flash burn. (Colour/Shutterstock)

is like a sunburn on the eye itself. Refer to **Figure 1-3**. It is usually not felt until many hours after the burn occurs. Flash burn makes the eye feel as though there is sand in it. Every time the eyelid comes down, there is a sensation of sandpaper being rubbed across the eye.

Ultraviolet light burns either the white of the eye or the retina of the eye. Burns on the retina can cause blindness, but are usually not very painful. However, a burn on the white of the eye is very painful. As with all burns, the major risk is infection of the burn area.

The dead cells in the burned area of the eye can easily promote bacterial growth because of the moisture. Always consult a doctor when an eye burn occurs.

Explosions

Welding on containers can be very dangerous. Proper precautions must be taken when welding on any previously used containers. If the container held a flammable liquid, it would need to be flushed with water, filled with water, or filled with an inert gas, depending on the size and shape of the container. Repairing a motorcycle gas tank, for example, requires removing the fill cap and running consistent air flow through the tank for a couple of hours and then purging it with argon. Gasoline and many other flammable liquids can catch fire, but the fumes are explosive.

Burns

Burns are one of the most common and one of the most avoidable injuries that occur while welding. Burns occur when the welder does not use the appropriate personal protective equipment (PPE). Being careless and not respecting the power of the radiation produced by the welding arc can be a tough lesson. Burns can be very dangerous. There is a risk of infection from the dead tissue. It is important to understand the three different types of burns and how to treat each type so you will be prepared in the event of a burn. See **Figure 1-4**.

First degree burns are identified by skin that is red, painful, and tender like a sunburn. There is no sign of any broken skin. When a first degree burn occurs,

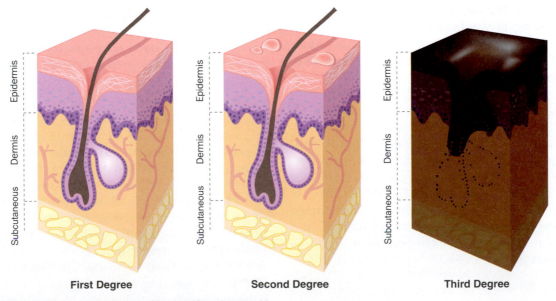

First Degree   **Second Degree**   **Third Degree**

Figure 1-4. The three types of burns. (Blamb/Shutterstock)

put the burned area under cool water or apply a cool washcloth or towel. In order to avoid possible allergic reactions, do not apply any ointments to the burn.

Second degree burns are more serious than first degree burns. They are identified by redness, blisters, and breaks in the skin. As with first degree burns, the first step is to place the burned area under cool water until the pain is somewhat relieved. Gently pat the burned area with a dry towel and cover it the best you can to help prevent infection. A second degree burn may require medical attention.

Third degree burns are the worst burns and always require professional medical attention. In a third degree burn, the surface and the tissue below the skin are white in color or charred and black in color. Strangely, these burns may not be as painful in the beginning because the nerves ends have probably been destroyed. *Never* remove clothing that might be stuck to the skin. Do *not* apply ice water or ice to the burn; doing so could cause shock and worsen the situation. The best response to a third degree burn is to get the victim immediate medical attention.

Heat Stress

The excessive heat produced by the welding process can lead to health and safety risks to the welder. These risks include heat exhaustion and heat stroke. Heat exhaustion is characterized by faintness, dizziness, and heavy sweating. Heat stroke symptoms include an elevated temperature, dry skin, high heart rate, and confusion. Heat stroke can lead to unconsciousness or even death, and requires immediate medical attention.

Precautions should be taken to avoid heat stress. The welder should drink plenty of water, and drink it frequently. Personal protective equipment, such as under-hood air circulators or water-cooled vests, can be worn to help dissipate heat. The *Occupational Safety and Health Administration (OSHA)* provides a publication on heat stress.

Electrical Hazards

Serious injuries, burns, and even death can result from electrical shocks. All shop tools, including grinders, drills, and cutting tools, can be hazardous if proper safety measures are not taken. Death can occur from equipment that operates on as few as 80 volts. Most equipment uses voltage ranges from 115 volts to 460 volts, and some equipment may require even higher voltages. Shock most frequently happens from bare or inadequately insulated conductors, loose connections, or damaged cables. Cables need to be checked frequently for damaged areas, loose connections, or frayed ends. Any problems must be corrected before the equipment can be put into use.

Special precautions should be taken when working in damp or wet conditions, including when sweating on a hot day. Welders need to wear dry gloves and rubber-soled boots. Before welding equipment can be worked on, the power needs to be turned off and locked out to prevent accidental electrocution. Primary current to an electrically-powered welding machine is usually 110 volts ac or more. This amount of voltage can cause extreme shock to the body and possible death.

Welding shops have a large number of electrical circuits, all of which require fuses or breakers. If you think your welding machine has blown a fuse, inform the instructor. *Never* try to find or change a fuse or reset a breaker. The instructor or a trained technician will first need to determine why the fuse or breaker failed before replacing it. *Never* replace a fuse with one that has a higher amperage rating. This will enable too much current to flow through the circuit and cause severe damage, a fire, or possibly electrocution. Electrical issues should always be handled by qualified personnel.

Welding machines have internal grounds to greatly reduce the chances of electrical shock. These grounds supplement the main grounding cable. A loose ground cable can cause many problems, including fire and electrical shock. It is very important to frequently check both ends of the grounding cable. First, make sure the end of the cable is secure in the brass fitting or clamping apparatus. Then make sure the bolts or nuts holding the cable ends are tight on the clamp and the machine. Loose ground cables will heat up severely, possibly causing a burn or fire. A bad ground can also have a negative effect on the welding arc or result in an electrical shock.

All electrical components must be installed by the instructor or by a qualified technician. All electrical work must be completed in accordance with the applicable codes to ensure proper electrical flow and to protect people and equipment. All electrical components must have good connections. All electrical connections should be tight and should be checked frequently. This applies to all connections, not just ground cable connections.

Unless you are a professionally-trained welding machine technician, *never* open any panels on a welding power supply, especially if it is powered up. *Never* switch the polarity switch when the welder is under a load. This can cause severe damage to the equipment.

Always lock out primary voltage switches when working on the interior components of any welding machine. Locking out refers to placing a paddle lock on the switch to prevent the power from being turned on before the machine is repaired. Also, be sure to remove the fuse. Welding current supplied by the welder has an open circuit voltage of 80 volts as a maximum. This is a fairly low voltage, and the possibility of shock is small, but it can still produce a significant shock.

Always keep the welder and the work area dry. Welding in damp or wet areas greatly increases the chances of shock. Welders sometimes need to weld in less than perfect conditions, so be aware of the added risk. Dry gloves and clothing will greatly reduce the chances of shock. Wear rubber boots to help insulate you from the water or damp area. It is also a good idea to check the connections on the ground and attach it as close as possible to the work, preferably directly on the work itself. This makes it easier for the current to pass through where it should.

Some GTAW machines use high-frequency voltage to initiate the arc without the electrode touching the base metal. A high-frequency generator can be either an added component on the machine or it can be built into the machine. Being able to start the arc without touching the metal is very advantageous for producing very clean welds without any tungsten inclusions. The high-frequency voltage also maintains a good arc when using alternating current. When normal frequency ac is used, the arc is actually extinguished when it reverses direction. High frequency maintains the arc in this area of the cycle. Voltage is significant with high frequency, but the amperage is very low. With such a low amperage, the current will usually not travel through the body. These factors make high-frequency ac less dangerous than the other currents.

Respiratory Hazards

Most of the gases used in GTAW are inert, colorless, and tasteless. Therefore, special precautions must be taken when using them. Nitrogen, argon, and helium are nontoxic; however, they can cause asphyxiation (suffocation) in a confined or closed area that lacks adequate ventilation.

Any atmosphere that does not contain at least 18% oxygen can cause dizziness, unconsciousness, or even death. The gases used in GTAW cannot be detected by the human senses and will be inhaled like air. *Never* enter any tank or pit where gases may be present until the area is purged (cleaned) with air and checked for oxygen content.

Ozone is a gas that is produced by the interaction of ultraviolet light emitted from the welding arc and oxygen in the atmosphere. The amount of ozone produced depends on the intensity of the ultraviolet energy. Ozone fumes can cause irritation of the eyes and throat with long term exposure. Ozone is generally not a concern with proper ventilation.

Phosgene is another gas that can be present during welding. Phosgene results from a chemical breakdown of degreasers, cleaning solvents, or other chlorinated hydrocarbon cleaning agents. The intense heat and UV radiation from a welding arc accelerates the production of phosgene gas. Excessive cleaning solvent residue should be removed from metal before it is welded.

Fumes produced by GTAW are minimal due to the cleanliness of the process. The base metal must always be free of paint, residues, and solvents to ensure a proper weld. With proper material preparation and ventilation, the risks from fumes produced from GTAW welding are minimal compared to those associated with other welding processes.

Safety in the Shop

GTAW is a skill that can be performed safely if the welder uses common sense and follows all safety rules. You should establish good safety habits as you work in this field. Check your equipment regularly and be sure your environment is safe. Safety concerns in GTAW include electrical current, inert gases, welding environment safety rules, personal protective equipment, MSDS sheets, and special precaution areas. Refer to the safety checklist in **Figure 1-5**.

Personal Protective Equipment

Working in a welding shop can be extremely hazardous unless the proper personal protective equipment is selected and worn at all times. Most accidents occur when a worker does not follow safety protocols. Welders should wear a cotton shirt with long shirt sleeves, leather arm sleeves, or a leather welding jacket. Shirt pockets should be covered so slag or hot particles cannot enter the pocket. Shirts or welding jackets must be buttoned to the top and should fit snugly around the neck area. One of the most easily burned areas is the lower neck area, which is covered by neither the shirt nor the welding hood. You usually cannot feel radiation burns until skin damage has already occurred.

Welding boots should always be worn. They should be made from leather and must cover the ankle.

Welding Safety Checklist		
Hazard	**Factors to Consider**	**Precautionary Summary**
Electric shock can kill	• Wetness • Welder in or on workpiece • Confined space • Electrode holder and cable insulation	• Insulate welder from workpiece and ground using dry insulation, rubber mat, or dry wood. • Wear dry, hole-free gloves. (Change as necessary to keep dry.) • Do not touch electrically "hot" part or electrode with bare skin or wet clothing. • If wet area and welder cannot be insulated from workpiece with dry insulation, use a semiautomatic, constant-voltage welding machine or stick welding machine with voltage-reducing device. • Keep electrode holder and cable insulation in good condition. Do not use if insulation is damaged or missing.
Fumes and gases can be dangerous	• Confined area • Positioning of welder's head • Lack of general ventilation • Electrode types, i.e., manganese, chromium, etc. • Base metal coatings, galvanizing, paint	• Use ventilation or exhaust to keep air-breathing zone clear and comfortable. • Use helmet and position of head to minimize fume breathing zone. • Read warnings on electrode container and material safety data sheet for electrode. • Provide additional ventilation/exhaust where special ventilation requirements exist. • Use special care when welding in a confined area. • Do not weld unless ventilation is adequate.
Welding sparks can cause fire or explosion	• Containers that have held combustibles • Flammable materials	• Do not weld on containers that have held combustible materials (unless strict AWS F4.1 explosion procedures are followed). Check before welding. • Remove flammable materials from welding area or shield from sparks, heat. • Keep a fire watch in area during and after welding. • Keep a fire extinguisher in the welding area. • Wear fire-retardant clothing and hat. Use earplugs when welding overhead.
Arc rays can burn eyes and skin	• Process: gas-shielded arc most severe	• Select a filter lens that is comfortable for you while welding. • Always use helmet when welding. • Provide nonflammable shielding to protect others. • Wear clothing that protects skin while welding.
Confined space	• Metal enclosure • Wetness • Restricted entry • Heavier-than-air gas • Welder inside or on workpiece	• Carefully evaluate adequacy of ventilation, especially where electrode requires special ventilation or where gas may displace breathing air. • If basic electric shock precautions cannot be followed to insulate welder from work and electrode, use semiautomatic, constant-voltage equipment with cold electrode or stick welding machine with voltage-reducing device. • Provide welder helper and method of welder retrieval from outside enclosure.
General work area hazards	• Cluttered area	• Keep cables, materials, tools neatly organized.
	• Indirect work (welding ground) connection	• Connect work cable as close as possible to area where welding is being performed. Do not allow alternate circuits through scaffold cables, hoist chains, ground leads.
	• Electrical equipment	• Use only double insulated or properly grounded equipment. • Always disconnect power to equipment before servicing.
	• Engine-driven equipment	• Use only in open, well-ventilated areas. • Keep enclosure complete and guards in place. • Refuel with engine off. • If using auxiliary power, OSHA may require GFCI protection or assured grounding program (or isolated windings if less than 5 KW).
	• Gas cylinders	• Never touch cylinder with the electrode. • Never lift a machine with cylinder attached. • Keep cylinder upright and chained to support.

Figure 1-5. This general safety checklist should be followed at all times.

Figure 1-6. Over-the-ankle leather boots protect the feet from many potential hazards in a weld shop. (Mark Prosser)

Figure 1-8. Cracks and spatter on the lenses from other welding processes can allow dangerous light to enter the welding hood. (Mark Prosser)

Refer to **Figure 1-6**. Steel toes are not absolutely necessary, but are a good idea when working with heavy parts. Jeans or work pants should be made of denim or other fire-resistant natural materials and should be worn over the boots. The pants should have no frayed edges, which are easily combustible.

Gloves used for GTAW are thinner and much more manageable than other welding gloves. These soft leather gloves, shown in **Figure 1-7**, allow the welder greater control and movement than traditional welding gauntlets or heavy leather gloves.

Helmets need to be checked for cracks and loose-fitting lenses. Refer to **Figure 1-8**. Small undetectable leaks can make the eyes feel irritated throughout the day. Lenses can be damaged from cracks or from spatter

created during a different welding process. Check the lenses to make sure they fit into the hood properly, have a good tight seal, and have clear plastic lenses on both the front and back. The welding hood should have the correct shade *filter lens* for the welding process being performed. The filter lens, also called a filter plate, is a protective lens that filters out harmful rays and intense light. See **Figure 1-9**. If overhead welding will be performed, a cotton or leather welding cap should be worn under the helmet. This will prevent hot metal from burning the welder's scalp.

Safety glasses should be worn in the shop at all times, even under your welding hood. You only have two eyes, and they need to be protected. Safety glasses with side shields are fine for general use, but a full face shield or goggles need to be worn in addition to the safety glasses for grinding or cutting operations. See **Figure 1-10**.

Welding situations can be very noisy. The decibel levels from some equipment can be damaging to the ear. Like eye damage caused by flash burn, hearing damage may not be noticed until it is too late. Earplugs and earmuffs prevent hearing damage and block foreign particles from entering the ear. Refer to **Figure 1-11**.

Much more information on personal safety equipment can be found in the AWS publication ANSI Z49.1 *Safety in Welding, Cutting, and Allied Processes*. This publication is available on the American Welding Society's website.

Figure 1-7. There are many different types of GTAW gloves. They are made of leather and provide protection against arc rays. These gloves provide little protection against hot materials. (Mark Prosser)

Guide for Shade Numbers				
Operation	Electrode Size 1/32 in (mm)	Arc Current (A)	Minimum Protective Shade	Suggested [1] Shade No. (Comfort)
Shielded metal arc welding	Less than 3 (2.5) 3–5 (2.5–4) 5–6 (4–4.8) More than 6 (4.8)	Less than 60 60–160 160–250 250–550	7 8 10 11	– 10 12 14
Gas metal arc welding and flux cored arc welding		Less than 60 60–160 160–250 250–550	7 10 10 10	– 11 12 14
Gas tungsten arc welding		Less than 50 50–150 150–500	8 8 10	10 12 14
Air carbon arc cutting	(Light) (Heavy)	Less than 500 500–1000	10 11	12 14
Plasma arc welding		Less than 20 20–100 100–400 400–800	6 8 10 11	6 to 8 10 12 14
Plasma arc cutting	(Light) [2] (Medium) [2] (Heavy) [2]	Less than 300 300–400 400–800	8 9 10	9 12 14
Torch brazing		–	–	3 or 4
Torch soldering		–	–	2
Carbon arc welding		–	–	14
PLATE THICKNESS				
	in	mm		

Gas welding
  Light
  Medium
  Heavy
	Under 1/8
	1/8 to 1/2
	Over 1/2
		Under 3.2
		3.2 to 12.7
		Over 12.7
				4 or 5
				5 or 6
				6 or 8

Oxygen cutting
  Light
  Medium
  Heavy
	Under 1
	1 to 6
	Over 6
		Under 25
		25 to 150
		Over 150
				3 or 4
				4 or 5
				5 or 6

[1] As a rule of thumb, start with a shade that is too dark, then go to a lighter shade which gives sufficient view of the weld zone without going below the minimum. In oxyfuel gas welding or cutting where the torch produces a high yellow light, it is desirable to use a filter lens that absorbs the yellow or sodium line of the visible light spectrum.

[2] These values apply where the actual arc is clearly seen. Experience has shown that lighter filters may be used when the arc is hidden by the workpiece.
Data from ANSI Z49.1–2005

Figure 1-9. Always use the proper shade lens to provide the correct amount of protection.

Fire Safety

Starting a fire with GTAW is much less likely than with GMAW, SMAW, or FCAW. This is simply because GTAW produces no flying sparks, slag, or spatter. However, you may not be the only one welding in the shop, and there may be other processes happening at the same time. Fire is always a potential hazard in a welding shop. Welding and grinding can produce sparks that fly 15′ (4.6 m) or more. There are several precautions that need to be taken in order to minimize your chances of starting a fire.

Every weld shop has potential fire hazards such as paper, wood, combustible liquids, and electrical wiring. Always keep combustible materials, such as rags and cleaning agents, out of the welding area to reduce the risk of fires and injury.

Every weld shop should have proper fire extinguishers. It is important for everyone working in the shop to understand the proper use of each of the

Figure 1-10. Different types and styles of face protection are available. (Mark Prosser)

Figure 1-11. Hearing and ears can be protected in various ways. Earplugs block sound and foreign material from entering the ear, while earmuffs provide additional protection to the outer ear. (Mark Prosser)

four different types of fire extinguishers. Each type of extinguisher fights certain classes of fires. Some extinguishers can be used on multiple classes of fires. Refer to the fire extinguisher classification chart in **Figure 1-12**.

Type A extinguishers are identified by a green triangle label with the letter A in the middle. These extinguishers are used on flammable solids such as clothing, wood, and paper.

Type B extinguishers are identified by a red square with the letter B in the middle. These extinguishers are used on flammable liquids such as gasoline, oil, paint thinners, and any other flammable liquids.

Type C extinguishers are identified by a blue circle with the letter C in the middle. These extin-

guishers are used on electrical fires such as those in welding machines, fuse boxes, or electrical motors on shop equipment and hand tools. Any fire that is associated with electricity requires a type C extinguisher.

Type D extinguishers are identified by a yellow star with a D in the center. These extinguishers are good for flammable metals such as magnesium or titanium. Catching solid metals on fire by welding is uncommon, but the dust of these metals created from grinding can cause flash fires.

Fire extinguishers should be placed around the shop, with each type of extinguisher in the area of the potential fire type. The extinguishers should have large red labels that are easily seen and spotted around the shop. Make sure extinguishers are always easily accessible. Keep materials, tools, and equipment away from extinguishers. Fire extinguishers are relatively inexpensive compared to the cost of not having one in an emergency.

It is not always possible to remove all potential fire hazards from a work area. In these cases, you should cover flammable materials with fire blankets and post a fire watch. The fire watch can be anyone who can sit and pay attention to the work area with the proper type of fire extinguisher ready for use. Using a fire watch is always a great idea when welding in a potentially dangerous area. Never sacrifice safety and caution for the sake of speed. Whatever welding you are doing is not worth a burned down shop, or worse. Fires can spread quickly, leaving you little time to react.

If there is fire in the shop, make sure you grab an extinguisher that is capable of putting out that type of fire. Using the wrong type of extinguisher can accelerate the fire. For example, using a water-type extinguisher on an electrical or gasoline fire can quickly compound your problems. Most shop extinguishers use foam, carbon dioxide (which removes oxygen), or dry chemicals. All of these remove one of the mandatory components of a fire: heat, fuel, and oxygen.

After the correct type of extinguisher has been selected, it must be used properly. When using a foam type extinguisher, aim the nozzle so the foam falls on the base of the fire. When using a carbon dioxide–type extinguisher, aim the nozzle at the edge of the fire and gradually move the spray toward the center of the fire. Carbon dioxide extinguishers will remove all of the oxygen in the area. If these are used in a confined space, they can suffocate the user. When using chemical extinguishers, always direct the spray at the base of the fire in a slow swooping action.

Fire extinguishers should be checked and charged periodically by professional technicians. If an attempt to extinguish a fire is unsuccessful, you should vacate the building and call the proper authorities.

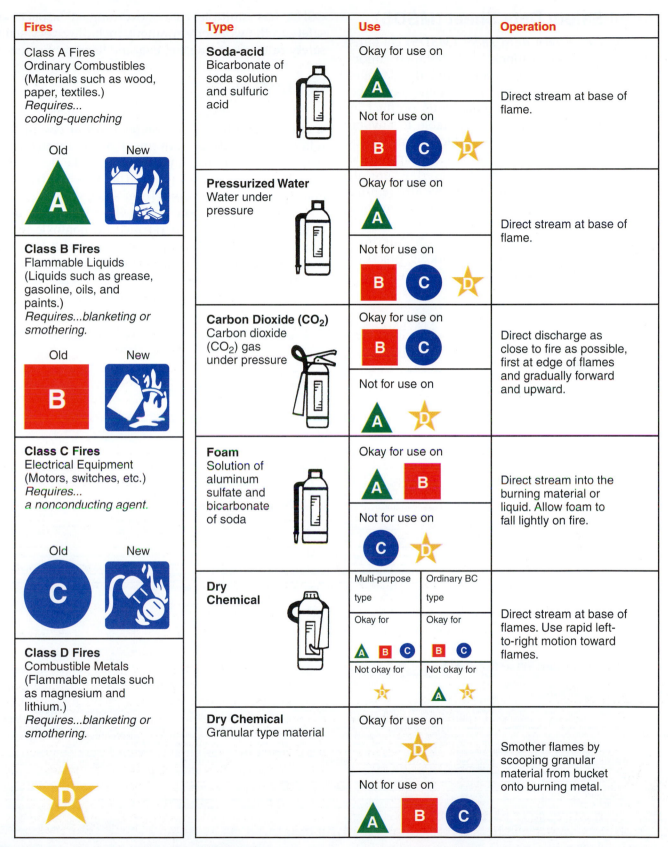

Figure 1-12. Fire extinguisher classification chart.

Material Safety Data Sheet (MSDS)

A *material safety data sheet (MSDS)* contains manufacturer's specifications and detailed information about a hazardous product and the dangers associated with it. See **Figure 1-13**. There should be an MSDS available for every hazardous product in the shop. These sheets must be available to anyone working with or around the product. Some states have right-to-know laws mandating that all applicable material safety data sheets be displayed where all workers can easily view them. These laws also require employers to provide proper training to anyone working with the hazardous materials.

The Occupational Safety and Health Administration (OSHA) monitors the safe working practices and conditions of companies to ensure worker safety.

OSHA has multiple rules governing all aspects of safety in the work environment, including very strict safety rules regarding ear, eye, and head protection.

Ventilation

The two different types of ventilation are natural ventilation and forced ventilation. *Natural ventilation* relies on the natural movement of air through the workspace to remove any fumes or smoke developed during welding. *Forced ventilation* uses electric fans and ducts to remove fumes from the work area.

Natural ventilation is usually adequate if only light welding is being performed on nontoxic materials in a wide open space. Work areas with tall ceilings and enough space per welder to provide natural

MATERIAL SAFETY DATA SHEET

For Welding Consumables and Related Products. Conforms to OSHA Hazard Communication Standard 29CFR 1910.1200. Standard must be Consulted for Specific Requirements

Date: 01/01/2011 MSDS No. 724

SECTION 1: IDENTIFICATION

Manufacturer/Supplier	Welding Material Sales Inc. 1340 Reed Road Geneva, Il 60134
Telephone Number	630-232-6421
Emergency Number	800-424-9300
Product Type	Pure Tungsten (EWP), 2% Ceriated Tungsten (EWCe-2), 2% Lanthanated Tungsten (EWLa-2),1% Thoriated Tungsten (EWTh-1), 2% Thoriated Tungsten (EWTh-2), 0.3% Zirconiated Tungsten (EWZr-1), 1.5%LanthanatedTungsten (EWLa-1.5), Rare Earth (EWG)
AWS Classifications	AWS A5.12

SECTION 2: HAZARDOUS INGREDIENTS

Important: This section covers the materials from which the product is manufactured. The fumes and gases produced during welding with the normal use of this product are covered under Section V. Thorium dioxide is subject to the reporting requirements of Section 313 of Title III of the Superfund Amendments and Reauthorization Act of 1986 (SARA) and 40 CFR Part 372.
*The term "HAZARDOUS MATERIALS" should be interpreted as a term required and defined in OSHA HAZARD COMMUNICATION STANDARD 29 CFR 1910.1200 however the use of this term does not necessarily imply the existence of any hazard.

PRODUCT	W	ZrO_2	ThO_2	LaO_2	CeO_2	Tip Color
Pure Tungsten	>99.5%	-	-	-	-	Green
0.3% Zircoriated Tungsten	>99.1%	0.15-0.4 %	-	-	-	Brown
1.5% Lanthanated Tungsten	>97.80%			1.3-1.70%		Gold
2% Thoriated Tungsten	>97.3%	-	1.7-2.2%	-	-	Red
Rare Earth (EWG)	>97.30%					Color determined by Manufacturer
2% Ceriated Tungsten	>97.3%	-	-	-	1.8-2.2%	Gray

Occupational Safety and Health Administration 28 CFR 1910.1000 Permissible Exposure Limit (PEL). American Conference of Governmental Industrial Hygienists (ACGIH) Threshold Limit Value (TLV[R]).

SECTION 3: PHYSICAL DATA

Melting Point: Approximately 3400°C   **Solubility in Water:** Insoluble
Boiling Point: Approximately 5900°C   **Specific Gravity (H_2O=1):** Approximately 19
Vap. Press: N/A at 25°C   **Appearance and Odor:** Gray, no odor
Vap. Density: N/A
Radioactive Isotope: Th-232

SECTION 4: FIRE AND EXPLOSION HAZARD DATA

NonFlammable: Welding arc and sparks can ignite combustibles. See Z49.1 referenced in Section 6.

SECTION 5: REACTIVITY DATA

Hazardous Decomposition Products
Welding fumes and gases cannot be classified simply. The composition and quantity of these fumes and gases are dependent upon the metal being welded, the procedures followed and the electrodes used. Workers should be aware that the composition and quantity of fumes and gases to which they may be exposed, are influenced by: coatings which may be present on the metal being welded (such as paint, plating or galvanizing), thenumber of welders in operation and the volume of the work area, the quality and amount of ventilation, the position of the welder's head with respect to the fume plume, as well as the presence of contaminants in the atmosphere (such as chlorinated hydrocarbon vapors from cleaning and degreasing procedure). When the electrode is consumed, the fumes and gas decomposition products generated are differer in percent and form from the ingredients listed in Section II,The composition of these fumes and gases are the concerning matter and not the composition of the electrode itself.
Decomposition products include those originating from the volatilization, reaction, or oxidation of the ingredients shown in Section II, plus those from the base metal, coating and the other factors noted above.

INGREDIENT	CAS No.	OSHA PEL	ACGIH TWA	ACGIH STEL
Tungsten (W)	7440-33-7	-	5mg/m^3	10mg/m^3
Thorium Dioxide	1314-20-1	-	-	-
Zirconium Oxide	1314-23-4	5mg/m^3	5mg/m^3	10mg/m^3
Cerium Dioxide	1345-13-7	-	-	-
Lanthanum Dioxide	1312-81-8	-	-	-

Gaseous reaction products may include carbon monoxide and carbon dioxide.
Ozone and nitrogen oxides may be formed by the radiation from the arc.
One method of determining the composition and quantity of the fumes and gases to which the workers are exposed is to take an air sample from inside the welder's helmet while worn or within the worker's breathing zone. See ANSI/AWS F1.1 publication available from the American Welding Society 550 N.W. LeJeune Road, Miami, FL 33126.

SECTION 6: HEALTH HAZARD DATA

Threshold Limit Value: The ACGIH recommended general limit for welding fume NOC (Not otherwise classified) is 5 mg/m^3. ACGIH -1985 preface states: "The TLC-TWA should be used as guides in the control of health hazardsand should not be used as fine lines between safe and dangerous concentrations." See Section V for specific fume constituents, which may modify this TLV.

Figure 1-13. A portion of the first page of an MSDS. (Welding Material Sales, Inc.)

ventilation may not need forced ventilation, unless there is evidence of fumes gathering.

Forced ventilation should be used if heavy welding is being performed, the air space in the work area is limited, or dangerous materials are being welded. Generally, the amount of hazardous fumes generated by gas tungsten arc welding of steel or aluminum is minimal compared to the other welding processes. However, if the fumes generated from the welding process are excessive, displace too much oxygen in the work area, or are toxic, a forced ventilation system, such as the one shown in **Figure 1-14**, must be used. A forced air ventilation system is required when welding is performed on materials that contain zinc, austenitic manganese, cadmium, mercury, lead, copper, beryllium copper, or any other material that gives off hazardous fumes. See **Figure 1-15**.

Forced ventilation can consist of a negative pressure that draws the fumes out of the welding area. Forcing clean air into the work area to create a positive pressure that displaces the contaminated air can also be considered forced ventilation. Fume extraction is a forced ventilation technique that relies on both positive pressure and negative pressure. Fumes are drawn out of the work area with negative pressure. After the fumes have been drawn from the area, they are filtered, and the clean air is returned to the welding area, creating a positive pressure. Generally, fume extraction systems remove contaminated air, filter the air, and return it to the area. Some local, state, or federal regulations require filtration of welding fumes to remove dangerous components before the fumes are released into the outside air.

Figure 1-15. A fume extraction system removes and filters harmful fumes and then returns the clean air back into the building. This particular unit uses 16 filters. (Mark Prosser)

Thermal Cutting

Thermal cutting of stainless steels produces toxic fumes from the chromium in the stainless. These fumes are known to cause several serious health problems. A welder can come into contact with chromium in several ways—through water, food, and materials—but the levels are low, and the types of chromium are different. Certain types of chromium can be dangerous, especially when cut by a thermal cutting process. Prolonged exposure to these types can cause health problems, such as respiratory problems, kidney and liver damage, weakened immune systems, skin rashes, upset stomach, and irritations to the nose, eyes, and lungs. Be careful when using degreasers because the welding arc can change the vapors from these chemicals into poisonous phosgene gas, which can be deadly. Always be aware of the applicable fume hazards of your welding materials and use the proper ventilation.

Welding in an Enclosed Space

Without proper ventilation, welding in confined spaces can be very dangerous and even deadly. Often, a permit is required in order to enter a confined space or vessel. The air must be monitored before anyone enters the space, and periodic tests need to be done to ensure the space remains safe to work in. The biggest danger when working in a confined area is oxygen depletion. Shielding gases and fumes can build up and quickly deplete oxygen levels, causing unconsciousness. Continuous ventilation should be in place to circulate fresh air, and oxygen should *never*

Figure 1-14. This is a flexible forced ventilation arm that has a powerful LED light in the hood. (Mark Prosser)

be used for this. All welding equipment and cylinders should remain outside the space. The worker should always wear a lifeline harness in case rescue by fellow workers is needed. Verbal contact should always be maintained with the worker to ensure cohesion.

Gas Cylinder Safety

Inert gases used in GTAW are available to the user in two forms—as high pressure gases and as liquids. All storage vessels used for inert gases are approved by the Department of Transportation and are so stamped on the vessel nameplate or the cylinder wall, as shown in **Figure 1-16**. High-pressure gas cylinders contain gases under a very high pressure of approximately 2000–4000 psi (13,800–27,600 kPa). These cylinders must be handled with extreme care.

The following sections explore the proper storage and handling of shielding gas cylinders. Following the guidelines presented in these sections can greatly reduce the risk of accidents and injuries associated with shielding gas cylinders.

Cylinder Storage

Cylinders should always be stored in the vertical position with the protective cap on. Secure all cylinders with safety chains or cables. Keep the cylinder's protective cap in place until the cylinder is ready to use. This cap prevents the cylinder valve from being accidentally opened or broken off. All cylinders should be marked for identification. This is done with a stencil, stamp, or, most commonly, a label that is not easily

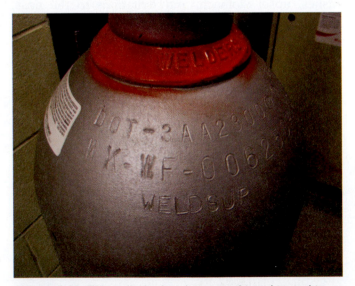

Figure 1-16. This cylinder has been made and tested to a Department of Transportation (DOT) specification. The letter T indicates the amount of gas that it will hold. The name imprinted is the owner. (Mark Prosser)

removed. This labeling identification method was established by the ANSI Z48.1-1954 document. The current document that details labeling criteria is the ANSI/CGA C-7 *Guide to the Preparation of Precautionary Labeling and Marking of Compressed Gas Containers*.

Cylinders are to be stored inside of buildings in a well-ventilated, dry, and protected location. The cylinders should be at least 20' (6.1 m) away from any highly combustible materials, such as oil or gasoline. Oxygen cylinders are to be stored separately from fuel gas cylinders and separated by at least 20' (6.1 m) or a 5-foot-tall (1.5 m) noncombustible barrier with a half-hour fire resistant rating. High pressure cylinders should not be exposed to temperatures over 120°F (49°C). Safety guidelines specifically for GTAW cylinder handling can be found in the ANSI Z49.1 *Safety in Welding, Cutting, and Allied Processes* publication.

Cylinder Handling

When moving gas cylinders, there are some very important precautions that must be taken. Do *not* use gas cylinders as rollers. Do *not* move a cylinder without its protective cap in place. When moving a cylinder, use a cylinder cart and place safety chains around the cylinder to prevent it from falling. See **Figure 1-17**.

Attaching a Cylinder

When preparing to connect a cylinder to the welding outfit, a welder must follow an additional set of safety rules. *Never* use oil or grease on any tanks, especially oxygen tanks. If gas from the cylinder comes in contact with lubricants, an explosion may result. If the cylinder's safety cap or outlet threads are severely rusted or damaged, return the cylinder to the supplier. Do *not* attempt to lubricate the threads or fix a damaged valve!

Know the contents of the cylinder before attaching a regulator. Make sure the cylinder is clearly labeled, **Figure 1-18**. If you are unable to identify a gas from cylinder labeling, do *not* attach the regulator. Instead, set the cylinder aside until the gas can be identified.

Check the outlet threads and clean the valve opening by *cracking* (partially opening and then immediately closing) the cylinder valve before attaching regulator, **Figure 1-19**. Cracking the cylinder valve blows out any particles or dirt in the valve opening. Select a regulator that is approved for use with the specific shielding gas being used. Be sure that the regulator is rated for the pressure found at the cylinder valve. Attach the regulator securely, and make sure the *flowmeter* (valve with a floating ball that is adjusted to set the shielding gas flow working pressures) is in the vertical position, as shown in **Figure 1-20**.

Figure 1-17. Secure all cylinders with safety chains. The cylinder in the foreground is chained to a cylinder cart for transport. The cylinders in the background are chained to the wall for long-term storage. (Mark Prosser)

Figure 1-18. The label indicates the type of gas stored, a caution, and the cylinder contents in cubic feet. The safety cap is kept in place until the cylinder is ready to use. (Mark Prosser)

Figure 1-19. Cracking a cylinder before attaching a regulator blows out any dirt in the valve opening. (Mark Prosser)

Pressure relief valve

Figure 1-20. This regulator and flowmeter have been properly attached to the cylinder. Note the pressure relief valve on the cylinder. This device releases excess pressure in the cylinder and should never be tampered with. (Mark Prosser)

Open the cylinder valve slowly. Always stand to the side of the regulator when opening a cylinder valve. *Never* tamper with or try to fix a leaky cylinder valve. Instead, return the cylinder to the supplier.

Be especially careful when using any gas mixture that includes hydrogen gas. Remember, hydrogen gas will explode and burn. Hydrogen is lighter than air, so it rises. Also, keep in mind that a hydrogen flame is almost invisible.

When a cylinder is empty and needs to be replaced, close the cylinder valve, remove the regulator, and replace the valve cap. Mark "MT" on the cylinder in chalk and store it separately from full cylinders.

Figure 1-21. A typical tank dolly used for transporting tanks. (Mark Prosser)

Dewar Flasks (Micro Bulk Systems)

Liquefied gas cylinders, commonly called *Dewar flasks* or *micro bulk systems*, are basically vacuum bottles. The gas in a Dewar flask has been reduced to a liquid at the supplier's plant. This greatly reduces the volume of the gas, making handling and shipping easier and reducing the space required for storage. The shielding gas in a Dewar flask is changed from liquid form to gas form as it passes through heat exchangers within or on the system. The following are safety issues to keep in mind when working with a Dewar flask:

- Cylinders must be kept in an upright position.
- Cylinders should be moved on special dollies as shown in **Figure 1-21**.
- Always use equipment designed for inert gases.
- Do *not* interchange equipment components.
- Liquid gases are extremely cold and can cause severe frostbite when exposed to the eyes or the skin. Do *not* touch frosted pipes or valves.

General Safety Rules

The following safety rules should always be observed:

- *Never* use oxygen in place of compressed air. Oxygen supports combustion and will make a fire burn violently.
- Power wire brushes are very dangerous because they expel broken pieces of wire. Always wear safety glasses and a clear plastic face shield when using power wire brushes.
- Grinders have a lot of torque and can kick back easily, especially when equipped with a wire wheel. Always wear safety glasses, gloves, and safety shields when working with grinders or power wire brushes.
- When working with mechanical, hydraulic, or air clamps on tools, jigs, and fixtures, be alert to the clamp operation. Serious injury may result if parts of the body are exposed to the clamp action.
- Ultraviolet light can be very dangerous because you cannot feel the burn happening. Make sure all skin is covered and the appropriate personal protective equipment is properly used.
- To avoid injury if a regulator malfunctions, always stand to the side of any tank regulator when opening.
- Do *not* weld with a lighter or matches in your pocket. A spontaneous fire or even an explosion may result.
- *Always* tag out any equipment being worked on or serviced.
- When a piece of equipment is broken, it must be fixed by qualified personnel only.
- *Never* use a piece of equipment you have not been trained on. Always ask for your instructor's help when performing a task you are unfamiliar with or unsure of.
- Make sure you always have proper skin protection.
- Be sure your helmet has the correct shade filter lens.

Summary

Gas metal arc welding has been used for over 60 years to join many different types of materials. GTAW is still the preferred welding process when high-quality and precision welds are desired.

Material set up and preparations are very important when creating a quality GTAW weld. As with any welding process, safety is the primary concern. Safety concerns associated with GTAW are addressed in the ANSI Z49.1 *Safety in Welding, Cutting, and Allied Processes* publication. Some GTAW safety issues include proper ventilation, personal protective equipment (PPE), tank safety, and dangerous fume potential. PPE equipment needed for welding includes a welding helmet with the correct filter lens, gloves, welding boots, and a long-sleeved shirt or welding jacket. Safety glasses should be worn at all times. Gloves and a face shield or goggles should be worn when grinding or cutting. Earplugs and earmuffs prevent hearing damage.

Review Questions

Write your answers on a separate sheet of paper. Do not write in this book.

1. GTAW was invented as an improved process to weld what two materials?
2. A(n) _____ gas is a gas that does not react chemically with other materials.
3. Before the term *gas tungsten arc welding (GTAW)* came into use, the American Welding Society called the Heliarc process _____.
4. List the three types of light radiation emitted from a welding arc.
5. Which type of light can cause temporary blindness but is otherwise harmless?
6. Which type of light is the most dangerous because it can burn the welder without being felt or seen?
7. List the three different degrees of burns.
8. Which type of burn requires immediate medical attention?
9. When working on the interior components of any welding machine, always _____ primary voltage switches and remove the _____.
10. The greatest potential danger of welding in confined spaces is _____ depletion.
11. What do the letters *PPE* stand for?
12. When should safety glasses be worn?
13. List the different classifications of fire extinguishers and the types of fires that each is used to extinguish.
14. What documents contain manufacturers' specifications and information about hazardous materials?
15. What are the two different types of ventilation?
16. Where should you stand when opening a cylinder valve?
17. Where should shielding gas cylinders be stored?
18. What action(s) should be taken if a cylinder valve is found to have rusty or damaged threads?
19. The shielding gas in a Dewar flask is changed from _____ form to _____ form as it passes through heat exchangers.
20. Never use _____ in place of compressed air.

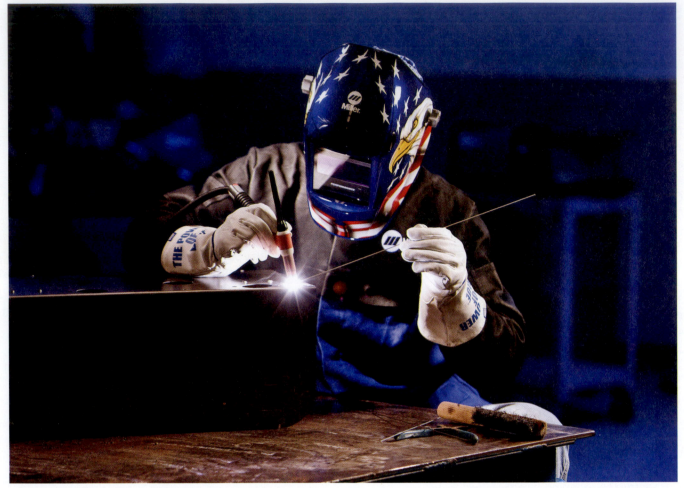

The proper personal protective equipment, including a helmet and leather gloves, must always be worn during GTAW operations. (Miller Electric Mfg. Co.)

Chapter

Power Sources

Objectives

After completing this chapter, you will be able to:

- ☐ Recall the GTAW power source specifications.
- ☐ Explain the purpose of a power source.
- ☐ Distinguish between the different types of welding current.
- ☐ Identify the commonly used methods to start an arc.
- ☐ Recall the meaning of *duty cycle*.
- ☐ Explain how a volt-ampere curve is used.
- ☐ Explain the differences between the various types of power sources.
- ☐ Recognize the different types of GTAW power source controls.
- ☐ Recall the factors that influence the selection of power sources.
- ☐ Recognize the correct procedures for the installation and maintenance of welding machines.

Key Terms

anode
arc plasma
auxiliary controls
cathode
closed-circuit voltage
constant-current power source
contactor
contaminate
direct current electrode negative (DCEN)
direct current electrode positive (DCEP)
direct current reverse polarity (DCRP)
direct current straight polarity (DCSP)
droopers
duty cycle
engine-driven welding generators
high-frequency arc stabilizer box
high-frequency arc starter
high-frequency generator
ions
open-circuit voltage
power source
variable-voltage power source

Introduction

In the welding industry, the term *power source* describes a machine for producing welding current. The machine may be a power supply only; a power supply with controls for manual welding; a power supply with additional equipment such as slopers, pulsers, or timers; or a power supply with additional equipment for pipe welding, tube welding, or spot welding. Power sources have evolved through the years to deliver a wide range of capabilities. Power source technology is continually evolving to create more powerful machines with smaller footprints. These machines have options and capabilities that were unavailable only a few years ago. The newest inverter technology has resulted in power sources that produce twice the power and are half the size and weight of some of the more traditional machines.

Power Source Basics

The GTAW power source must be able to convert the incoming power into usable power. It must produce sufficient amperage (current) to provide the heat necessary for melting metal at a low voltage. Different types of machines accomplish this in different ways. The length of time a machine can operate continuously, output ratings, and performance (minimum and maximum current) are significant concerns when selecting power sources.

Guidelines established by the National Electrical Manufacturer's Association (NEMA) under the title Specification EW-1, *Electric Arc-Welding Power Sources* cover power sources using utility-supplied electricity. NEMA EW-1 is a specification that defines the performance characteristics and rating procedures for ac/dc arc-welding equipment. This specification also recommends installation and test procedures for high-frequency arc-welding machines.

The electrical power received from the utility line is high-voltage, low-amperage alternating current and is called *primary power* or *primary current*. A power source changes this current to low-voltage alternating current or low-voltage direct current. The major components of a typical system are shown in **Figure 2-1**.

Power sources that are specially designed and manufactured for use in GTAW must produce a variable low voltage and a constant electrical current to the welding arc. The various types of power sources used in GTAW are shown in the chart in **Figure 2-2**.

Power sources produce *open-circuit voltage* when the power switch is turned on, but before the arc is struck. The machines will produce welding current when an arc is struck. When the electrode is moved close enough to the base metal, the welding current

Welding Current Output			
Machine Type	**AC Only**	**DC Only**	**AC or DC**
Transformer	X		
Transformer-rectifier		X	X
Motor generator	X	X	X
Motor alternator	X	X	X
Inverter		X	X

Figure 2-2. Five types of power sources are used in gas tungsten arc welding.

will flow and create the arc. When the arc is struck, the voltage drops and forces the current to jump the gap between the electrode and the base metal, sustaining the arc. *Closed-circuit voltage* is the voltage that is present while the weld is being made. A *contactor* switch prevents accidental shortage of the welding current to the workpiece prior to the start of the weld. The contactor turns the flow of welding current from the power source to the torch on and off.

The contactor is located in the power source and opens and closes the power circuit from the power source to the torch. Switches located on the main power source select the desired modes (foot or machine) for amperage control and contactor operation. Operation of the hand or foot control energizes the contactor and allows the arc to start.

When using a power source, observe the following rules:
- *Never* use the power source above the rated capacity.
- *Never* change an output lead during welding.
- *Never* change the current type (polarity) while welding.

A GTAW arc is produced when welding current jumps the gap between the negatively charged electrode

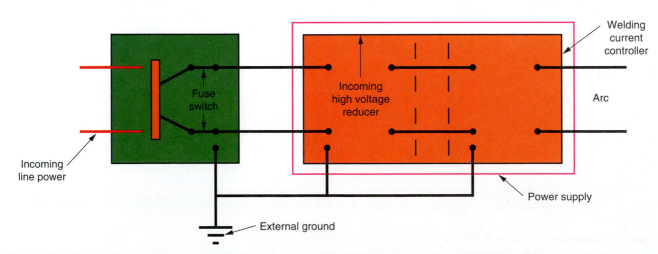

Figure 2-1. The basic components required to change utility electrical power to welding current.

(*cathode*) and the positively charged workpiece (*anode*), creating a small amount of gaseous particle ionization, or *arc plasma*. (This description of the *cathode* and *anode* apply only to DCEN setups. DCEN and DCEP will be discussed later in the chapter.) The positively charged gaseous atoms are attracted to the electrode, where they give up their kinetic energy (energy of motion) in the form of heat. This heat is enough to bring about electron emission (flow) from the tip of the electrode.

The emitted electrons are attracted to the positively charged workpiece. They raise the temperature of the shielding gas atoms through collisions with them. Repeated collisions of these extremely hot particles cause the atoms to change their makeup. They pick up or lose an electron, and they become *ions*. This happens to more and more atoms, and they form a plasma zone. These gaseous atoms, or ions, are also attracted to the electrode. This movement creates the required heat to sustain electron emission and thus the flow of welding current. Electrical current produced by the power source is a flow of negative charges (electrons). The amount of current flowing in the circuit is measured in amperes (A).

In GTAW, the arc voltage is a measurement of electrical pressure. Voltage controls the distance (arc gap) that the electrons can jump to form the welding arc. Higher voltages can jump a larger gap distance between the end of the electrode and the workpiece. **Figure 2-3** shows arc voltage meter readings with the corresponding tungsten-to-work gap settings.

Types of Welding Current

The GTAW process uses three types of current. Each is needed to weld different types of metals. The three types of current are direct current electrode negative (DCEN), direct current electrode positive (DCEP), and alternating current (ac). Each complete cycle of alternating current contains a half cycle of straight polarity and a half cycle of reverse polarity. In the *straight polarity (SP)* portion of the cycle, the welding current flows from the electrode to the workpiece. In the *reverse polarity (RP)* portion of the cycle, the welding current flows from the workpiece to the electrode. **Figure 2-4** shows a sine waveform of one complete cycle of alternating current.

Direct Current

The American Welding Society has defined current flow as moving from the negative (–) terminal to the positive (+) terminal. If the current continually flows from the welding machine negative terminal to the electrode holder, across the arc gap to the workpiece, and back to the welding machine, the circuit is known as **direct current electrode negative (DCEN)**. Another term used to describe DCEN is **direct current straight polarity (DCSP)**. Many of the older welding machines in the field use the DCSP markings, while newer equipment identifies the current as DCEN.

If the current continually flows from the welding machine to the workpiece, across the arc gap to the electrode, then back through the lead to the welding

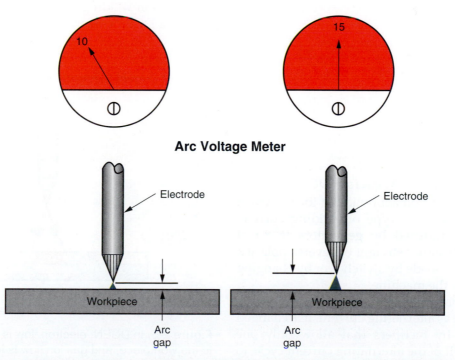

Arc Voltage Meter

Figure 2-3. Arc voltage meters measure the gap between the electrode tip and the weld on the workpiece.

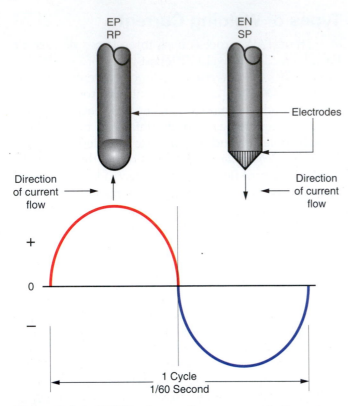

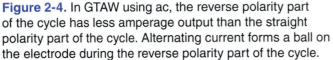

Figure 2-4. In GTAW using ac, the reverse polarity part of the cycle has less amperage output than the straight polarity part of the cycle. Alternating current forms a ball on the electrode during the reverse polarity part of the cycle.

machine, the circuit is known as *direct current electrode positive (DCEP)* or *direct current reverse polarity (DCRP)*.

Note

Since the American Welding Society standard identifies the terms *direct current electrode negative (DCEN)* and *direct current electrode positive (DCEP)* as standard terminology, this book will use these two terms. However, *direct current straight polarity (DCSP)* and *direct current reverse polarity (DCRP)* should be recognized as interchangeable terms with DCEN and DCEP respectively.

Direct Current Electrode Negative (DCEN)

DCEN is the common abbreviation for the direct current electrode negative type of welding current. Welding current produced by generators is direct current and may be either straight or reverse polarity. Polarity changes are made by switches or by changing the output leads of the machine.

Welding current produced by transformer-rectifiers starts out as alternating current. It is changed to direct current by rectifiers that allow only one polarity, or one-half of the alternating current cycle, to pass through as welding current.

With DCEN polarity, the electron flow is always from the electrode to the workpiece. The gas ion flow is always from the workpiece to the electrode. In DCEN mode, the tungsten stays sharp, or pointed, and most of the heat is generated in the workpiece. See **Figure 2-5**.

Direct current electrode negative (DCEN) is used to weld steels, stainless steels, nickel, titanium, and many other materials. DCEN creates a deep, penetrating, narrow weld by concentrating about two-thirds of the heat from the arc onto the workpiece and one-third of the heat input onto the electrode. In some automatic welding of aluminum and magnesium, DCEN is used with helium shielding gas.

Direct Current Electrode Positive (DCEP)

DCEP is the common abbreviation for the direct current electrode positive type of current. DCEP is produced in the same way as DCEN. However, the electron flow is always from the workpiece to the electrode. The gas ion flow is always from the electrode to the workpiece.

In DCEP mode, the electrode receives most of the heat. A large electrode is required. A ball will usually form on the tip of the electrode. **Figure 2-6** shows the electron and gas ion flow for DCEP.

DCEP has limited use in GTAW, as most of the heat in the arc is directed to the electrode instead of the workpiece. About two-thirds of the arc heat is concentrated onto the electrode, and one-third of the heat is concentrated onto the workpiece, creating a wide and shallow weld bead profile. DCEP polarity is usually used only on materials needing shallow penetration. Flat position welding is almost always a requirement for using DCEP current.

Direct Current Electrode Negative (Straight Polarity)

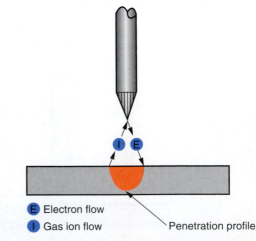

Figure 2-5. In DCEN, electron flow is from the electrode to the workpiece, and gas ion flow is from the workpiece to the electrode.

Direct Current Electrode Positive (Reverse Polarity)

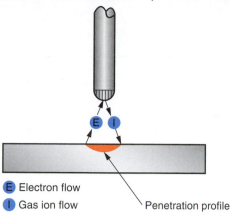

Ⓔ Electron flow

Ⓘ Gas ion flow   Penetration profile

Figure 2-6. In DCEP, electron flow is from the workpiece to the electrode, and gas ion flow is from the electrode to the workpiece.

Alternating Current

Alternating current (ac) is generally used for manual and semiautomatic welding of aluminum and magnesium. AC concentrates about one-half of the arc heat onto the work and about one-half onto the electrode. In ac welding, the straight polarity (electrode negative) portion of the ac cycle provides better penetration characteristics, and the reverse polarity (electrode positive) portion of the ac cycle provides a necessary cleaning action. The number of times the current switches from straight polarity to reverse polarity in one second is referred to as *frequency*. In the United States, the electrical current cycles at 60 hertz (Hz), or 60 times per second.

Power companies supply high-voltage and low-amperage current. The transformer of an alternating current welding machine changes the current to lower voltage and higher current (amperage) for welding. The frequency remains the same at 60 cycles per second, or 60 Hz.

Alternating Current, High-Frequency (ACHF)

The letters *ac* indicate the welding current is alternating current. The letters *hf* indicate that a high-frequency voltage is used to maintain the alternating current arc. These letters are often used in specifications, blueprints, and fabrication orders throughout the industry.

Normally, the current is unbalanced in GTAW. The imbalance happens because hot tungsten can emit (send out) electrons better than molten metal. The current flows more readily in one direction than the other. As a result of this imbalance, current flow during the reverse polarity part of the cycle will not equal current flow during the straight polarity part of the cycle.

During each part of the current cycle, electrons will carry heat across the arc gap. Gas ions are formed from the inert gas. The welding current flows through the ionized path that develops. **Figure 2-7** shows the

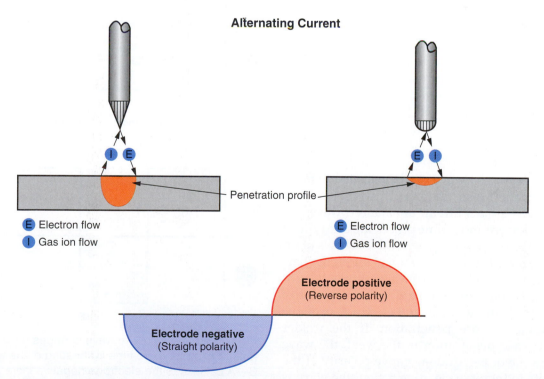

Alternating Current

Ⓔ Electron flow

Ⓘ Gas ion flow

Penetration profile

Ⓔ Electron flow

Ⓘ Gas ion flow

Electrode positive
(Reverse polarity)

Electrode negative
(Straight polarity)

Figure 2-7. Penetration into the base material is greater during the straight polarity part of the alternating current cycle. Electron flow is in the opposite direction of gas ion flow.

electron and gas ion flow for one complete cycle of alternating current.

High-frequency voltage plays an important role when used in conjunction with alternating current. High-frequency voltage maintains and stabilizes the arc during the transitions from one polarity to the other. When the arc switches from positive to negative, which occurs 60 times a second, the arc is extinguished right at the zero point. The superimposed high-frequency voltage maintains the arc at this transition point.

In the straight polarity part of the cycle, the electrons leave the electrode tip and strike the metal. Enough heat is created to melt the workpiece. The reverse polarity part of the cycle removes oxides from the surface of the material as the electrons flow from the surface of the metal to the electrode. This part of the cycle will also cause the tip of the electrode to melt. The tip will then become rounded because most of the heat is directed to the electrode.

Alternating Current, High-Frequency Square-Wave

The square-wave designation indicates that the alternating current produced by the machine has a square waveform rather than the sine wave normally associated with alternating current. See **Figure 2-8A**. The modification of the waveform reduces the transition period from one polarity to another, which increases the actual weld time at the desired welding current.

The square-wave power source can produce changes in the output current of each polarity, allowing for the selection of the time period for each cycle of current polarity. The ac waves may be balanced or unbalanced, as shown in **Figure 2-8B**.

A *balanced wave* occurs when an equal amount of time is spent on each side of the wave during each cycle. This means that for any given minute, the power source spends as much time producing electrode positive current as electrode negative current.

In many cases, the wave can be adjusted to favor one type of current over another. This is referred to as an *unbalanced wave*. With an unbalanced wave, the power source spends more time producing one type of current than the other during each cycle. This is often done because each type of current has different characteristics that the welder may want to take advantage of for a certain job. Electrode-positive polarity provides more cleaning action and electrode-negative polarity provides more penetration. If the welder wants to increase penetration in the weld, the wave is adjusted to favor electrode-negative polarity. If the welder wants to increase cleaning action, the wave is adjusted to produce more electrode-positive current.

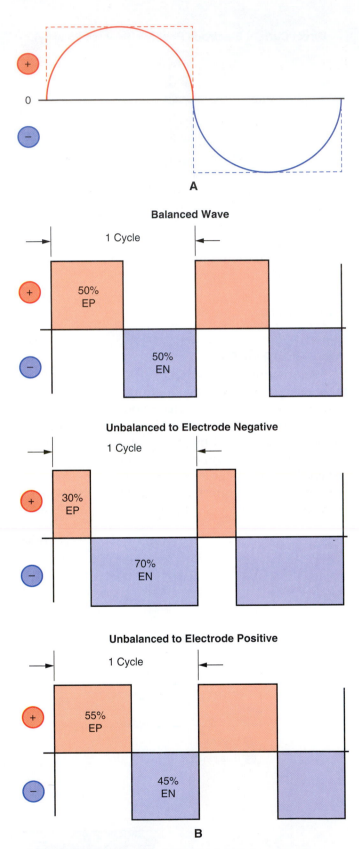

Figure 2-8. A—A square wave changes polarity very rapidly, allowing more time for heating of the weld zone. B—A balance to the electrode negative increases penetration. A balance to the electrode positive increases the cleaning of the metal surface.

The current from a square-wave machine can be adjusted to provide excellent cleaning action of the base material, a much more stable arc, and uniform heating of the material. Square-wave power sources accomplish this with minimal heating of the machine.

High-Frequency Generator

When the cycle of alternating current crosses the zero amperage point, it changes polarity. The arc will go out at this point. In the straight polarity portion of the cycle, electrode emission of the electrons is sufficient to ignite the arc. All of the electrons are concentrated at the electrode tip. However, on the reverse polarity portion of the cycle, the electron flow is very poor because the electrons are scattered all over the workpiece surface. As a result, the arc may not ignite during that half-cycle. This condition may last for one or more cycles. Reverse polarity has a cleaning effect on the weld because the arc flows away from the weld, carrying contaminants with it. Without the reverse polarity portion of the cycle, oxides will be absorbed into the weld. Welds containing oxides are usually porous and weak.

The *high-frequency generator* was developed to provide the voltage necessary to ignite the arc during the reverse polarity part of the cycle and to stabilize the arc. The high voltage of the current generated by the high-frequency generator causes the arc gap to become ionized, thus becoming a path for the electrons to flow. This produces a continuous steady arc as the output current constantly changes polarities.

Starting the Arc

The method used to start the arc in the GTAW process depends on the type of power source being used and how the machine is set up. Three commonly used methods are scratch or tap start, high-frequency start, and pilot arc start.

Scratch or Tap Start

The welder scratches or taps the workpiece, creating a short circuit and an arc, allowing the current to flow. This method is commonly used with motor generators or direct current rectifiers. This type of arc start may contaminate the electrode or weld area if the electrode sticks. (To *contaminate* means to make unfit for use because undesirable material has entered.) Therefore, this method must be used with extreme care.

High-Frequency Start

A high-frequency voltage of 3000–5000 volts with a very low current flow is used to generate a spark when the electrode nears the workpiece. The high-frequency generator may be integral to (part of) the welding machine, or it may be an accessory wired into the system. With this system, the high-frequency spark will jump a gap of about one-half inch. Since the electrode does not touch the workpiece, contamination of the workpiece and the electrode is avoided. **Figure 2-9** shows a high-frequency circuit. Note the main power cable that carries the high-frequency voltage.

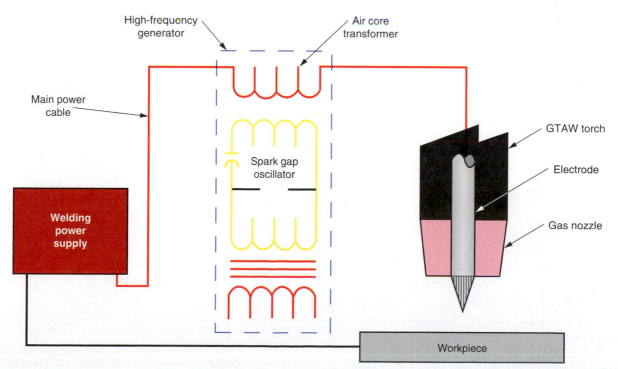

Figure 2-9. The main power cable of the GTAW torch carries the high-frequency voltage from the generator to the electrode.

Figure 2-10. This high-frequency arc starter can be used in conjunction with an inverter power source. (Miller Electric Mfg. Co.)

A *high-frequency arc starter*, also referred to as a *high-frequency arc stabilizer box*, is shown in **Figure 2-10**. These add-on boxes superimpose a high frequency above the electrical current that establishes the arc without touching the electrode to the base metal. They also maintain the arc stability when welding with ac. When the current switches back and forth between DCEN and DCEP, there is a spot at the zero point of the cycle where the arc actually extinguishes. High-frequency stabilizer boxes keep the arc consistent during this part of the cycle.

Inverter machines are equipped with a touch start system that requires only the tungsten tip to be in contact with the workpiece for the arc to fire. Additional components may be added to the basic machine for high-

frequency arc starting. These machines are very small and are used in shops and industrial areas where high amperage or long duty cycles are not required.

Pilot Arc Start

A special pilot arc power source is used to create a very small arc between the electrode and an anode ring inside the insulated gas nozzle. The small arc heats the electrode just before the starting of the main arc, ensuring positive starting when the main arc is desired. This type of arc start is generally used in GTAW spot welding. **Figure 2-11** shows a pilot arc circuit.

Open-Circuit Voltage (OCV)

Manual GTAW power sources can produce a maximum of 80 open-circuit volts, according to NEMA specifications. Semiautomatic and automatic power sources may have a maximum open-circuit voltage (OCV) rating of 100 volts. These voltage ratings are required to protect a welder working on the secondary (output) side of the welding machine from the high primary (input) voltage.

The OCV readings are measured across the power source terminals (output electrode and ground) while the machine is turned on but not welding. As shown in **Figure 2-12**, machines equipped with voltmeters display OCV when the machine is operating without welding current being produced. When the arc is started, the voltmeter will register *actual* arc voltage, as shown in **Figure 2-13**. This is the voltage across the gap between the electrode and the workpiece.

Most power sources designed for ac/dc output operate near the maximum OCV allowed by NEMA. The high OCV is required for good arc characteristics during the reverse polarity part of the ac cycle. These

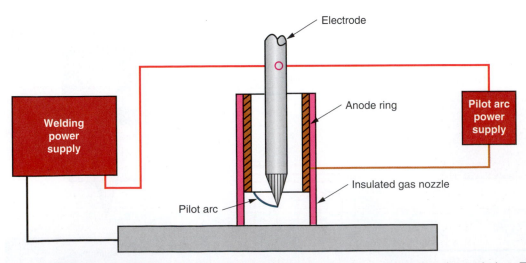

Figure 2-11. The pilot arc generator produces voltage for maintaining a pilot arc to keep the electrode hot. The main arc is very easy to start when the electrode is hot.

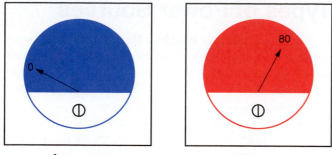

Figure 2-12. A welding machine operating without producing welding current. The voltmeter registers the maximum OCV allowed by NEMA for manual GTAW power sources.

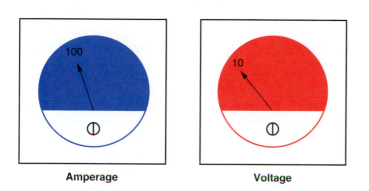

Power supply — On
Machine under load
Amperage registers
Actual arc volts registers

Figure 2-13. A welding machine operating under load. The voltmeter registers the actual arc voltage in the gap between the electrode and the work.

machines have no external method for changing the OCV. The maximum OCV for a welding machine is shown on the machine's data label.

Rectifier welding power sources do not have the problems associated with ac current, since they produce only direct current. However, some allow the welder to adjust the OCV. By lowering the OCV, the welder gains additional current control.

Duty Cycle

Welding machines are rated for the amount of current and length of time they can operate without damage to the power source. This rating is called the machine's *duty cycle*. Welding machines should never be operated beyond these limits.

The duty cycle of a power source determines the maximum length of time the machine can be continuously operated at a given amperage. Every type of

power source has an established duty cycle. Operation within these ratings will greatly extend the life of the equipment. The basic design of the power source determines the duty cycle. Design features that affect the machine's duty cycle include the size and quality of the internal components, size of the wiring, amount of insulation, and effectiveness of the cooling system.

Power sources are manufactured with duty cycles ranging from 20% to 100%. Each 10 percentage points represents one minute of operation in a 10-minute period. **Figure 2-14** shows the cycle and time period for various types of GTAW power sources. For example, a 300-ampere-rated power source with a 60% duty cycle could be used at 300 A for a six-minute period. The welder must then allow four minutes for the machine to cool. If the machine's duty cycle is exceeded, the machine will overheat. Overheating will result in an unstable arc and premature equipment failure.

Some power sources have a thermal (heat) overload protection thermostat to guard against overheating. The thermostat will automatically shut down the welding circuit until the machine cools. Machines that do not have thermal overload protection may overheat, causing damage to the unit. If you are unsure whether a machine is equipped with thermal overload protection, you should assume that it is *not*.

Volt-Ampere Curve

The volt-ampere curve, which shows the relationship between arc voltage and current, is determined and provided by the machine manufacturer. **Figure 2-15** shows a typical volt-ampere curve for a constant-current power source. The point at the very top of the arc represents that the machine produces approximately 80 open-circuit volts.

The current for a given arc voltage can be determined by drawing a horizontal line from the arc voltage to the curve and then reading the amperage value at that intersection. The arc voltage is determined

Duty Cycle	Number of Minutes Machine Can Be Operated at Rated Load in a 10-Minute Period
100%	Full Time
60%	6
50%	5
40%	4
30%	3
20%	2

Figure 2-14. Power source duty cycles limit the number of minutes the unit may be operated at the rated load.

by the arc length. If the welder maintains a constant arc length, the welding current will remain the same. For this reason, power sources used for GTAW are called *constant-current power sources*. If the welder raises or lowers the torch height, a new voltage intersection line is obtained. This movement of the torch raises or lowers the current output by the power source. Notice the curve is not straight, but drooping. Because of this characteristic, these machines are also called *droopers*.

As shown in **Figure 2-15**, the welder has an adjustment range of about 11 amperes of welding current that can be controlled by adjusting the distance between the torch and the workpiece. This allows the welder to vary the welding current by changing the arc gap. Welders can raise and lower the arc as needed to bridge gaps, achieve deeper weld penetration, and adjust welding speed.

GTAW power sources that have a selection of current ranges, or steps, change the volt-ampere curve at each setting. This change makes the volt-ampere curve more flat. The flatter the curve, the more the amperage will change as the welder varies the arc length.

To determine the actual variance in the amperage change, consult the machine instruction book. In many instances, the manufacturer will have the volt-ampere curve determined by factory test and illustrated in the book.

Because the welding arc voltage can be adjusted by changing the arc gap during the welding operation, a GTAW power source is also termed a *variable-voltage power source*.

Types of Power Sources

Technology has enabled the creation of power sources that are smaller and more powerful, with arc adjusting capabilities that were unavailable in older machines. Power sources may be alternating-current, direct-current, ac/dc, or inverter types.

Welding facilities that do not require welding machines with high amperage ranges or duty cycles often use small welding power sources. Such a unit is shown in **Figure 2-16**. These units will only weld lighter gauges of aluminum and steel. If the machine is overheated, it will automatically shut down, preventing damage to the equipment. These units usually have a maximum current of approximately 150 A with a 20% duty cycle.

Alternating-Current Power Sources

Alternating current power sources are normally single-phase transformers that use alternating current from the incoming (primary) power line. High-voltage and low-amperage current is then changed (transformed) into low open-circuit voltage and variable-amperage current for welding power. When used for ac GTAW, these machines require high frequency for arc starting and solenoids for gas control. Welding amperage is controlled by a panel rheostat. The unit shown in **Figure 2-17A** is a basic transformer power source.

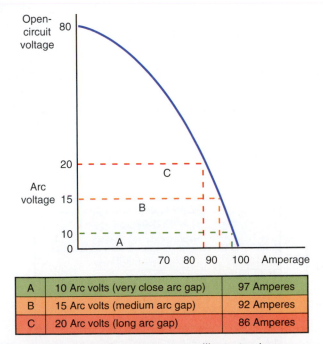

A	10 Arc volts (very close arc gap)	97 Amperes
B	15 Arc volts (medium arc gap)	92 Amperes
C	20 Arc volts (long arc gap)	86 Amperes

Figure 2-15. The volt-ampere curve illustrates how a welder can select output amperage by adjusting the arc voltage (gap). The manufacturer develops the volt-ampere curve for each machine.

Figure 2-16. A typical ac/dc power source that can be used for small welding jobs. (Miller Electric Mfg. Co.)

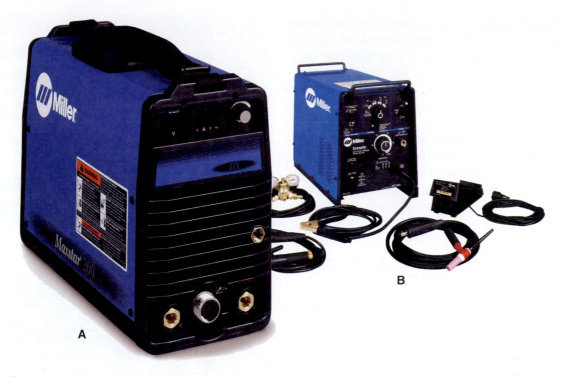

Figure 2-17. AC power sources. An ac/dc power source allows a welder to select the type of current required for the weld from a single machine. (Miller Electric Mfg. Co.)

With the development of ac/dc power sources in a single unit, like the one shown in **Figure 2-17B**, the ac-only units are being used only in isolated cases. Rather than investing in a new ac/dc welding machine, some shops may use cheaper ac-only machines for welding of aluminum or magnesium.

Direct-Current Power Sources

Direct-current rectifiers are machines that use a transformer to convert high-voltage single- or three-phase primary power to low-voltage ac power, which is then rectified to produce direct-current welding power, as shown in **Figure 2-18**. The polarity for GTAW will be DCEN for steels, stainless steels, and

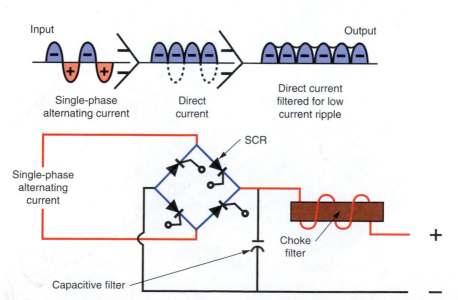

Figure 2-18. Silicon-controlled rectifiers (SCRs) convert the positive current part of the ac cycle into a negative current part of the cycle. This operation is called *rectification*. After being rectified, the current flows only in one direction (direct current). The filters smooth the ripple to make the current flow evenly.

most other metals. Due to its lack of penetration characteristics, DCEP is used in very limited and specific areas such as die rebuilding, where penetration is not important.

The machine shown in **Figure 2-19** also provides direct current power. Because this machine is used for other welding processes, the controls must always be set before welding, with the amount of welding power controlled by a rheostat on the machine face. These types of machines are often employed where multiple processes are used. They require additional auxiliary equipment for high-frequency arc starting and shielding gas control.

Alternating-Current/Direct-Current Power Sources

The *alternating current/direct current* type of power source (commonly called an ac/dc welding machine) is very useful because of the dual current selection available from a single machine. An ac/dc welding machine produces alternating current or direct current with either straight polarity or reverse polarity. A machine of this type is shown in **Figure 2-20**. This

Figure 2-19. This DC power source can be used for GTAW or SMAW. (Miller Electric Mfg. Co.)

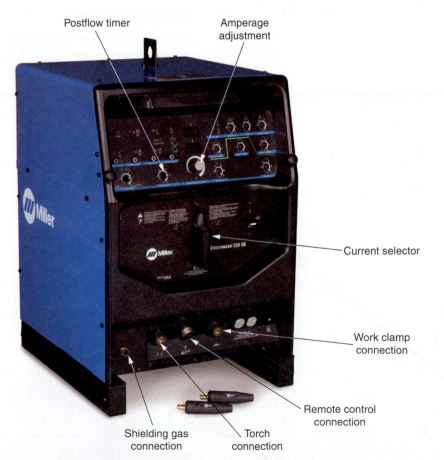

Figure 2-20. This ac/dc welding machine provides the welder with the choice of alternating current or direct current (straight or reverse polarity). (Miller Electric Mfg. Co.)

machine can be used for various thicknesses of material and different types of metal. It is capable of remote control of the entire welding operation. Note the connections on the bottom front of the machine for the GTAW torch, shielding gas, remote control, and the work clamp. An adjustable postflow timer switch is provided to shield the electrode with argon gas after the welding stops. These machines are manufactured for various amperage capacities. They are usually rated at a 60% duty cycle.

Another version of the ac/dc power source, shown in **Figure 2-21**, allows for control of the ac waveform during welding. This control is called *Syncrowave®, balanced wave,* and *square wave* by different manufacturers. It is used to change the pattern of the alternating current output with additional electrical components. The output may be balanced wave (normal) or unbalanced wave. The sine wave can be adjusted to provide either greater penetration or more cleaning action. This adjustment is called an unbalanced wave.

The waveform control is not used to select the amount of the output current, but rather the percentage of time spent in each polarity. Rheostats or programs selected from the machine panel give the operator complete control of the waveform desired. Machine programs are stored in the computer of the

machine and are accessed from the machine's control panel. The machine shown in **Figure 2-22** has independent current control for both the electrode positive and electrode negative parts of the cycle, controls for adjusting the frequency of EP to EN alternations, and balance control to regulate the time in each polarity. **Figure 2-23** defines each control's uses in the welding cycle.

Inverter Power Sources

Inverter power sources for welding are actually several converters in one machine. They convert the high-voltage (120 V or 240 V), low frequency (60 Hz) ac power supplied to the machine into lower voltage, higher-frequency ac power. The machine then uses that intermediate, high-frequency ac power to produce either alternating or direct current for welding. Inverter power sources operate at frequencies from several thousand cycles per second (kilohertz or kHz) to 100 kHz. Because they operate at higher frequencies, the inverter power sources perform better than the older, transformer-type machines. Transformer size and weight are reduced considerably, response time improves, and noise is reduced. A block diagram

Figure 2-21. The current waveform can be controlled by the welder using this ac/dc power source. (Miller Electric Mfg. Co.)

Figure 2-22. Complete control of each cycle of welding current can be programmed into the machine prior to starting the welding operation. (Miller Electric Mfg. Co.)

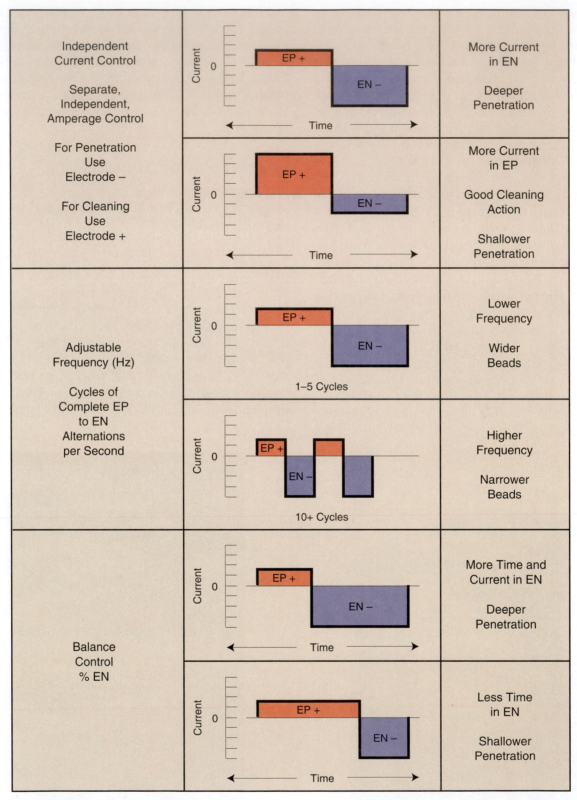

Figure 2-23. Adjustment of each control separately, or in any combination, can be made to shape the desired weld bead. (Miller Electric Mfg. Co.)

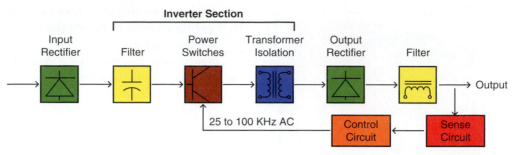

Figure 2-24. A block diagram of an inverter power source.

of a dc inverter operation is shown in **Figure 2-24**. A machine of this type is shown in **Figure 2-25**.

An additional benefit derived from the use of inverters is that incoming power type and quality is less critical than when using older-type power sources. Either 50 Hz or 60 Hz power can be used. The power source output is very smooth, which improves the quality of the final weld. Some inverter machines produce only DCEN and DCEP welding current. Other inverter machines produce both alternating and direct current. A dc-only machine is much less expensive than one that produces both ac and dc, and may be a cost-effective choice if you are welding only steels. An auxiliary high-frequency arc starter can be used in conjunction with inverter power sources.

Inverter technology has made the physical size of these welders very small with adequate power. The power sources are lightweight, portable, and use a variety of types of input power. See **Figure 2-26**.

Welding power sources are often made for use with several welding processes. The machine shown in **Figure 2-27** can be used for GTAW, SMAW, and other processes using different controls supplied with the machine or added as accessories. Controls that do not have any function with the GTAW process are not in the system when the machine is set up for GTAW.

Multi-Operator Welding Systems and Controls

Multi-operator welding systems are used in schools, in nuclear plants, on construction sites, and

Figure 2-25. The fiberglass case on this dc inverter makes it light enough to be carried. Note the handles on top of the machine. (Miller Electric Mfg. Co.)

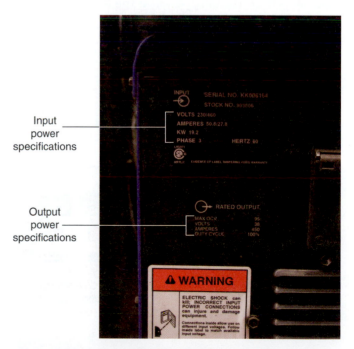

Figure 2-26. This constant-current dc inverter arc welder can operate on either 230 V or 460 V input power. Note that the duty cycle, amperage, and maximum OCV are listed in the lower left corner of the front panel. (Mark Prosser)

Figure 2-27. This ac/dc inverter power source has a waveform control and the versatility to be used for both GTAW and SMAW. (Photo used with the permission of Lincoln Electric, Cleveland, OH)

at other locations where large numbers of welders are trained or employed. Two types of multi-operator welding systems are currently in use—master power sources and rack systems.

A master power source consists of a single, centralized power source that supplies welding power to multiple welders. See **Figure 2-28**. Cables carry the welding power to control units at the individual work stations. The control units allow each welder to adjust the welding current to match the job being performed at that work station.

If anything happens to a master power source, any welders supplied welding current from the power source are unable to work. This is one of the reasons master power sources are being replaced by rack systems with a separate power source for each welder. Rack systems are much lighter, more portable, more versatile, and more dependable than master power sources. A rack system is a centralized system that has an individual power source for each welder, as shown in **Figure 2-29**. Notice the hooks on the frame that make it easy to move around a construction site. Rack systems have a heavy duty framework for durability. The main advantages of rack systems over master power sources are lower operating cost, portability, and lower maintenance cost. Another advantage is the ability to remove individual machines from the system for repair without affecting the remaining power sources.

Multi-operator welding systems were originally designed for shielded metal arc welding (SMAW), which required starting the arc by scratching the electrode to the work. Stopping the arc was done by withdrawing the electrode from the workpiece. Multi-operator welding systems set up for gas tungsten arc welding must include additional equipment. The additional equipment includes GTAW torches and solenoid valves in the gas lines to control the flow of inert gas to individual welders. The additional

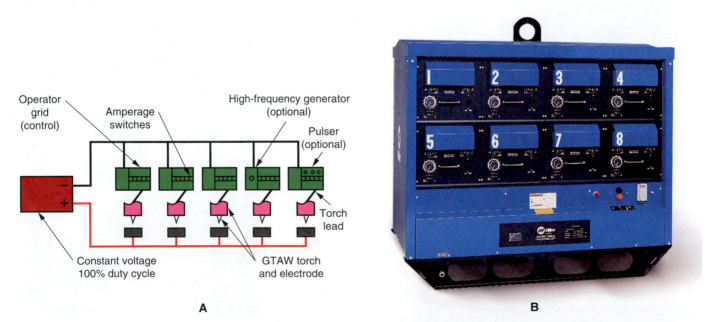

A

B

Figure 2-28. Multi-operator power sources allow the use of several separate grids with only one master power source. A—Diagram of a multi-operator system. B—This power source provides welding current to eight individual grids. Since the grids will be operating at less than full power, the power source does not need to provide the total ampere capacity of the collective grids. (Miller Electric Mfg. Co.)

Hooks

Figure 2-29. Current is supplied to eight individual welders by this rack system. (Miller Electric Mfg. Co.)

Figure 2-30. An engine driven ac/dc power source is useful when utility hookup is not an option. (Photo used with the permission of Lincoln Electric, Cleveland, OH)

equipment may also include a water cooling system, a high frequency generator for arc starting, slopers for starting and ending the welding current, and pulsers to control weld penetration. This equipment will be described in greater detail later in the book.

Portable Engine-Driven Welding Generators

Engine-driven welding generators are often required for welding in the field when utility power is not available. These machines are designed for GTAW and other welding processes. Engine-driven welding generators also provide power for the various power tools that may be required for the welding project. The units provide ac and dc welding current for GTAW. Machines like the one shown in **Figure 2-30** can be moved around the work site in the back of a truck, on a trailer, or by crane. These types of machines are typically capable of generating several hundred amps of welding current with a 100% duty cycle.

Power Source Controls

Controls for a power source range from very simple to very complex. The simplest control is a tap connection for changing the amount of welding current. One of the most complex is a series of controls that fully sequence the welding current throughout the entire operation.

Industrial-rated power sources generally have two methods of controlling the arc welding current. The first is a machine-mounted rheostat, which is preset to the desired amperage prior to starting the weld. This type of amperage control is usually associated with the scratch starting technique.

The second type of amperage control is a remote hand- or foot-operated rheostat. This type of control is adjusted by the welder during the welding operation to set the desired current level. These controls are very useful to the welder because the current may be started and ended at a very low level. This reduces the possibility of melt-through and craters at the start and end of the weld. For these reasons, remote rheostats are often preferred over machine-mounted rheostats. A remote amperage foot control used with a power source is shown in **Figure 2-31**.

The number of built-in controls on a power source determines the degree of the welder's involvement in the welding process. When using a machine with the minimum number of controls, the welder must perform some procedures manually, such as controlling the gas flow, controlling the water flow, and scratch-starting the arc. A machine with a wide range of built-in controls, such as the one shown in **Figure 2-32**, may require the welder only to set up

Figure 2-31. The welder can use a foot pedal to control amperage. (Mark Prosser)

Figure 2-32. This is a manual ac/dc inverter power source. The ac output is a square wave with a wave balance control. (Photo used with the permission of Lincoln Electric, Cleveland, OH)

and initiate the weld. The machine would then carry out the welding process according to preprogrammed settings.

Amperage Controls

Industrial-rated welding machines control the amperage output in steps, or ranges. Three or four steps for control of welding current may be provided. Industrial-rated welding machines deliver precise amperage control with minimal drift, or variation, during the welding operation.

One of the two types of current control (machine or remote) is selected depending on the mode in which the machine is to be used. Machine control can be used if amperage variation is not required during the welding process. For example, machine control is suitable when many parts are welded at a set amperage. On the other hand, if a welder is not working near the power source but must make adjustments during the operation, a remote circuit must be included for current control.

Amperage can be controlled by changing a rheostat connected to a dial on the machine front, the foot pedal, or the dial on the torch if you are using a scratch start. Amperage can also be controlled remotely by an electronic circuit. Electronic circuits are used to operate the main contactor, thus allowing control of the amperage by remote hand or foot controls.

Remote electronic controls are used to start the arc at very low levels of current, to raise the amperage to the desired setting, and to reduce the current level to a low setting at the end of the weld. Amperage controls are also used by the welder to increase current for deeper penetration, to reduce current when dealing with poor fit-ups, or when bridging gaps. The amperage controls also help the welder reduce melt-through on thin materials. In short, they give the welder the control needed to perform the intricate welds associated with GTAW welding.

Auxiliary Controls

Auxiliary controls are pieces of equipment that can be built into or added onto a power source to give the welder greater control of the welding process than is possible with the amperage control alone. Examples of auxiliary controls include pulsers, slopers, and timers. Many power sources do not include these units built-in, but have the required circuitry included so the auxiliary controls can be added at a later time.

Today, many of the auxiliary controls are solid-state modules. These auxiliary controls plug into a port or connect to a wiring stripboard within the machine. In either case, installation is completed using the manufacturer's instructions. **Figure 2-33** shows a power source with an auxiliary pulse control.

Figure 2-33. Auxiliary control modules have been added to this power source for pulse weld control. (Mark Prosser)

Power Source Selection

Selecting a power source requires research. You must determine exactly what you need in a power source to perform the type of welds desired. Determine the material thickness range you will mostly use to determine the amount of power you will need.

Major points to consider are the ability of the power source to produce the necessary current for the desired welding processes, machine current loads (amperage ranges), duty cycles, and cost. Choose the power source that best fits your application and budget.

When selecting a power source, consider the auxiliary controls you are likely to need. Take into account the controls built into the power source, as well as the cost and ease with which auxiliary controls can be added to the machine to meet present and future needs.

Comparison of various power source types can be made using the following factors:

- Current type
- Amperage range
- Duty cycle
- Overload-prevention features
- Open-circuit voltage
- Amperage controls
- Auxiliary controls
- Cost
- Method of installation

Power Source Installation

Power sources should be installed in an area free of dust, dirt, and fumes. Also, the area must allow machine heat to dissipate (escape). Dirt and heat will cause a power source to produce an output below its rated load. Continued exposure to dirt and heat will ruin the machine. The area selected for installation must be free of objects blocking the flow of air into and out of the machine. The machine should not be exposed to moisture, which may cause electronic components to fail.

Power sources that plug into shop power are designed to operate on 110 V, 220 V, and/or 440 V ac power, either single- or three-phased. These power sources operate on 60 Hz power unless otherwise specified. Machines that operate on a power frequency other than 60 Hz can be special-ordered from the factory for use outside North America.

The required fuse size for the incoming power is always shown on the power source's data label. The fuse panels should be close to the power source for

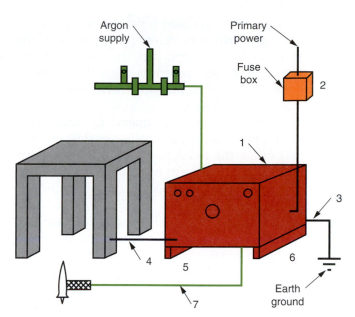

1. Locate power supply away from walls for proper airflow.
2. Locate primary fuse box near power supply.
3. Connect power supply frame to earth ground.
4. Connect work ground with 2/0 min–4/0 max cable.
5. Keep power supply dry.
6. Keep area near power supply clean.
7. Connect torch cable and shielding gas hose.

Figure 2-34. A typical power source installation includes a fuse box mounted near the welder. An earth ground from the power source protects the welder from primary line high voltage.

safety reasons. A typical power source installation is shown in **Figure 2-34**. Machines that do not match the utility power voltage can be used if a step-up or step-down transformer is installed between the power main and welding machine. The manufacturer's instruction manual should always be consulted when the utility voltage does not match the machine requirement. Using a power source voltage other than the one stated on the power source data label may nullify the manufacturer's warranty.

Primary power supplied to homes for appliances and other uses is not suitable for powering most welding machines. Some light duty welding machines operate on 110 V power, but most welding machines require 220 V or higher. Check the machine manufacturer's recommendations for power requirements.

Power Source Maintenance

Given reasonable care and routine maintenance, a welding machine will operate satisfactorily for many hours before repairs are required. Modern methods of insulating transformers, the use of solid-state components, and good design have extended the service life of power sources. A qualified repair person is always required to perform repairs on power sources and auxiliary controls. However, periodic inspection and maintenance of the equipment is often the responsibility of the operator. The manufacturer's instructions should always be followed.

Most manufacturers recommend blowing out the power source periodically. This removes dust buildup that could cause the machine to overheat. Use only dry, filtered, compressed air; nitrogen gas; or a can of compressed gas specifically designed for electronics to clean these machines.

All terminals on the machine and the torch should be checked for damage and looseness. Loose connections can cause overheating and electrical problems with the arc. Check a power source's high-frequency points for excessive pitting and proper gap every six months or when operation is faulty. Replace pitted points, if required. Regap points according to the manufacturer's instructions. A typical high-frequency point assembly is shown in **Figure 2-35A**. Too small a gap will cause improper high-frequency operation. Too large a gap will cause high-frequency radiation loss, which results in radio and television interference. Proper adjustment can be accomplished by following the steps shown in **Figure 2-35B**. It is usually not necessary to lubricate the fan and motor bearings on most power sources because they are sealed units. However, the bearings in certain power sources may

require periodic lubrication. Always follow the manufacturer's maintenance recommendations.

Check the mechanical linkages and switches for freedom of movement. Mechanical joints can be lightly greased, if required. The terminal blocks for cable connections should be tight and clean. Corroded cable connections will increase resistance to current flow, which can result in a drop in welding current, overheating in the connection, and possibly a fire. The connections can be cleaned by brushing or wiping with a cloth. Check motor generator brushes and replace them when they are worn beyond the manufacturer's tolerance. Worn brushes cause rapid armature wear. Worn armatures must be sent to a qualified shop for reconditioning, resulting in the power source being out of service for an extended period.

Lubrication, inspection, and adjustment of portable gasoline and diesel power sources require close attention to the manufacturer's instructions. This applies to both the engine and the power source. Because these units usually operate under very adverse conditions, improper maintenance will drastically reduce the machine's effectiveness and service life. Always use authorized factory replacement parts. Use extreme caution when working with primary power. Never perform maintenance or repairs on a power source that is operating. Be careful when fueling a gasoline or diesel engine. Spilled fuel will create a fire hazard.

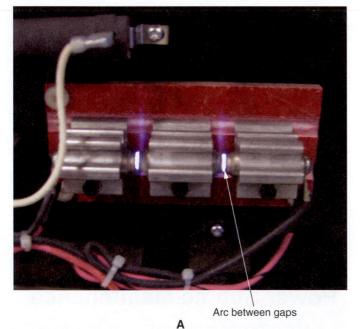

A

Arc between gaps

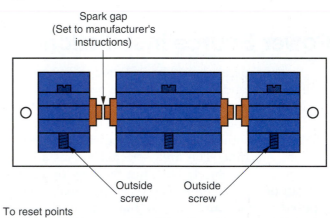

Spark gap
(Set to manufacturer's
instructions)

Outside screw   Outside screw

To reset points
1. Loosen outside screws only. (Do not loosen center screw).
2. Insert proper feeler gauge between contact points.
3. Press point to feeler gauge.
4. Tighten set screw.
5. Check other set screw and reset as required.

B

Figure 2-35. Primary power connections are made to boards within the power source. A—A high-frequency spark is generated in the unit shown. Note the blue arc. (Mark Prosser) B—A typical procedure for setting point gap. The proper gap setting on high-frequency points should always be maintained.

Summary

GTAW power sources convert incoming power to produce a variable low voltage and a constant electrical current to the welding arc. The GTAW process uses three different types of current—direct current electrode negative (DCEN), direct current electrode positive (DCEP), and alternating current (ac).

Three commonly used methods for starting the arc are scratch or tap start, high-frequency start, and pilot arc start. The method used depends on the type of power source used and how the machine is set up.

GTAW power sources produce open-circuit voltage (voltage produced when the power switch is turned on, but before the arc is struck) and actual arc voltage (voltage across the gap between the electrode and the workpiece). The duty cycle of a power source determines the length of time the machine can be continuously operated at a given amperage. The volt-ampere curve, provided by the manufacturer, shows the relationship between arc voltage and current.

Some power sources allow the welder to switch between ac output and dc output. These machines are commonly referred to as ac/dc welding machines, provide dual current selection from a single machine. Inverter power sources convert incoming utility power alternating current into high-frequency voltage, which is then used to produce either alternating or direct welding current.

There are two types of multi-operator welding systems—master power sources and rack systems. A master power source is a single, centralized power supply that provides welding power to multiple welders. A rack system is a centralized unit that contains an individual power source for each welder.

Engine-driven welding generators are used in the field when utility power is not available. Because they often operate in extreme conditions, proper maintenance is critical.

Industrial-rated power sources control the arc welding current with either a machine-mounted rheostat or a remote hand- or foot-operated rheostat. Auxiliary controls can be added at a later time.

Welding power sources vary in size, complexity, and capabilities. The ways in which the machine can control and manipulate current and the resulting arc are very important to understand when selecting the power source. When selecting a power source, make sure to research the type of machine needed to perform the type of welds desired. Determine the material thickness range you will mostly use to determine the amount of power you will require.

Care must be taken to install power sources in an area that is free from dust, dirt, fumes, and moisture. The area must provide enough air space around the machine to effectively dissipate heat. Routine maintenance, according to the manufacturer's instructions, is required to keep welding machines in good operating condition.

Review Questions

Write your answers on a separate sheet of paper. Do not write in this book.

1. Which specification defines the performance characteristics and rating procedures for ac/dc arc-welding equipment?
2. The GTAW power source provides electrical current, which is also called _____.
3. Primary power is _____-voltage, _____-amperage alternating current.
4. What type of voltage do welding machines produce after they are started and before the arc is made?
5. When welding using DCEN, the current jumps the gap between the negatively charged _____ and the positively charged _____.
6. What does the welding machine contactor do?
7. Alternating current contains a half cycle of _____ polarity and a half cycle of _____ polarity.
8. In the _____ portion of an alternating current cycle, the welding current flows from the electrode to the workpiece.
9. When using DCEN, how much of the heat from the arc is concentrated onto the workpiece?
10. Shallow penetration welds are made with _____ current.
11. Steels, stainless steels, and titanium are typically welded using _____ current.
12. What does a high-frequency generator do? Why is it required when using alternating current?
13. The _____ start method of starting an arc must be used with care to avoid contaminating the electrode or weld area.
14. What is the machine duty cycle?
15. What design features affect duty cycle ratings for power sources?
16. Power sources are manufactured with duty cycles ranging from _____% to 100%.
17. How is the volt-ampere curve used?

18. The gap between the electrode and the work-piece is called _____.
19. The welder can change the _____ being delivered to the workpiece by changing the arc length.
20. What is a multi-operator welding system?
21. Name two types of amperage control.
22. According to NEMA specifications, manual GTAW power sources can produce a maximum of _____ open-circuit volts.
23. What is the typical duty cycle of an ac/dc welding machine?
24. What factors influence where power sources should be installed?
25. What should be used to blow out welding power source equipment?

Chapter

Auxiliary Equipment and Systems

Objectives

After completing this chapter, you will be able to:

- ❏ Explain the function of and proper procedure for installing auxiliary high-frequency generators.
- ❏ Recall the various methods used to control GTAW current, including pulsers, slope controllers, timers, and remote controls.
- ❏ Differentiate between the different types of GTAW torches.
- ❏ Recall the purpose of a gas lens.
- ❏ Recall the tasks involved in proper torch maintenance.
- ❏ Recall the steps needed for correct installation of torches.
- ❏ Identify the various types of tungsten electrodes.
- ❏ Recall the finishes, color codes, and sizes applicable to tungsten electrodes.
- ❏ Use proper methods to prepare electrodes.

Key Terms

air-cooled torches
arc wandering
ceriated tungsten
collet
collet body
downslope timer
emissivity
end caps
final current
gas-cooled torches
gas lens
gas nozzles
initial current
initial time period
laminar flow
lanthanated tungsten
main welding current timer
postflow timers
pulser
pure tungsten
remote arc start switch
slope controllers
spitting
tapering down time
thoriated tungsten
torches
upslope timer
water-cooled torches
whiskers
zirconiated tungsten

Introduction

Various types of equipment are used in GTAW to assist in the arc starting process and improve control of the welding current. These systems also help to maintain the arc gap, pulse the welding current, time the weld cycle, and program the weld operation. High-frequency generators, pulsers, slope controllers, timers, and remote controls are available options on new machines, or can be purchased as auxiliary systems to add to an older machine. Welding torches, electrodes, and some special welding systems are also termed auxiliary equipment, even though they attach to the basic power source.

Auxiliary High-Frequency Generators

An auxiliary high-frequency generator is designed for use on power sources that do not have a high-frequency generator built in. Some auxiliary high-frequency generator units have only high-frequency voltage controls. See **Figure 3-1**. Others could have a combination of a high-frequency voltage control and preflow and postflow gas controls.

When the high-frequency generator is in the *Off* position, it is not operating. It will not interfere with the normal operation of the power source and can remain connected to the system. When the high-frequency generator is in the *On* or *Continuous* position, it will provide continuous operation for welding with alternating current. When the high-frequency generator is in the *Start Only* position, it is used to start the arc only when direct current is being used.

Postflow timers allow the shielding gas to flow for an adjustable period of time after the arc is broken.

Figure 3-1. This auxiliary high-frequency generator has a high-frequency intensity control. Start or continuous options can be selected, depending on the application. (Mark Prosser)

This gas flow shields the electrode and the weld from air and contamination. The postflow time is variable, depending on the size of the electrode.

The *remote arc start switch* is attached to the torch or foot control. When the welder presses the remote arc start switch, the control sequence starts. Once the arc is established, the switch can be released, and the arc will continue until it is withdrawn from the work.

Radiated Interference

The spark gap oscillator in a high-frequency generator is similar to a radio transmitter. Improper installation can result in radio and TV interference or problems with nearby electronic equipment.

Radiated interference can develop in four different ways. Interference can be caused by radiation from the high-frequency generator itself, from the welding leads, or from feedback into the power lines. Interference can also be caused by reradiation of "pick-up" by metallic objects. The radiated interference from the high-frequency generator induces currents in nearby metallic objects, which then generate their own interference. Keeping these four contributing factors in mind and installing equipment according to the following instructions can greatly reduce the possibility of interference.

Installation

Auxiliary high-frequency generators must be installed according to the manufacturer's instructions to ensure radiation protection and operator safety. Failure to observe the following recommended installation procedures can result in radio or TV interference problems. Unsatisfactory welding performance can also result due to lost high-frequency power.

Try to keep the work lead and electrode lead as short as possible and as close together as possible. Usual lengths should not exceed 25′ (7.5 m). Tape the leads together whenever practical. Be sure the rubber coverings of torch and work cables are free of cuts and cracks that allow high-frequency leaking. Cables with high natural rubber content resist high-frequency leakage better than neoprene and other synthetic rubber insulated cables.

Keep the torch in good repair. Make sure all connections are tight to retard high-frequency leakage. The work terminal must be connected to a solid ground at the welding power source or to a water pipe that enters the ground within 10″ (3 m) of the welding machine. Ground the connection using cable of the same size as, or larger than, the work cable. Grounding to the building frame or a long pipe system can result in reradiation, effectively making these members radiating antennas. A solid ground is one that is driven into the ground, or earth.

When the high-frequency generator is in operation, keep all power source and generator panels and covers securely fastened in place to minimize radiated interference. All electrical conductors within 50′ (15 m) of the welder should be enclosed in grounded rigid metallic conduit or equivalent shielding. Flexible helically wrapped conduit is generally not suitable.

If the welding operation takes place in a metallic building, several good electrical grounds should be driven into the earth and connected to the metallic walls all around the building. When high frequency is being used, the spark intensity should be kept at the lowest setting that produces acceptable results. This will minimize the level of radiated interference.

Pulsers

A *pulser* continuously switches between a low and a high amperage level to reduce the overall heat input used to make a weld. Each time the output current switches from the low amperage level to the high amperage level and back to the low level is a pulse. The number of pulses per second (frequency) and the duration of each pulse are set by the welder. Increasing the duration of a pulse increases the amount of heat added to the weld. In addition to control of overall heat input, GTAW pulsing aids in penetration control, distortion control, and melt-through control when thin materials are being welded.

Pulsing systems are used with DCEN and alternating currents. They can be installed on power sources as a plug-in module or as an auxiliary control. When a pulsing system is used with manual welding power sources, the maximum current level desired is set using the amperage control on the power source. The low level current (background current) is then set on the pulser. The welder uses a remote foot control with a power source upslope control to start the sequence and a downslope control to end the operation without craters. Upslope and downslope controls will be discussed in the next section.

The basic controls for regulation of a pulsing sequence are shown in **Figure 3-2**. *Peak Amperage*, which is the high-level amperage, is set a little higher than a normal GTAW amperage setting. *Background Amperage*, which is the low-level amperage, is set at a value lower than the peak amperage.

Pulses per Second refers to the number of times per second the welding current achieves peak amperage,

Figure 3-2. Each of the required controls for starting, sequencing, and ending the operation are located on this panel. (Mark Prosser)

and the *Percent On* time is the pulse peak duration as a percentage of the time for one cycle. The *Percent On* time controls the length of time the peak amperage level is maintained before it drops to the background amperage level.

For manual GTAW, .2–10 pulses per second are used. Higher pulse rates are used for automatic welding. In some cases, a welding job will require a pulse-type weld, but the available power source does not have a pulse controller installed. Manufacturers have designed pulsers that can be added to nearly any GTAW power source. An accessory pulser is shown in **Figure 3-3**.

One way a pulser affects output is by establishing a background current at a fixed amperage, as shown in **Figure 3-4**. The peak current can vary, but the background current will not. This setting is normally used

Figure 3-3. Pulsers of this type control all of the pulsing operations for both ac and dc welding. (Mark Prosser)

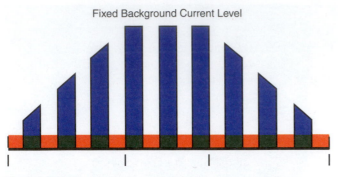

Fixed Background Current Level

Figure 3-4. Upper limit control by the operator allows changing maximum amperage levels to compensate for thickness changes and gaps. The background current level remains constant.

for manual welding on all types of metals. On older equipment, the pulsation is controlled or adjusted with a dial. On newer machines, these settings are displayed on a digital readout and are controlled by a computer program. The operator can adjust the settings in the program by using buttons or dials on the control panel. These programs can be updated by the manufacturer as new programs are developed.

A different program, which varies the background current at a preset ratio to changes in the peak current level, can be run on the pulser. See **Figure 3-5**. This type of program is normally used for automatic welding and to produce rapid pulsing to agitate molten metal for outgassing (removing gases).

Pulsers with pulse frequency control are used in many applications. Frequency control gives the operator the ability to pulse the welding current at a range of frequencies, from a low number of pulses per second to a very high number of pulses per second. This type of pulser generally has additional controls, such those

that enable pulsing of the welding wire. Pulsing the welding wire can be an advantage on different types of joint setups, especially open root butt joints. Pulsing can help with heat control and outgassing.

Pulsers with pulse frequency control also have sequencers for programming the sequence of various functions performed during the welding operation. These sequenced functions can include switching the pulsed current during the welding process. For example, a sequencing program can instruct the machine to weld the root pass with high-frequency pulsing, automatically reduce the pulsation frequency for the next two weld passes, and then increase the pulse frequency again when necessary. Pulsers with sequencing capabilities use solid-state circuitry, operate on 60 Hz 110 V ac, and are relatively trouble free. They are very effective for creating high-quality, repetitive welds.

Slope Controllers

Slope controllers are used in both manual and automatic GTAW. They control the rise and fall of the welding current at the start and end of the welding operation. These devices might be called *slopers*, *programmers*, *electroslopes*, or *sequencers*. A typical sloper/sequencer is shown in **Figure 3-6**. These units are designed to perform the following functions:

- control the amount of initial current (and possibly the *initial time period*)
- control the upslope time (initial current to welding current)
- control the length of welding current time (optional)

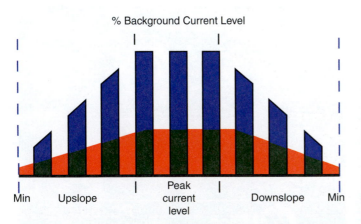

% Background Current Level

Min Upslope Peak current level Downslope Min

Figure 3-5. Hydrogen gas has the opportunity to escape from the molten metal when the background current changes during the pulse agitation. This reduces the risk of porosity.

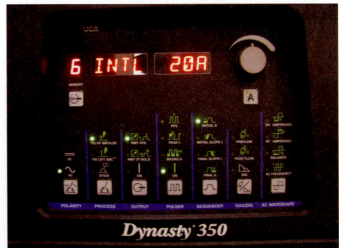

Figure 3-6. This sequencer (programmer) gives the operator complete control over the entire welding operation. (Mark Prosser)

- control the downslope time (welding current to final current)
- control the final current for a prescribed period of time

The *initial current* and the initial time period are generally set at levels that prevent melt-through on thin materials. The initial time period also gives the welder time to position and adjust the welding torch before full current is applied. Welding wire, if required, can be brought into proper position at this time.

The *upslope timer* determines the amount of time the power source takes to transition from producing initial current to producing welding current. During this time period, the welder may obtain penetration through the base material. Welding wire, if required, is added when the upslope times out. The *main welding current timer* is set for the length of time welding current is desired.

The *downslope timer* determines the length of time the power source takes to transition from producing welding current to producing final current. During this period of time, called *tapering down time*, the addition of filler metal is gradually reduced. *Final current* is set at a low amperage to prevent craters at the end of the weld.

Timers

Timers used in sequencing circuits can be grouped into three categories: electrical-mechanical, electronic-potentiometer, and electronic-digital.

Electrical-Mechanical Timers

Electrical-mechanical timers are electrically operated after the desired time period is set mechanically. When the circuit is activated, the timer will automatically reset to the time period desired and operate for the set length of time. Timers of this type are generally used to control the preflow and postflow of gas and water.

Electronic-Potentiometer Timers

Commonly called *pots*, these timers are used to control all types of welding operation times. Time periods are set for seconds, minutes, or cycles. These timers are easy to set and easy to adjust through the entire range. When replacing these timers, be sure to use an exact replacement, as each type of potentiometer has a separate range of operation within the circuit.

The knobs on a potentiometer can loosen, and calibration or original settings can be lost. If the knob on the potentiometer slips, the value indicated by the dial will not match the actual values achieved during operation. This will make it nearly impossible to properly set the equipment's controls. In such cases, the potentiometer must be recalibrated.

Calibrate the potentiometer in this manner:
1. With the knob removed, turn the potentiometer shaft as far to the left as possible.
2. Align the knob so that its indicator mark aligns with the first mark on the scale on the case. (Note: In some cases, the scale may be on the knob and the indicator mark may be on the case.) Install the knob while holding the potentiometer shaft to ensure it does not move.
3. Tighten the knob-holding screw.
4. If the knob comes loose again at a later date, repeat the previous steps to recalibrate the potentiometer.

This procedure will recalibrate the potentiometer. However, if welding machine with the slipping potentiometer knob is the same machine used to develop the welding procedure, the calibration may have been off when the welding procedure was created. If that is the case, it will be necessary to create a new welding procedure to account for properly calibrated potentiometer settings.

Electronic-Digital Timers

Electronic-digital timers are the most recent type developed and have several advantages over the older types of timers:
- Time is kept more accurately.
- The readings can be seen as individual numbers.
- No calibration is required.
- The numbers can be reset by pressing the push button controls until the proper number is showing.

Remote Controls

Control of welding sequences and welding current is often accomplished by remote controls, such as foot- or hand-operated switches and potentiometers. Foot controls and hand controls are designed for specific machines and, in most cases, are not usable with other types and models.

There are different types of foot controls. The most basic type includes the foot pedal with only a microswitch. Another type consists of a foot pedal and a potentiometer. The third type combines the foot pedal, a microswitch, and a potentiometer. Each type has slightly different capabilities.

A foot control with *just a microswitch* has three basic purposes: to start the high frequency, start the cover gas and coolant flow, and energize only the welding current contactor. A foot switch equipped with *just a potentiometer* is used to control only welding current within the range established on the power source. A foot switch equipped with a microswitch and potentiometer is used to start high frequency, start gas flow (and in some cases, water flow), and energize the welding current contactor. The potentiometer is then used to increase current (upslope) to welding current level and decrease current (downslope) to final current level. A basic foot control is shown in **Figure 3-7**.

Hand controls operate in the same manner as foot controls. They can include a microswitch, potentiometer, or both. Mounted on the torch for accessibility, they can be located according to the user's preference and are usually taped or bolted on. Some require constant pressure on the switch for continuous operation. Others can be released during welding, and then reoperated to energize the postflow operation. **Figure 3-8** shows a torch-mounted control with a microswitch to turn on the cover gas and a dial to adjust the current.

A welder normally uses a foot control when working at a table or in an area where the welder can sit down. Since one foot is required to operate the unit, out-of-position and long welds become very difficult as the welder tires from standing on one foot. For these types of welds, a hand-operated control allows

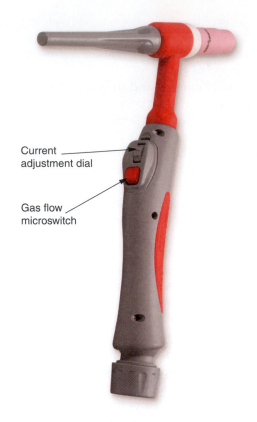

Current adjustment dial

Gas flow microswitch

Figure 3-8. Welding current is controlled by rotating the dial on the torch handle. (Weldcraft Co.)

the welder more freedom of movement and still offers complete control of the welding operation.

GTAW Torches

Electrode holders are commonly called *torches*. They are available in many styles and types for varied GTAW welding tasks. Various types of GTAW torches are illustrated in **Figure 3-9**.

A GTAW torch has several basic functions:
- grips and holds the electrode
- provides electrical current to the electrode
- conducts heat away from the electrode
- provides inert gas to the weld area
- protects the welder from heat and electrical shock

The electrical components in GTAW torches are made from high-conductivity copper or brass. The internal connections are silver brazed. The outer insulation covering is either phenolic rubber or silicone rubber. Torches are designed for either manual or machine welding, and can be gas- or water-cooled.

Figure 3-7. Depressing this foot control energizes all of the welding sequence circuits. The operator raises or lowers the pedal for more or less welding current. Releasing the foot control energizes the afterflow circuits. (Mark Prosser)

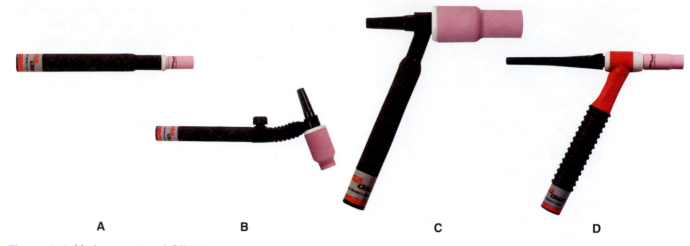

A **B** **C** **D**

Figure 3-9. Various types of GTAW torches. A—A pencil-type, gas-cooled manual torch is used for light duty work. B—A flexible head, gas-cooled, manual torch is useful in limited-access situations. C—This liquid-cooled torch has a large cup for increased gas coverage. D—Cooling water circulates through passages in an angle head water-cooled torch to remove heat. As a result, the duty cycle is greatly increased over gas-cooled torches. (Weldcraft Co.)

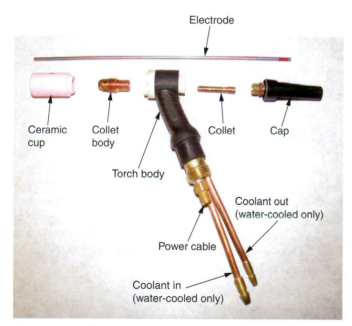

Electrode

Ceramic cup   Collet body   Collet   Cap

Torch body

Coolant out (water-cooled only)

Power cable

Coolant in (water-cooled only)

Figure 3-10. Manual torch components. (Mark Prosser)

Figure 3-10 shows the main components of a GTAW torch:
- torch body
- collet body
- collet
- cap or cover for electrode
- ceramic cup

All GTAW torches are rated for the maximum amperage the torch can be continuously operated at without overheating. Manufacturers have designed accessories such as valves, nozzles, and adapter accessories for each type of torch. Accessories from one manufacturer might not fit another manufacturer's torch.

Gas-Cooled Torches

Gas-cooled torches (commonly called *air-cooled torches*) are cooled by the flow of the inert shielding gas through a passage in the combination cable (cable that carries the gas flow and the electrical current). The gas carries away heat generated in the power cable. These types of torches are generally used for light duty and field work, where cooling water is not available.

Shut-off valves are required to be turned on before striking an arc to avoid electrode contamination and must be shut off after welding to avoid wasting shielding gas.

Water-Cooled Torches

Water-cooled torches are cooled by the passage of water through the torch head. The water then exits the system through the power cable. **Figure 3-11** shows a water-cooled torch with connections for water inlet, water outlet, and power cable with gas inlet.

Depending on the manufacturer's design, torches may have O-rings to seal the electrode caps. Nylon or Teflon® insulators are used to insulate the nozzle end of the torch body, as shown in **Figure 3-12**.

The *collet* and *collet body* are sets of electrical contacts that are designed to grip and provide current to one diameter of electrode and cannot be used with any other size, **Figure 3-13A**. Using the wrong size collet and collet body can result in damage to the torch, and the electrode will not be properly gripped. Install and tighten collet bodies by hand, or use a special chuck wrench. Failure to properly tighten the collet body damages the end cap threads when

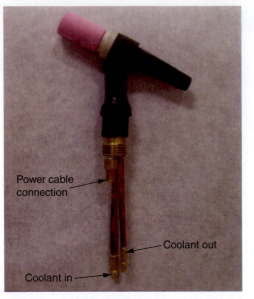

Figure 3-11. Connections for gas, water out, and power are silver brazed into the manual torch body. (Mark Prosser)

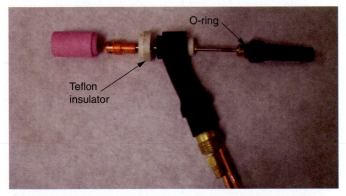

Figure 3-12. Nylon or Teflon® insulators shield the high-frequency spark. The O-ring seal is used to prevent gas leakage. (Mark Prosser)

attempting to grip the electrode. A single-piece collet/collet body is available, **Figure 3-13B**. According to the manufacturer, the new design makes changeover from one size electrode to another simpler and faster, while reducing spare part inventories.

End caps are designed in varying lengths to match standard electrode lengths and adapt the torch for welding in small areas. Screwing the end cap into the torch head forces the collet into the collet body. This action positions the electrode and causes the collet to grip it. *Never* overtighten the end cap. If the end cap is overtightened, the brass threads of the torch can easily strip, ruining the torch. If the electrode does not stay in position, the cap threads might be stripped, or the size of the collet might be wrong. The collet must be the same size as the electrode. The end of the collet may be squished or damaged, making it unable to properly grab the electrode. Other possible problems are damaged threads where the collet body screws into the torch, or a loose collet body. After checking all of these components for proper function and replacing any damaged components, you should not have a problem with electrode adjustment.

Gas nozzles are usually designed for installation into a particular type of torch and generally do not adapt to another make or model. Gas nozzles come in many sizes, shapes, and materials. Typical nozzle configurations are shown in **Figure 3-14**.

The appropriate gas nozzle shape depends on the type of base metal and the type of joint. For example, more gas flow is required on stainless steel, so a larger nozzle is needed. It is better to use the next larger size nozzle than to turn the working pressure up. A long, narrow nozzle is useful for a deep weld joint. Always check the manufacturer's recommendations when choosing a gas nozzle.

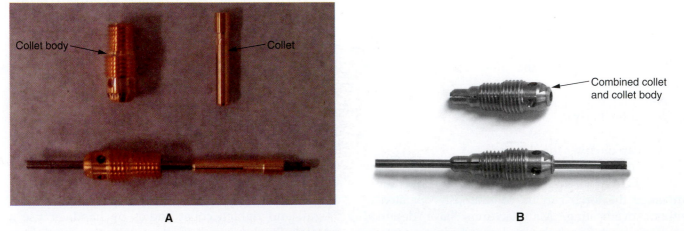

Figure 3-13. Collet types. A—A traditional two-piece unit consists of a collet that fits into a collet body, which then screws into the torch. (Mark Prosser) B—A single-piece unit combines the collet and collet body. Either type requires a different size collet for each electrode diameter. (Weldcraft, Inc.)

Figure 3-14. A variety of nozzle shapes and sizes are shown here.

Gas nozzles are relatively inexpensive and should simply be replaced when they become unusable. Discard a nozzle that has chips, cracks, or a metal buildup on the outlet end. These defects can alter the gas flow pattern from the nozzle, resulting in contamination of the weld metal.

Nozzles are identified by the size of the orifice (opening) and the length of the nozzle, as shown in **Figure 3-15.** Each torch manufacturer assigns part numbers to the various nozzles for individual types of torches, and these numbers must be used when ordering replacement nozzles.

Gas Lens

A *gas lens* allows the welder to use a longer electrode extension than would be used with a standard nozzle. A gas lens is a special collet body with a series of stainless steel wire mesh screens to make

Size (diameter)		No.
3/16	=	No. 3
4/16	=	No. 4
5/16	=	No. 5
6/16	=	No. 6
7/16	=	No. 7
8/16	=	No. 8
9/16	=	No. 9
10/16	=	No. 10
11/16	=	No. 11
12/16	=	No. 12
Lengths		
Short Regular Long Extra long Special		

Measure number of 1/16th inches

Figure 3-15. Standard nozzle diameters and lengths.

the column of shielding gas flowing through the collet body and out of the nozzle less turbulent and more streamlined. A gas flow with these characteristics is referred to as a *laminar flow.* Because a laminar flow is more concentrated than a turbulent flow, it aids in maintaining a blanket of inert gas around the electrode and over the weld area. This is very helpful when wind or drafts are present, or when the electrode must be extended due to the location of the weld area. **Figure 3-16A** shows how the column of gas is formed when a regular collet body is used. The column of gas is somewhat erratic and cone-shaped. **Figure 3-16B** shows how the gas column is formed when a gas lens is used. The column is uniform, cylindrical, and provides heavier gas coverage than would be possible with a regular collet body.

Cables, Hoses, Connectors, and Valves

Power cables for GTAW torches are usually made of copper wire enclosed in a vinyl tube. Either gas or water is used to carry away the heat generated in the torch and the power cable. There are three tubes in the handle of the torch—one provides cover gas, one is the coolant feed (in) line, and one is a coolant return (out) line. See **Figure 3-17.** Special connectors are used at each end of the cable assembly for the gas or water to flow

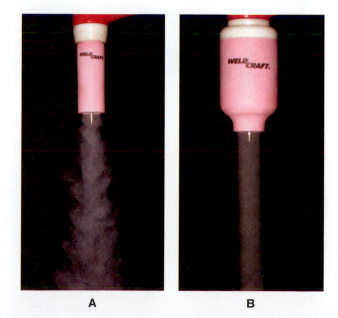

A **B**

Figure 3-16. A gas lens can improve gas flow. A—Gas flow through a normal collet body. Note the inconsistency in density and the conical shape of the gas flow. B—When a gas lens is used, the gas flow is cylindrical and consistent. (Weldcraft Co.)

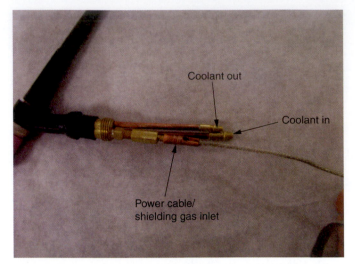

Figure 3-17. The power cable for the torch is contained inside a vinyl tube. Gas or water passes around the cable for cooling. (Mark Prosser)

through. **Figure 3-18** shows various types of connectors for attachment to the power source. Standard cable lengths are 12-1/2′ (3.8 m) and 25′ (7.5 m). Couplings can be used to increase the length beyond these dimensions.

Some torches use a separate power cable and do not use coolant to carry away heat, as shown in **Figure 3-19**. With this type of assembly, repairs to the torch cable are easily made and are generally less expensive. However, the cable assembly is heavier due to the added weight of the electrical cable and insulation.

Manual shielding gas shut-off valves can be mounted on the torch body or on the inlet gas hose. These are generally used in field type work or where timers, circuits, or solenoids are not included in the system. Several types of gas shut-off valves are in use. These include a threaded screw type that installs in the torch body, a push-pull type that installs in the torch body, and a push-pull type that installs in the shielding gas line. See **Figure 3-20**. Remember, these

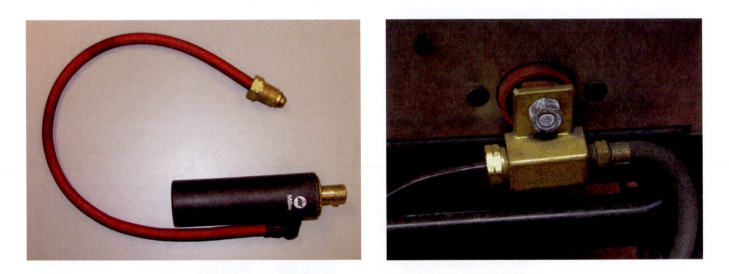

Figure 3-18. A variety of power cable adapters are used for connecting the cable to water or gas and welding machine terminals. (Mark Prosser)

Combined shielding gas/power
cable connection

Shielding gas
connection

Power cable
connection

Figure 3-19. External-type power cables do not require cooling. Note that the shielding gas line combines with the power cable at the base of the torch fitting. (Mark Prosser)

shut-off valves operate manually. If the valve is left open after the operation is completed, the shielding gas will continue to flow as long as gas is available from the main supply.

Torch Maintenance

Daily maintenance of the welding torch and torch components can greatly reduce potential problems, such as contaminated welds. Proper maintenance will save money by reducing shielding gas leakage and can even increase welder safety. Proper maintenance requires you to perform the following tasks:

- Inspect the welding outfit.
- Replace burned or cracked insulators.
- Replace burned or cracked O-rings.
- Test all water and gas lines daily for leaks.
- Repair or replace leaky tubing.
- Keep all connections tight on the torch.
- Keep coolant water clean.
- Use recommended coolant water pressure levels.

Torch Installation

For a torch to operate properly at its rated capacity, it must be installed correctly. Always follow the manufacturer's instructions for installation. If you do not understand the instructions, stop the installation and contact the manufacturer or representative for information. Various torch installations are shown in **Figures 3-21** through **3-23**. Note that the torch in **Figure 3-23** is attached with a separate gas hose and power cable. As you attach equipment, keep in mind that gas fittings use right-hand threads on the nuts and adapters. Also, *never* use vinyl tubing that has contained water for argon gas lines.

Water-Cooled Torches

Some GTAW power sources are equipped with built-in water coolers. See **Figure 3-24**. These machines are used in conjunction with water-cooled torches to provide extended duty cycles for heavy welding operations. Some water-cooling systems, like the one shown in **Figure 3-24**, pump coolant from a reservoir to the

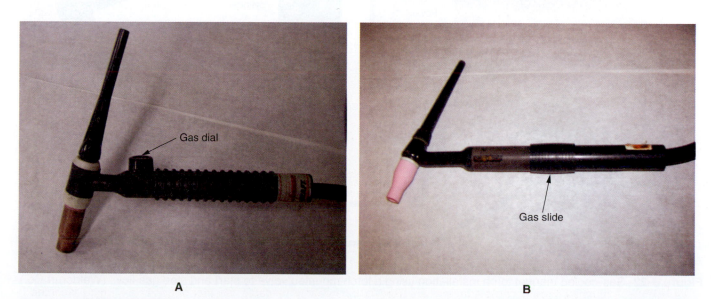

Gas dial

Gas slide

A

B

Figure 3-20. Gas shut off valve configurations. A—Gas dial on the torch body. B—Gas slide on the torch body. (Mark Prosser)

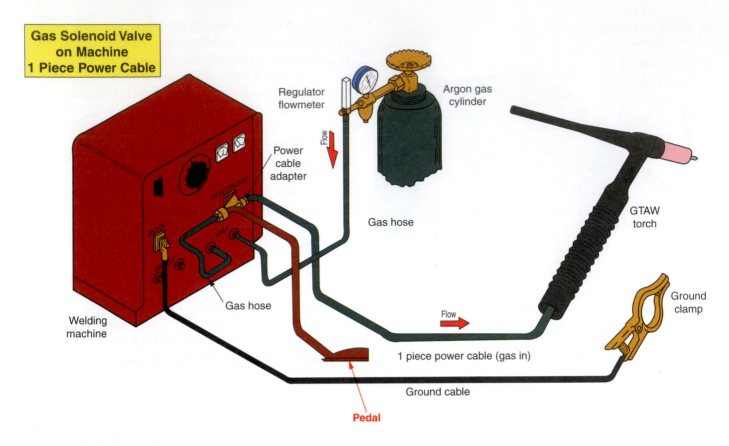

Figure 3-21. A gas-cooled manual torch installation requires a switch on either the torch or foot control to activate and deactivate the gas solenoid, which controls gas flow. (Weldcraft Co.)

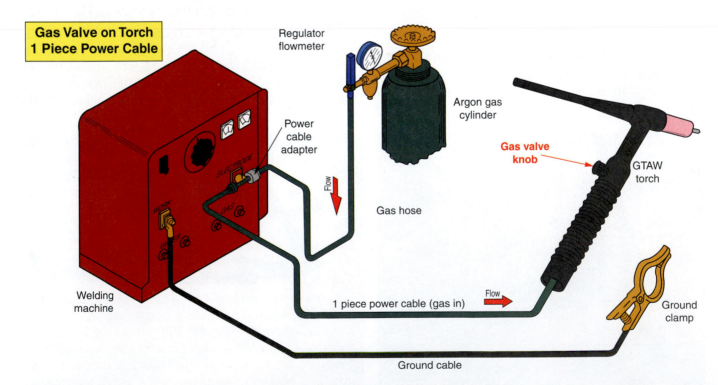

Figure 3-22. Gas-cooled manual torch installation using a torch mounted valve to start and stop gas flow. (Weldcraft Co.)

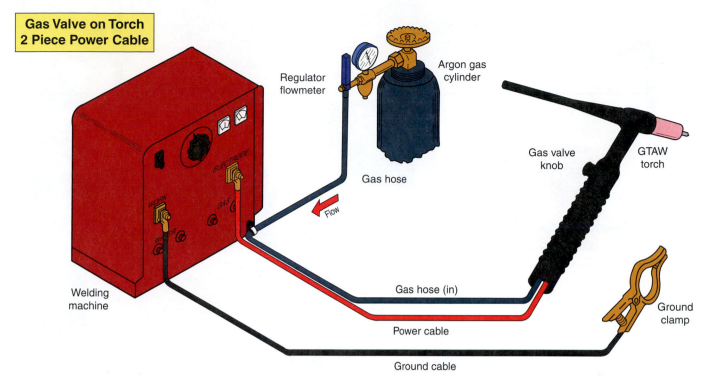

**Gas Valve on Torch
2 Piece Power Cable**

Regulator flowmeter

Argon gas cylinder

Gas valve knob

GTAW torch

Gas hose

Flow

Gas hose (in)

Ground clamp

Welding machine

Power cable

Ground cable

Figure 3-23. A manual torch installation with a separate external power cable and gas hose. (Weldcraft Co.)

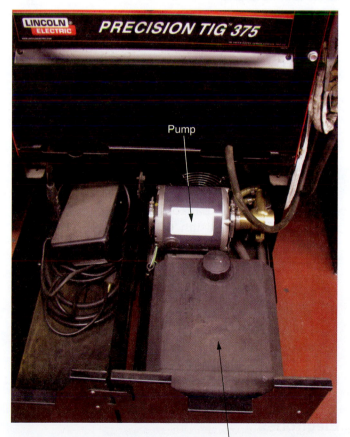

Pump

Coolant reservoir

Figure 3-24. Cooling systems are an integrated part of some machines. Note the coolant reservoir and pump on this Lincoln machine. (Mark Prosser)

torch, where the coolant picks up heat. From the torch, the warm coolant flows to a cooler, where it is cooled and then recirculated to the reservoir, **Figure 3-25**.

Auxiliary torch coolers can be added to a power source for heavy duty welding, **Figure 3-26**. **Figure 3-27** shows a typical auxiliary cooler installation.

Always use a filter on all water lines into the torch to reduce the possibility of contaminating the coolant. These filters are usually located inside the power source and need to be changed on a routine maintenance schedule. Use a regulator and pressure gauge to determine pressure in the torch. *Never* exceed the manufacturer's recommendations for pressures. When using a water cooler, check the flow rate at the water return line outlet. Use a rust inhibitor in water-cooled systems. Always use special left-hand nuts and adapters for water connections, and always connect water lines to the torch so the water will enter the torch and exit the torch properly.

Torch Cable Cover Jacket

After the torch installation is complete and the system is checked for proper operation, a torch cable cover jacket can be installed to protect the vinyl tubing from heat, abrasion, and puncture from sharp objects. These covers are available in leather and heavy vinyl and are made in standard torch cable lengths. A typical cover jacket is shown in **Figure 3-28**.

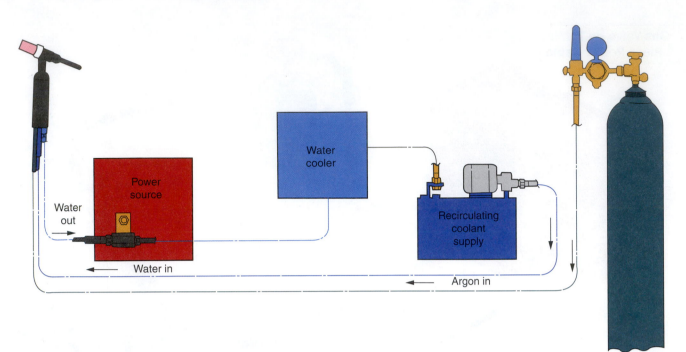

Figure 3-25. Recirculated water system with a water cooler installed in the system. (CK Worldwide, Inc.)

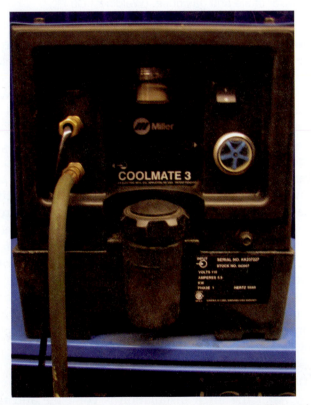

Figure 3-26. Small coolers of this type are very efficient in lowering water coolant temperature. (Mark Prosser)

Electrodes

The electrodes used in GTAW are manufactured to meet the requirements of American Welding Society Specification A5.12M/A5.12. There are several different types of electrodes used for different applications. The chemical requirements for various types of electrodes are shown in **Figure 3-29**.

Electrode Types and Classifications

Pure Tungsten (EWP) Electrodes

Pure tungsten electrodes achieve good arc stability and a clean balled end for ac welding, with reasonably good resistance to contamination. Pure tungsten has the lowest resistance to heat of all the tungsten electrodes and therefore is not recommended for use with inverter-type power sources.

Alloys have been developed that have the favorable characteristics of pure tungsten but are more durable. These alloys also have much higher current-carrying capacities than pure tungsten. The trend in industry is to replace pure tungsten with alloys, even for welding with ac current on aluminum or magnesium.

Thoriated Tungsten (EWth-1 and EWth-2) Electrodes

In *thoriated tungsten* (EWth) electrodes, thorium is added to tungsten to improve the arc-starting characteristics. EWth-1 electrodes have 1% thorium added and EWth-2 electrodes have 2% thorium added. Thoriated

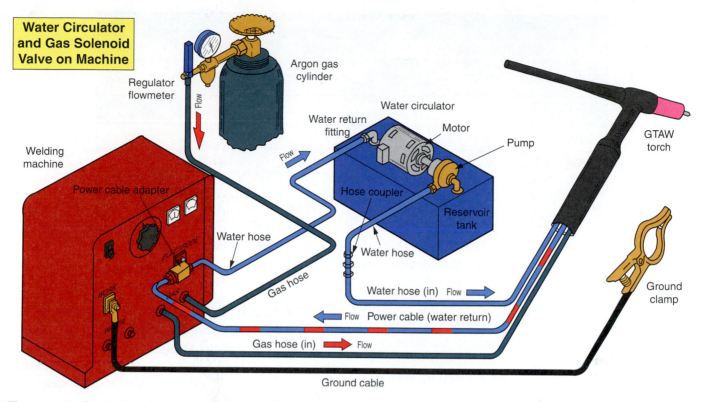

Figure 3-27. Recirculated water-cooled manual torch installation. (Weldcraft Co.)

tungsten electrodes are good for wider current ranges because their tips do not become molten as easily. If used properly, thoriated tungsten electrodes have less of a tendency to stick, or freeze, to the work and have good resistance to contamination with either ac or dc current.

Both 1% and 2% thoriated tungsten electrodes have a long life due to the thorium content. These electrodes start an arc easily and have high current capacity and increased resistance to weld pool contamination. Overheating thoriated tungsten electrodes can cause spitting

Figure 3-28. This light tan cable cover jacket protects the torch cables from damage. (Mark Prosser)

of vaporized droplets of tungsten into the weld, which can make a weld rejectable. Thoriated tungsten produces radioactive dust during the grinding process.

Zirconiated Tungsten (EWZr) Electrodes

Zirconiated tungsten (EWZr) electrodes contain 1/4% to 1/2% zirconium. EWZr electrodes combine the desirable characteristics of a pure tungsten electrode with the capability and starting characteristics of a thoriated tungsten electrode. An EWZr electrode is normally used for ac welding, when tungsten inclusions are not tolerated. Zirconiated tungsten electrodes are preferred for welding aluminum and magnesium.

Ceriated Tungsten (EWCe-2) Electrodes

Ceriated tungsten (EWCe-2) electrodes contain 2% cerium. EWCe-2 electrodes possess very good starting characteristics, have excellent performance in low amperage ranges, and can be used continuously for extended periods with either ac or dc current. This type of electrode is very well suited for thin sheet metal and delicate precision work.

Lanthanated Tungsten (EWLa-1, EWLa-1.5, and EWLa-2) Electrodes

Lanthanated tungsten (EWLa) electrodes contain 1%, 1.5%, or 2% lanthanum, which is a non-radioactive, "rare earth" material. The numbers at the end of

Chemical Requirements					
AWS Classification	Tungsten, Minimum Percent (by difference)	Thorium Percent	Zirconium Percent	Total Other Elements (maximum percent)	Color
EWP	99.50	—	—	0.5	Green
EWTh-1	98.50	0.8 to 1.2	—	0.5	Yellow
EWTh-2	97.50	1.7 to 2.2	—	0.5	Red
EWLa-1	—	—	—	1.0	Black
EWZr	99.20	—	0.15 to 0.40	0.5	Brown
EWCe-2	97.30	—	—	0.5	Orange

Figure 3-29. Tungsten electrode chemical requirements and color codes.

the designations indicate the percentage of lanthanum added to the tungsten electrode. EWLa electrodes are good general shop electrodes that have the same operating characteristics as ceriated tungsten electrodes. Lanthanated tungsten maintains a sharp tip very well and does not ball on ac current.

Electrode Finishes

Tungsten electrodes are manufactured with surfaces free of impurities, undesirable films, foreign inclusions, pipes, seams, or slivers. This level of manufacturing quality is necessary to ensure the correct operation of welding equipment with no adverse effect on the weld deposit. Tungsten electrodes are available in a clean finish (chemically cleaned to remove surface impurities) or ground finish (ground to a uniform shape with a polished surface). The ground finish electrode is more expensive.

Electrode Color Codes

Each electrode type is identified by a different color code that is painted on one end of the electrode. Some manufacturers use this same color code to identify the electrode type on the electrode's package. Electrode color codes are shown in **Figure 3-29**.

Electrode Sizes

Electrodes are made in 3-1/2″ (89 mm), 7″ (178 mm), and longer lengths. Electrodes are also available in diameters that range from .010″ (.3 mm) to .25″ (6.5 mm). It is normal practice to order the 7″ (178 mm) length and use a long cap on the torch for the tungsten cover. The tungsten shortens when it is ground or tapered. Electrode sizes and amperage ranges for the various diameters are shown in **Figure 3-30**.

Electrode Preparation

Always handle electrodes carefully to prevent breakage. Wear clean gloves to avoid contaminating the electrode with oil or grease. If contamination occurs, remove the oil or grease with a degreaser before using the electrode.

The electrode's point can be deteriorated by excessive currents, metal contamination, splintering or cracking of the electrode during preparation, or by using continuous high frequency. Always remove all contamination from the electrode when grinding new tapers, and inspect the point for cracks and splinters.

An arc can be hard to start if a pure tungsten electrode is being used with DCEN polarity. This is due to the low electron emissivity of pure tungsten. (*Emissivity* is the relative power of a surface to emit heat by radiating electrons.) A scratch start is usually required to start the arc. If the scratch start blunts the tip, then a thoriated electrode must be used. The arc can also be started by holding the electrode against the work. If continuous high frequency is available and not detrimental to the weld operation, it can be used as an arc starter.

Tapering and Balling

For use with direct current, a tungsten electrode is normally ground to a taper point, which helps to direct and control the arc stream during welding. When an electrode has a taper point, a lower amperage rating must be used to compensate for the metal removed to make the taper. If the current used is too high, the thinner metal at the tip could overheat.

Electrodes can be tapered using an electrode grinding machine (**Figure 3-31**), sanding belts, or a chemical powder. Since tungsten is very brittle, extreme care must be taken when grinding a new taper, or the

Typical Current Ranges for Tungsten Electrodes*						
Electrode Diameter (inches)	Straight Polarity Direct Current, amps	Reverse Polarity Direct Current, amps	High-Frequency Unbalanced Wave AC, amps		High-Frequency Balanced Wave AC, amps	
	EWP EWTh-1 EWTh-2	EWP EWTh-1 EWTh-2	EWP	EWTh-1 EWTh-2 EWZr	EWP	EWTh-1 EWTh-2 EWZr
0.010	up to 15	**	up to 15	up to 15	up to 15	up to 15
0.020	5–20	**	5–15	5–20	10–20	5–20
0.040	15–80	**	10–60	15–80	20–30	20–60
1/16	70–150	10–20	50–100	70–150	30–80	60–120
3/32	150–250	15–30	100–160	140–235	60–130	100–180
1/8	250–400	25–40	150–210	225–325	100–180	160–250
5/32	400–500	40–55	200–275	300–400	160–240	200–320
3/16	500–750	55–80	250–350	400–500	190–300	290–390
1/4	750–1000	80–125	325–450	500–630	250–400	340–525

* All values are based on the use of argon as the shielding gas. Other current values may be used depending on the shielding gas, type of equipment, and application.
** These particular combinations are not commonly used.

Figure 3-30. Current ranges for tungsten electrodes.

electrode can break. Use a fine grit wheel or belt, and use only moderate pressure on the tungsten point. Always rotate the electrode evenly during manual grinding to maintain the tip in the centerline of the electrode.

To prevent loss of identification of the tungsten type, do *not* grind the painted end. Once the painted end is ground and the paint removed, identification of the tungsten type is very difficult.

Shop belt sanders, **Figure 3-32**, are sometimes used to reestablish the point on an electrode. The electrode is held at a shallow angle to the belt to set the angle of the point. The point is centered in the electrode by rotating the electrode in your fingers while sanding.

If a belt sander is used to sharpen the electrode, the grinding marks must run with the length of the tungsten after the point is established. See **Figure 3-33**. The electrode is turned to align the grinding marks with the length of tungsten. The belt on the sander must have a very fine grit, or the tungsten will break during the tapering operation.

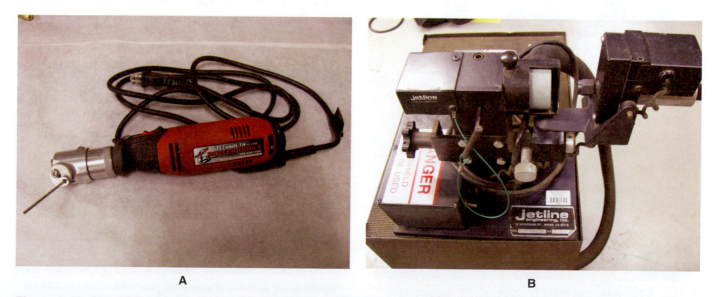

A

B

Figure 3-31. Special tungsten grinding machines prepare precision ground electrodes. Wear safety glasses and a full face shield when grinding electrodes. A—Handheld tungsten grinder. B—Bench-type tungsten grinder. (Mark Prosser)

Figure 3-32. The electrode is held against the belt at the desired angle and rotated during sharpening to establish the point. Wear safety glasses and a full face shield while grinding electrodes. (Mark Prosser)

Figure 3-33. After the point is established, reposition the electrode so the grind marks will run parallel to the electrode length. (Mark Prosser)

The grinder or belt sander should be dedicated to tungsten grinding only. Grinding tungsten on the same wheel or belt used to grind other materials can cause tungsten contamination. Always wear a face shield when grinding or using a belt sander.

If a chemical powder is used to create the taper, as shown in **Figure 3-34**, the electrode must be red-hot when it is dipped into the powder. Chemical powders produce a very good taper angle. These powders are mainly used in the field when a grinding instrument is not available. The tungsten is heated to a red-hot condition and dipped into the chemical powder. The tungsten is pushed into the powder about 1/2″ (12.7 mm) and pulled back out. If the electrode is hot enough, it can be dipped several times to achieve the desired effect. The chemical powder causes a chemical reaction that removes layers of the electrode, creating a pointed end. Gloves and safety glasses should be worn when this operation is performed. The welder should always wear proper personal protective equipment when using any sharpening technique.

The method used to prepare the electrode is not as critical for manual welding applications as it is for automatic welding. In manual welding, the welder can make adjustments such as heat inputs, torch angle, and travel speed as required for the operation. In automatic welding, the welding parameters are locked in at the start of the operation, and are not generally adjustable. Therefore, electrode preparation is much more critical.

Figure 3-35 shows some common electrode taper angles. The 30° taper is most often used by manual welders; however, the point is the most fragile of all the angles, and resharpening is often required during the operation.

The electrode for ac welding can be balled by grinding a radius on the end of the electrode. Next, an ac arc is started on a test piece of steel or copper with a

Figure 3-34. Chemical powders produce sharp points on electrodes. The electrode must be dipped into the powder while red-hot. (Mark Prosser)

low current, and then the current is increased to form the desired ball diameter. Application of too much current during balling or during the welding operation will cause the ball to dance and possibly separate from the electrode. If the ball falls into the weld joint, it will contaminate the weld metal. If the ball dances during welding, change to a larger diameter electrode that will carry more amperage.

Arc Wandering

Arc wandering is the erratic movement of the arc around the end of the electrode. A wandering arc will not focus in one concentrated area. Arc wandering is usually caused by poor preparation of the electrode taper, or by an electrode that is bent or not centered in the gas nozzle. The taper should be carefully ground so the tip is in the centerline of the electrode. Improperly ground electrode tips are shown in **Figure 3-36**. To correct the condition, the bent portion of the electrode should be removed and the electrode reground. Otherwise, the electrode should be discarded.

Arc wandering is sometimes caused by an electrode with the grinding marks across the centerline, as shown in **Figure 3-36**. To prevent this condition, grind the electrode so the grinding marks are parallel to the centerline of the electrode, as shown in **Figure 3-35**.

Contamination

Contamination of the electrode can be caused by contact between the electrode and the weld metal or filler metal, insufficient shielding gas during or after the weld operation, or expelled metal from the molten weld pool attaching to the electrode tip. **Figure 3-37** shows examples of contaminated electrodes.

Contaminated electrodes introduce contamination into the molten weld metal, causing porosity and cracking. If contamination occurs, remove the electrode, break off the entire contaminated area, and

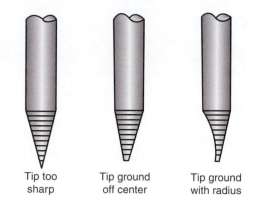

Tip too sharp | Tip ground off center | Tip ground with radius

Figure 3-36. Note the horizontal grinding marks and incorrect geometry of each tip. Incorrectly ground electrodes will cause many problems during welding.

retaper or ball the tip before reusing the electrode. Foreign material that is not removed from the electrode tip will burn away as the tungsten becomes hot. This residue will cause smoke fumes to further contaminate the molten metal.

Spitting occurs when the electrode slowly dissolves and begins to release droplets of tungsten into the weld. Excessive spitting from the electrode taper point is usually caused by electrode contamination. Spitting can result in significant weld defects. If excessive spitting occurs, remove the electrode from the torch and regrind it.

Tungsten *whiskers* forming near the electrode tip indicate a possible moisture contamination in the shielding gas. Tungsten whiskers are microscopic crystalline filaments that develop on the surface of tungsten when it is exposed to the right conditions, like moisture. Under magnification, the end of the electrode has small whiskers on the tip, which allow the arc to wander. Check the shielding gas supply, torch, and all connections for leaks of moisture or air. If whiskers form on the electrode, it must be removed and reground.

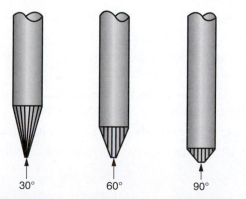

30° | 60° | 90°

Figure 3-35. Electrode tapers used in manual and automatic gas tungsten arc welding.

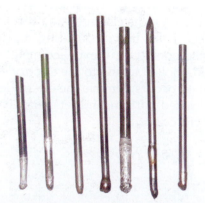

Figure 3-37. These electrodes have been contaminated by metal pickup or lack of inert gas shielding during the cooling period after the arc was extinguished. (Mark Prosser)

Summary

Some auxiliary high-frequency generator units have only high-frequency voltage controls, while others have a combination of a high-frequency spark control and preflow and postflow gas controls. Auxiliary high-frequency generators must be installed according to manufacturer's instructions in order to avoid radiated interference and poor welding performance due to loss of high-frequency power.

Auxiliary systems, such as pulsers, slope controllers, timers, and remote controls, assist in the arc starting process and improve control of the welding current. These systems also help to maintain the arc gap, pulse the welding current, time the weld cycle, and program the weld operation.

GTAW torches can be gas-cooled (air-cooled) or water-cooled. In a typical water-cooled torch, there are three tubes in the handle of the torch—one provides cover gas, one is the coolant feed line, and one is a coolant return line. Proper torch maintenance can reduce contaminated welds, reduce shielding gas leakage, and increase welder safety. Torches must be installed according to manufacturer's instructions in order to operate properly at their rated capacity.

GTAW electrodes are made of pure tungsten or tungsten alloys. Thorium, zirconium, cerium, or lanthanum is added to tungsten to create alloys that are more durable and have higher current-carrying capacities than pure tungsten. Each electrode type is identified by a different color code that is painted on one end of the tungsten.

An electrode grinding machine, sanding belts, or a chemical powder can be used to taper electrodes for dc welding. The tapering helps to direct and control the arc stream during welding. The electrode is balled for ac welding.

Contamination of the electrode can be caused by contact between the electrode and the weld metal or filler metal, lack of shielding gas during or after the weld operation, or expelled metal from the molten weld pool attaching to the electrode tip. If contamination occurs, remove the electrode, discard the contaminated portion, and retaper or ball the electrode before reusing it.

Review Questions

Write your answers on a separate sheet of paper. Do not write in this book.

1. What position should the high-frequency generator be set to when welding aluminum with alternating current?
2. What position should the high-frequency generator be set to when welding ferrous materials with direct current?
3. The postflow timer control on the high-frequency generator is adjusted for a period of time to prevent _____ of the electrode and the weld.
4. Improper installation of the high-frequency generator can result in _____ and _____ interference.
5. When installing an auxiliary high-frequency generator, keep the power source lines as _____ as possible.
6. What is the purpose of a pulser?
7. What does a slope controller do?
8. List the three basic types of timers used in sequencing circuits.)
9. Foot and hand controls contain a(n) _____, a(n) _____, or both.
10. List five main functions of a welding torch.
11. GTAW torches can be either _____-cooled or _____-cooled.
12. What is the purpose of a gas lens?
13. What are the two standard GTAW torch cable lengths?
14. When connecting an argon gas fitting onto a machine or torch, the threaded connection is always a(n) _____ hand thread.
15. List two advantages of using electrodes made of tungsten alloys rather than pure tungsten.
16. Which type of tungsten is radioactive?
17. Tungsten electrodes are available in what two finishes?
18. Which type of electrode has difficulty in starting the arc when DCEN is used? Why?
19. List three methods used to taper electrodes.
20. What causes excessive spitting of the electrode?

Chapter

GTAW Shielding Gases and Filler Metals

Objectives

After completing this chapter, you will be able to:

☐ Recall the various GTAW shielding gases and their properties.

☐ Recognize different types of shielding gas cylinders.

☐ Recognize proper procedures for storing gas cylinders.

☐ Explain how shielding gases are distributed to the welding area.

☐ Recall the equipment used to regulate shielding gas pressure in a GTAW system.

☐ Recall the safety guidelines relating to working with shielding gases.

☐ Recall the factors that influence the selection of filler metal.

☐ Explain the purpose of filler metal specifications.

☐ Distinguish between the various forms and sizes of filler metals.

☐ Summarize measures to be taken to avoid contamination of filler metals.

Key Terms

cast
Certificate of Conformance
certified chemical analysis report
filler metal
general use wire
helix
inert gases

Introduction

Shielding gases are used in gas tungsten arc welding to protect the electrode and the molten weld metal from atmospheric contamination. Shielding gases also help transfer the heat from the electrode to the metal. In addition, shielding gases can assist in arc starting and arc stabilization. Different shielding gases have very different effects on the welding arc and the weld itself and can aid in controlling the bead contour, the penetration, and the final appearance of the weld, especially on stainless steels.

Filler metal is additional metal added to build up the weld. Filler metal selection is a critical aspect of making a quality weld. Many different types and sizes of filler metals are available for the different types and thicknesses of materials commonly welded using GTAW. Filler metals must be compatible with the base metal to ensure proper metallurgical consistency.

Shielding Gases

The shielding gases used for GTAW include argon, helium, hydrogen, and nitrogen. Argon and helium are *inert gases* (stable gases that will not easily react with other materials). In their pure form, they are colorless, odorless, and tasteless. Inert gases will not contaminate the weld. The table in **Figure 4-1** lists commonly used shielding gases, their applications, and the advantages they provide.

Argon

Argon is the most commonly used shielding gas for GTAW. Argon is separated from the atmosphere during the production of oxygen and is readily available at a reasonable cost.

Of all the shielding gases, argon has the greatest tolerance for arc gap variation. Argon's lower arc voltage tolerance permits arc length gap variations with less influence on the arc power and bead shape. In addition, argon has better arc starting properties than helium. For welding with alternating current, argon is a better shielding gas than helium. It provides greater cleaning action and results in a higher quality weld with a better finished appearance.

Argon, which is heavier than air, falls downward as it leaves the gas nozzle. As a result, if the weld is in the flat position, a relatively low flow rate would sufficiently shield the weld area. However, if the weld is in the overhead position, the flow rate must be substantially increased to compensate for the shielding gas falling away from the weld.

Metal	Shielding Gas	Advantages
Aluminum	Argon	Better arc starting, cleaning action, and weld quality; lower gas consumption
	Helium	High welding speeds possible
	Argon-helium	Better weld quality, lower gas flow than required with straight helium
Magnesium	Helium	Metal thickness 0"–1/16" — Controlled penetration
	Argon	Metal thickness 1/16" + — Excellent cleaning, ease of pool manipulation, low gas flows
Mild Steel	Argon	• Better pool control, especially for position welding • Metal thickness 0"–1/8" — Ease of manipulation, freedom from overheating • Spot welding — Generally preferred for longer electrode life, better nugget contour, ease of starting, lower gas flows
	Argon-helium	Helium addition improves penetration on heavy gauge metal
Stainless Steel	Argon	Permits controlled penetration on thin gauge material (up to 14 gauge)
	Argon-helium	Higher heat input, higher welding speeds possible on heavier gauges
	Argon-hydrogen (65%–35%)	Prevents undercutting, produces desirable weld contour at low current levels, requires lower gas flows
	Helium	Provides highest heat input and deepest penetration
Copper and Nickel Cu-Ni Alloys (Monel and Inconel)	Argon	Ease of obtaining pool control, penetration, and bead contour on thin gauge metal
	Argon-helium	Higher heat input to offset high heat conductivity of heavier gauges
	Helium	Highest heat input for high welding speed on heavy metal sections
Titanium	Argon	Low gas flow rate minimizes turbulence and air contamination of weld, improved metal transfer, improved heat-affected zone
	Helium	Better penetration for manual welding of thick sections (inert gas backing required to shield back of weld against contamination)
Silicon Bronze	Argon	Reduces cracking of this "hot short" metal
Aluminum Bronze	Argon	Less penetration of base material

Figure 4-1. Common base metals and the recommended shielding gases for each.

The opposite is true if helium (which is lighter than air) is used instead. Helium rises as it leaves the gas nozzle. As a result, if a weld is being performed in the flat position, a high gas flow is required to replace the gas that is rising away from the weld area. In general, the volume of helium required to properly shield a weld is greater than the volume of argon that would be required to shield the same weld.

Helium

Helium gas was the first shielding gas used with GTAW. Helium is separated from natural gas produced from wells. Because helium is not as widely available as argon, it is more expensive. The higher flow rate required for proper shielding using helium compared to argon also contributes to the increased cost. The volume of helium used for a given job is greater than the volume of argon that would be used for the same job.

Pure helium offers advantages that might make it the preferred shielding gas in some circumstances. Helium creates a hotter arc than argon, and therefore allows deeper penetration. The use of helium also increases welding speed. Helium is the preferred shielding gas for welding high-heat conductivity metals (copper, aluminum, and refractory metals) and for welding aluminum with DCEN polarity.

Starting an arc in a pure helium shield is very difficult. Therefore, many automated welding applications use an argon gas starting system. After the arc is started using argon, the operation shifts over to using helium shielding gas. Manual welding is seldom done with pure helium shielding gas because the arc length (gap) tolerance is so critical that the welder cannot usually maintain the arc.

Hydrogen

Hydrogen is not an inert gas, but it can be added to argon in small amounts (usually less than 3%) to increase welding speeds and penetration. Shielding gas containing hydrogen is used for welding stainless steels that are sensitive to oxygen. Hydrogen causes porosity in aluminum welds and can cause porosity and embrittlement in carbon steel welds. Use of hydrogen is usually limited to welding of stainless steel and special purging applications.

Nitrogen

Nitrogen is not an inert gas. For shielding gas use, it is used only in small quantities and is mixed with argon. Like hydrogen, nitrogen is used in specific situations to help improve shielding of the weld. Nitrogen is added to argon to increase penetration when welding copper. Nitrogen is also commonly used as a backing gas for welding austenitic stainless steel because of its low cost compared to the other backing gas options. (A backing gas is a gas supplied to the back side of a weld joint to shield the portion of the weld area that the shielding gas from the torch cannot reach.) Nitrogen can also be used as an inexpensive purge gas for the initial purging of tanks, vessels, and pipelines prior to the admission of the inert gas for final purging.

Argon-Helium Mixtures

Argon and helium are mixed to provide the benefits of both gases. Argon-helium mixtures are used to increase control of the voltage and heat input of the arc, while maintaining the favorable characteristics of argon. Helium-rich mixtures are preferred in order to achieve significant benefits from the helium. Additions of less than 50% helium have little influence on the arc characteristics. **Figure 4-2** shows the substantial increase in heat input and weld penetration resulting from the use of an argon-helium mixture instead of pure argon.

The following are the commonly used argon-helium mixtures:

- **75He-25Ar**—The most commonly used mixture contains 75% helium and 25% argon. 75He-25Ar can improve the speed or quality of ac welding of aluminum. The cleaning action is almost as good as with pure argon. 75He-25Ar is sometimes used for manual welding of aluminum pipe and mechanized welding of butt joints in aluminum sheet and plate.

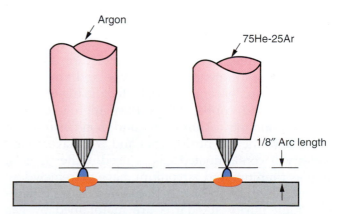

Figure 4-2. The thermal conductivity of helium is much higher than that of argon. Helium transfers more heat from the arc region to the weld pool.

- **90He-10Ar**—This mixture contains 90% helium and 10% argon. 90He-10Ar is used occasionally for dc welding when it is desirable to obtain heat input characteristics approaching those of helium while maintaining the good arc-starting behavior of argon. It is much easier to start an arc in an argon environment than a helium environment. In some cases, an argon-rich mixture is used while the arc is being started, and then a helium-rich mixture is used to make the weld.

Argon-Hydrogen Mixtures

Argon-hydrogen mixtures are used in very specific situations, such as in mechanized welding of light gauge stainless steel. The 15% hydrogen mixture, commonly called H-15, is the most popular argon-hydrogen mixture used for GTAW. H-15 is most often used for welding tight butt joints in stainless steel up to 1/16" (1.6 mm) thick at speeds comparable to those attained using helium. H-15 allows a 50% speed increase over argon. Faster speeds can be obtained by increasing the amount of hydrogen added to the argon. However, excessive hydrogen will cause porosity. A hydrogen content of 5% is sometimes preferred for manual welding to obtain welds with a cleaner appearance.

Shielding Gas Purity

Shielding gases for welding are refined to high purity specifications. Cylinder argon has a minimum purity of 99.996% and contains a maximum of about 15 parts per million (ppm) of moisture (a dew point temperature of –73°F [–58°C] maximum). Helium is produced to a minimum purity of 99.995% and generally contains less than 15 ppm of moisture. At these purity levels, differences in the type and amount of impurities usually cannot be detected during welding.

Steels and copper alloys have high tolerances for contaminants. Aluminum and magnesium are sensitive to the gas purity level and will have severe porosity if contaminated gas is used. Other metals, such as titanium and zirconium, have extremely low tolerances for contaminants in the inert gases. Therefore, high purity standards are maintained by gas suppliers to ensure that the shielding gases used will be more than adequate for even the most demanding application.

Shielding Gas Supply

Shielding gases are supplied in cylinders of various sizes for shop use. A Dewar flask (liquefied gas container) is shown in **Figure 4-3**. Dewar flasks are transportable on a truck due to their construction characteristics. The tanks are basically a tank inside a tank with protection rods that separate the inner and outer shells. This design enables a Dewar flask to meet U.S. Department of Transportation safety regulations.

The newest gas supply system, called a *perma-cell* or *micro bulk system*, is shown in **Figure 4-4**. Like Dewar flasks, these supply systems are liquefied gas containers. However, permacell or micro bulk systems are not transportable. They have a much larger capacity for their physical size. The stationary containers are filled by a truck that drives into the shop and hooks up a filler hose.

All shielding gas is sold by units of volume. One unit of gas is equal to the amount of gas that would fill one cubic foot of space at atmospheric pressure (14.7 psi) and a standard temperature of 70°F. The first time the customer purchases gas from the supplier, the supplier provides a cylinder filled with the gas. The

Figure 4-3. A Dewar flask containing argon. (Mark Prosser)

Figure 4-4. A micro bulk or permacell system is permanently installed on the premises and is filled by a truck when needed. (Mark Prosser)

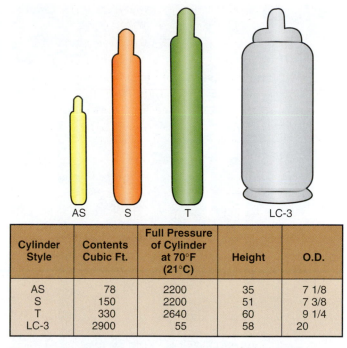

Cylinder Style	Contents Cubic Ft.	Full Pressure of Cylinder at 70°F (21°C)	Height	O.D.
AS	78	2200	35	7 1/8
S	150	2200	51	7 3/8
T	330	2640	60	9 1/4
LC-3	2900	55	58	20

Figure 4-5. Types and capacities of cylinders and Dewar flasks supplied to industry. Note that the LC-3 style cylinder is designed for liquefied gas and holds a greater volume of gas at a lower pressure than the other cylinder types.

supplier will charge the customer for the volume of gas contained in the cylinder and a one-time or recurring charge for leasing the cylinder (demurrage fee). When the customer runs out of gas, he or she simply returns the empty cylinder to the gas supplier and exchanges it for an identical cylinder prefilled with the desired gas.

Some customers own the cylinders they use rather than lease them from the supplier. In such cases, they may have to wait a day or more to have the cylinder refilled. If they exchange their cylinder for an identical prefilled cylinder, they should have the gas supplier provide a receipt acknowledging that the cylinder is owned by the customer. In most cases, leasing cylinders is more economical than owning the cylinders outright because the customer does not need to pay to have the cylinder periodically recertified.

If the customer uses large quantities of gas frequently, purchasing liquefied gases in Dewar flasks may be more economical than leasing a large number of smaller capacity cylinders. Various types and sizes of gas cylinders are shown in **Figure 4-5**.

Shielding Gas Storage

Storage of gas cylinders and containers should be rigidly controlled to prevent the incorrect use of shielding gases. Always store gas cylinders and containers in an upright position in a well-ventilated area separate from the weld shop. Make sure each container is labeled to properly identify the type or mixture of gas contained. Follow all safety precautions to avoid injury to the user. Remember, inert gases do not contain oxygen and therefore will not support life. You cannot see, smell, or taste inert gases.

Shielding Gas Distribution

Shielding gases can be distributed to the welding area in several ways. **Figure 4-6** shows a bank of gas cylinders that have been connected together. The gas cylinders can be located in a convenient area, with the gas piped to the welding area.

A distribution system must be leak-free to maintain the purity and inertness of the gas being used. Therefore, all of the cylinder fittings must be cleaned before installation and properly seated into the regulators. Protect high-pressure connectors, tubing, and pipe connectors to prevent foreign materials, such as water and oil, from entering. Plastic thread covers or tape can be placed over any unused openings for this purpose.

Figure 4-6. Individual cylinders are connected to a supply manifold by short lengths of high-pressure copper tubing, often referred to as *pigtails*. (Mark Prosser)

Figure 4-7. A single Dewar flask connected to a manifold that can supply up to six welding stations. (Mark Prosser)

Manifolds

Manifolds or pipelines are often used to distribute gases to the welding area from the supply area. The use of manifolds reduces the number of individual cylinders required at the welding station. More than one station can be supplied from a single cylinder or bank of cylinders connected to a manifold. See **Figure 4-7**.

Manifolds are generally constructed from rigid copper pipe. The joints are either soldered or silver brazed. The joints must be sound, and the manifold must be tested for leaks at a pressure above the normal operating pressure before the system is placed into use. During the leak test, the manifold is charged to the testing pressure and then the manifold supply valve is closed. The manifold is kept sealed off for several hours and then the manifold pressure is measured. A drop in pressure indicates that leaks are present and must be repaired.

Prior to using the system, the lines must be purged with argon or nitrogen to remove any foreign vapors and moisture. Connect the gas to the manifold and set the flow rate at 5 cfh. Purge the system until an analyzer test shows it is clear of oxygen. Most manifold systems are constructed out of copper tubing. If nitrogen is used to initially purge the manifold, the manifold must be purged a second time with argon. It should then be tested to make sure it is oxygen-free before being used.

The effectiveness of the purge can be evaluated by connecting a GTAW welding machine to the purged argon manifold and using the torch to make a spot weld on titanium. Hold the torch over the weld to supply postflow shielding until the weld cools below 700°F (371°C). If the weld is silver in color, the system is clear. Blue or gray colors indicate that the system is still contaminated and further purging is required. If the distribution system undergoes repair, disassembly, or modification, the entire system should be repurged and tested before being put back into use.

Shielding Gas Pressure Regulation

The pressure of shielding gases can be regulated with several types of special equipment. Gases distributed through a manifold require high-pressure regulators, shown in **Figure 4-8**, to reduce the pressure from the cylinder to the desired manifold pressure level. This pressure is usually set from 20 to 50 psi. A flowmeter is then used at the welding station to regulate the gas flow to the welding torch. See **Figure 4-9**.

If a cylinder is used at the welding station, a regulator/flowmeter is used to reduce the pressure from the cylinder and regulate the flow. See **Figure 4-10**. When installing a regulator/flowmeter or flowmeter with a ball tube, the ball tube should always be in the vertical position for proper operation. The amount of flow is always indicated by the position of the top of the ball unless otherwise indicated.

Regardless of the type of gas supply (cylinder, Dewar, or manifold), when the gas flow valve is opened for welding, a surge of gas will exit from the gas nozzle. This is due to the pressure buildup when the gas is not flowing.

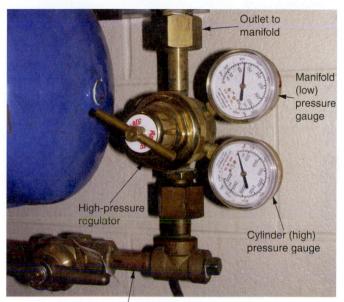

Figure 4-8. This manifold is used for high-pressure gases. A regulator reduces the pressure of the gas to the desired manifold pressure. (Mark Prosser)

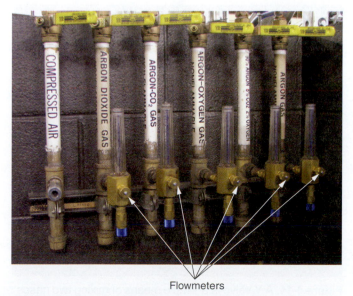

Figure 4-9. Station flowmeters of this type operate on manifolds at approximately 50 psi. (Mark Prosser)

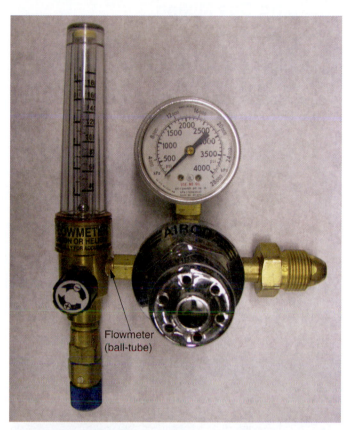

Figure 4-10. Two styles of regulators are shown here. A—Single-cylinder regulator/flowmeters have a gauge that shows cylinder pressure. Gas flow to the torch is adjusted by turning the knob on the attached flowmeter and reading the ball position in the vertical tube. B—A dial gauge is used with this regulator to read gas flow in cubic feet per hour (cfh). The desired gas flow is obtained by turning the adjustment knob on the front of the regulator. (Mark Prosser)

This surge of gas will last for several seconds until the excess pressure is reduced. To eliminate this condition, a specially designed surge protection valve may be used, **Figure 4-11**. Surge protection valves may be added to a standard regulator or they may be built into the regulator.

Shielding Gas Mixing

Premixed shielding gases can be purchased from gas suppliers, or the component gases can be mixed by the end user as they are needed. Different types of gas mixers are available. The type of mixer shown in **Figure 4-12** is used when large volumes of gas are

Figure 4-11. Surge protection valves eliminate surging of gas from the nozzle at the start of the operation. By eliminating the surge, surge protection valves readily pay for themselves in gas savings. (The Harris Products Group)

Figure 4-12. The large blue tank on the bottom of the mixer serves as a mixing and storage chamber. (Mark Prosser)

required. Smaller gas mixers, like the one shown in **Figure 4-13**, are often used in single-station operations. Another type of single-station mixer, like the one shown in **Figure 4-14**, can be used in manual applications. The gas is mixed within the Y valve. These Y valves are installed on the outlet side of the flowmeter, with the gas metered by the two separate

Figure 4-13. Small proportional mixers of this type are used for mixing at individual stations. They provide the proper mixture of gas at the desired flow rate. (Mark Prosser)

Figure 4-14. A Y valve is a simple means of mixing two gases in proportions different from those provided commercially. (Mark Prosser)

gas flowmeters. The mixture is changed by adjusting the valve on each branch of the Y valve.

To prevent backflow of the gases and improper mixing, install a reverse flow check valve, **Figure 4-15**, at the flowmeter outlet. **Figure 4-16** shows a reverse flow check valve installed on a row of flowmeters.

Purity Testing

A gas analyzer can be used to test for proper gas mixtures at individual welding stations. See **Figure 4-17**. These instruments can also be used for manifold leak testing and testing for adequate purging of pipes and vessels prior to the weld. A sample of

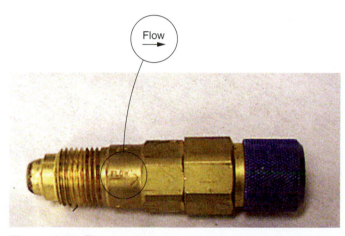

Figure 4-15. Reverse flow check valves prevent mixing of gases in the supply line. Flow direction through the valve is identified by a "flow" direction stamping, as seen here. (Mark Prosser)

Figure 4-16. Reverse flow check valves are installed on all of the flowmeters in this system. (Mark Prosser)

the purge gas is introduced into the analyzer and compared to a sample of reference gas. The machine analyzes the gas and gives an extremely accurate digital reading of the gas percentages. Gas analyzers are very portable and are available in handheld and tabletop versions.

Shielding Gas Safety

The following are important safety measures to keep in mind whenever you are working with compressed gas. Following these safety guidelines will help you avoid accidents and injuries.

Use only cylinders that have clearly-identified contents. If the wrong type of shielding gas is used for a welding operation, the arc may be difficult to manage, or the weld may develop defects that make it unacceptable. In extreme cases, a fuel gas or oxygen could be mistaken for shielding gas, creating the potential for fire and injury. If a cylinder is not clearly labeled, return it to the gas supplier for replacement. *Never* use it.

Install regulators and hoses that are designed for high pressures. Remember that cylinders contain highly pressurized gas. If a low-pressure hose is mistakenly used, it could easily rupture. Similarly, if the regulator is not designed to handle the pressure it is exposed to, it can burst or fail.

Always crack the cylinder valve to clean the cylinder valve outlet port before installing the regulator. Be sure that there is nothing in front of the cylinder valve outlet when you crack the cylinder valve. Any debris inside the outlet will be blown out of the cylinder outlet with enough force to cause injury. When installing the regulator, make sure it is firmly attached. However, do *not*

Figure 4-17. Portable analyzers of this type measure the percentages of the individual gases. (Mark Prosser)

overtighten it. Regulators are made of brass and can be damaged by overtightening.

To turn on the flow of shielding gas, begin by opening the cylinder valve. Always make sure the regulator pressure adjusting screw is fully closed before opening the cylinder valve. Open a cylinder valve slowly at first so the regulator is not immediately hit with the full cylinder pressure. Then open the valve the rest of the way. When you first open a cylinder valve, there is a sudden surge of high pressure at the regulator inlet. Open the regulator and adjust it to the desired pressure. When you open a regulator, always stand to one side or the other, but never directly in front of it. If the regulator is damaged, it may explode, blowing the adjusting screw out of the front of the regulator and shattering the glass in the pressure gauge.

Never use water, oil, Vaseline™, or any other lubricant on high-pressure tanks, regulators, or connectors in any part of the system. If the cylinder valve threads are damaged or rusty, return the cylinder to the supplier.

When a cylinder is empty, close the valve and replace the protective cap. Mark the container "empty" to avoid confusion.

High-pressure systems require the use of high-pressure hoses; you cannot use standard rubber hoses for high-pressure lines. If hoses become damaged, they must be replaced or repaired by a qualified repair technician. Torch lines must be kept off the floor and away from heat. A protective covering (sheath) should be installed over the torch lines. Do not allow the torch lines to be bent or kinked; too much pressure will build up in the gas and water hoses and can cause the hoses to burst. To limit gas leakage, always close the flowmeter valves when you are done welding. When you have finished the welding session, make sure the cylinder valves are completely closed and the regulator pressure adjusting screws are loosened until there is no tension in them.

Filler Metals

Filler metals used in GTAW must be of the highest quality possible in order to make acceptable welds. To ensure high quality, filler metals are manufactured using specialized machines, processes, and inspections.

Manufacturing

Several factors influence the selection of alloys for filler metals. These factors include chemical composition, mechanical properties, notch toughness, grain sizes, internal defects, and impurity level limits.

Once the alloy is selected, it is hot drawn through reducing dies to a predetermined size and cleaned. The wire is then allowed to cool before it is reduced to its final size. The wire is annealed and cleaned again as it is drawn down to its final size.

Various types of lubricants are used in the draw dies during the drawing process. These lubricants reduce wear on the dies by decreasing the friction between the wire and the die and carrying away the heat generated in the die. At the completion of the drawing process, the wire is cleansed of all surface impurities and residual lubricants. The wire is then processed for shipment to the user.

Specifications

Many quality specifications are used for the manufacture, testing, inspection, and packaging of various filler metals. **Figure 4-18** lists common filler metal specifications, as well as some of the AWS specifications for filler metals used in GTAW. A specification can include requirements for the classification and usability of the filler, manufacturing methods, acceptability testing,

Specifications	Material Types
American Welding Society	All types
ASTM International	All types
A151/SAE	Low-alloy steel
MIL-E-23765	Steel
Aeronautical Material Specifications	All types

A

Metal	AWS Specification	Classification
Copper alloys	A5.7	ERCu, ERCuSi-A, ERCuAl-A1
Stainless steel	A5.9	ER308, ER309, EC409, ER2209
Aluminum alloys	A5.10	ER1100, ER2319, ER4043, ER5356
Nickel alloys	A5.14	ERNi-1, ERNiCr-3, ERNiCrMo-3
Titanium alloys	A5.16	ERTi-1, ERTi-5, ERTi-6ELI, ERTi-15
Carbon steel alloys	A5.18	ER70S-2, ER70S-3, ER70S-6, E70C-3M
Magnesium alloys	A5.19	ERAZ101A, ERAZ61A, EREZ33A
Zirconium alloys	A5.24	ERZr-2, ERZr-3, ERZr-4
Low-alloy steel alloys	A5.28	ER80S-B2, E80C-B2, ER80S-D2

B

Figure 4-18. Common filler metal specifications. A—Common welding wire specifications used throughout the welding industry to establish quality. B—AWS specifications for GTAW filler metals.

chemical composition, reliability, sizes and lengths, finish and temper, packaging, and the filler guarantee.

In some cases, the user may add some additional requirements to the basic specification, such as requests for specific chemistry, weld testing, identification, or packaging. Each additional requirement will add cost to the filler metal.

Filler metal manufacturers guarantee only that their product meets the requirements of the specification. The manufacturers do not guarantee that every weld made with the metal will be satisfactory, because they cannot control the way the filler metal is used.

The filler metal available from manufacturers can be classified into three major categories. The first category is *general use wire*. Wire in this category meets a specification requirement, but no record of chemical composition or strength level is submitted to the user.

The second category of filler metals are those that conform to specific fabrication specifications. Sometimes customers require more rigid control over the filler metal than can be attained by using general use wire. In these cases, the customer can request a Certificate of Conformance with the purchase of the material. A *Certificate of Conformance,* shown in **Figure 4-19,** is a statement that the filler metal meets

---

CERTIFICATION OF QUALITY CONFORMANCE TESTS

Manufacturer or Distributor _____

Address_____

Customer's Name _____

Date_____

Specification MIL- _____

Type MIL-_____

Diameter & Length_____

Inspection Level_____

Lot No. _____

Customer's Order No. _____

Core Wire Heat No. _____

Lot Identification MIL- Para._____

Wet Batch No. _____

Chemical Analysis
(Complete)

Carbon_____

Manganese _____

Silicon _____

Phosphorous _____

Sulfur_____

Chromium _____

Nickel_____

Molybdenum _____

Vanadium_____

Chemistry was taken from: Chem Pad ☐   Groove Weld ☐

X-Ray Results _____

Concentricity (%)_____

Covering Moisture _____

Grinding During 8a Test Plate Preparation 3/

	Mechanical Test	
Yield Strength (0.2% offset method)	As-Welded	Stress-Relieved
	_____	_____
Tensile Strength	_____	_____
Elongation (%)	_____	_____
Reduction of Area (%)	_____	_____

Charpy Impacts

1. _____   1. _____
2. _____   2. _____
3. _____   3. _____
4. _____   4. _____
5. _____   5. _____

Groove Weld Test

Test No.   3   8a   Chem Pad

Amperage   _____   _____   _____

Operator Error (Layer Nos.)_____

We hereby certify that the above material has been tested in accordance with the listed specifications and is in conformance with all requirements.

_____

Figure 4-19. Certificate of Conformance form. (Techalloy Maryland, Inc.)

all the requirements of the material specifications. All of the material will be identified by heat numbers, lot numbers, or code numbers located on the package. On some work, the buyer may require that these numbers be recorded along with information about where the material is used in a job.

The last category is filler metal that has undergone a detailed chemical analysis. This type of filler metal is accompanied by a ***certified chemical analysis report***, shown in **Figure 4-20**, which lists the result of a chemical analysis of the individual heat or lot of filler metal. The test, which is made on a spectrometer machine, ensures the highest quality filler metal for critical welding operations on missiles, nuclear reactors, and pressure vessels. These operations usually require very close control of the actual filler metal chemistry. Records are maintained showing the specific filler metals used during the fabrication cycle and where they were used. Because of this detailed recordkeeping, if a joint fails, either because of the filler metal or base metal, other joints in the system can be located and evaluated for possible replacement.

Reid-Avery

Dundalk, Baltimore, Md. 21222

QUALITY ASSURANCE
TEST REPORT

DATE: ______________

P.O. NO.: P.O. NO.:

SPECIFICATIONS:

ITEM	POUNDS	SIZE	TYPE	LOT NO.	HEAT NO.
1.					
2.					
3.					
4.					
5.					
6.					

CHEMICAL ANALYSIS OF WIRES OR WELD METAL

ITEM	C	Mn	P	S	Si	Cr	Ni	Mo				
1.												
2.												
3.												
4.												
5.												
6.												

ADDITIONAL TEST RESULTS

State of ______________________________
City of ______________________________
Subscribed and sworn to before me this______day
of______________________________20____.

____
Notary Public
My commission expires______________________

I certify that the chemical analysis and physical or mechanical test results reported above are correct as contained in the records of the company.

____
QUALITY ASSURANCE DEPARTMENT

Figure 4-20. Certified chemical analysis form. (Techalloy Maryland, Inc.)

Filler Metal Forms

Filler metals used in GTAW are available in various forms and sizes. These include spools, coils, straight length rods, and different types of inserts.

Spools and Coils

Spools and coils are used in automatic or semi-automatic welding applications. The spools or coils adapt to a standard hub attachment made of wood or plastic. The filler metal is wound around these spools or coils which, depending on the size of the wire, contain varying amounts of material. To prevent wire feeding problems in automatic welding, all spool wire must meet cast and helix requirements. *Cast* is the diameter of one complete circle of wire as it lies on a flat surface. *Helix* is the maximum height of any point of this circle of wire above the flat surface. **Figure 4-21** indicates tolerances. Wire that has incorrect cast and helix will cause excessive wear of the liner and guide tip of the wire feeder.

Spools or coils of wire are identified with labels or tapes on the outside of the spool or flange. A spool with a label identifying the material is shown in **Figure 4-22**.

Straight Length Rods

Straight length rods are used in manual welding. The welder feeds the material into the pool as required. Straight length rods are made in a drawing mill and are available in varying sizes and lengths. The standard length is 36″ (91 cm).

Straight length rods are identified by numbers stamped into the end of the filler rod, as shown in **Figure 4-23**. Consumable inserts are identified by a stamped number on the insert itself.

Inserts

Inserts are generally used in manual, semiautomatic, and automatic welding of pipe. The AWS A5.30/A5.30M *Specification for Consumable Inserts* lists the material and design requirements for the various types of inserts shown in **Figure 4-24**.

Figure 4-22. The label identifies this particular material as 4043 aluminum. The lot number is 4270. The wire is 3/64″ diameter. The weight of the welding wire on the spool is 13.0 lbs. (Mark Prosser)

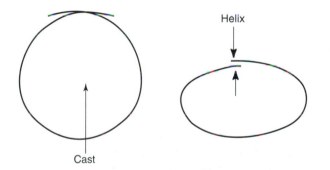

Standards for Cast and Helix						
		Cast			Helix	
Spool Size	Wire Types	Min. in	Min. mm	Max. in	Max. mm	Max. in	Max. mm
4″ (100 mm)	Low-alloy, stainless and nickel alloy	6*	150	9	230	½	13
	Aluminum	†		6	150	1	25
8″ (200 mm)	Low-alloy, stainless and nickel alloy	15*	380	30	760	1	25
12″ (300 mm)	Aluminum	†		15	380	1	25

* Measured on outside strand of full spool
† Diameter of wire level from which sample is taken

Figure 4-21. Standards for cast and helix. These terms define the characteristics of any form of continuous wire as it comes from the spool or coil.

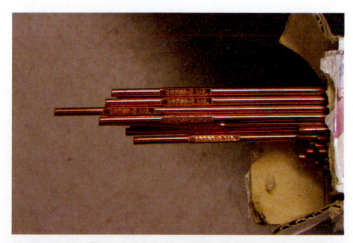

Figure 4-23. Filler metals have an identifying stamp on the end of the straight rod to identify the material. (Mark Prosser)

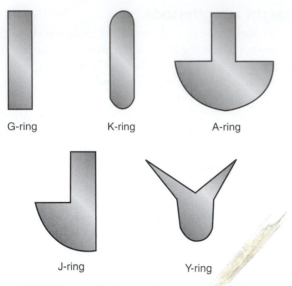

Figure 4-24. These drawings show several of the available profiles of pipe welding inserts.

Filler Metal Packaging

Manufacturers of filler metal use a variety of packaging methods to protect the material while it is being shipped. Wires or rods for general use are usually supplied in heavy-duty cardboard boxes. Wires or rods for very high-quality welds are packaged in hermetically-sealed plastic bags, which are airtight bags that protect the wire from the atmosphere. The shipping boxes also have labels to identify the filler metal contained in the box. See **Figure 4-25**.

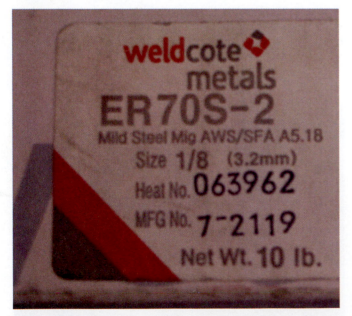

Figure 4-25. General use rods are packaged in sturdy cardboard packages with identification tags on the end or the side of the package. (Mark Prosser)

Filler Metal Use in the Welding Shop

Filler metals are packaged to prevent contamination when they are shipped to the user. Preventing contamination of the material after the package is opened is the user's responsibility.

Filler metals are easily contaminated by oil, moisture, grease, smoke, soot, and salt. Salts and oils on the hand, fingers, or gloves will also contaminate filler metals. Dirty work areas, tables, and rags, can easily contaminate any filler metal that comes into contact with them. These foreign materials often cause defects, such as porosity and cracks, in the weld metal. A rejected weld that requires rework to eliminate the defect adds cost to the job.

The simplest method of avoiding contamination of the filler metal is to keep it clean. This is done by keeping the material packaged for as long as possible. Once the seal on the filler metal package is broken, the filler metal can be stored in a heated cabinet to extend its shelf life.

Filler metals should be stored and used in clean and dry areas. Handle filler metals as little as possible, and then only with clean gloves. Before using a filler metal, wipe it down with a clean rag soaked in alcohol, acetone, or a similar general-purpose degreaser. If spools of filler metal are left on the welding machine, cover the spools to reduce exposure to dust, air, and moisture.

Snip off any contamination on the end of used filler metal wire prior to reusing the wire. See **Figure 4-26**. Welding filler metal removed from the inert gas shield during the welding operation is hot and readily absorbs oxygen, nitrogen, and hydrogen.

Quality welds cannot be made using contaminated filler metal. Requiring the filler metal manufacturer to clean, inspect, and package the material to a specification does little good if the material is contaminated prior to use by improper handling.

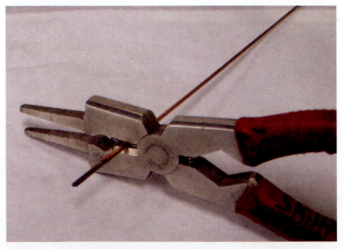

Figure 4-26. The dark, discolored area on the end of the wire indicates contamination. The area at the end should be cut off before the rod is reused. (Mark Prosser)

Summary

Shielding gases used for GTAW include argon, helium, hydrogen, nitrogen, argon-helium mixtures, and argon-hydrogen mixtures. Shielding gases protect the electrode and the molten weld metal from atmospheric contamination, help to transfer the heat from the electrode to the metal, and aid in arc starting and arc stabilization. Shielding gases can affect the look of the weld, the penetration of the weld, and even the quality of the weld. The type of shielding gas used depends on the type of material being welded and the type of welding current.

Shielding gases are supplied in transportable cylinders. Liquefied shielding gases are stored in Dewar flasks or gas supply systems called *permacells* or *micro bulk systems*, which are stationary. Manifolds or pipelines are often used to distribute gases to the welding area from the supply area. Prior to using the shielding gas distribution system, the lines must be purged with argon or nitrogen to remove any foreign vapors and moisture. Gas cylinders and containers should be stored in an upright position in a well-ventilated area.

It is important to follow safety guidelines when working with compressed gas. Only cylinders that have clearly-identified contents should be used. Regulators and hoses designed for high pressures are required. The cylinder valve should be cracked before installing the regulator. When opening a regulator, always stand to one side or the other. Water, oil, Vaseline™, or other lubricants should never be used on high-pressure tanks, regulators, or connectors.

The proper filler metal for a job depends on the type and thickness of the material being welded. Filler metals must also be compatible with the base metal. Filler metals used in GTAW are available in forms that include spools, coils, straight length rods, and inserts.

Many quality specifications are used for the manufacture, testing, inspection, and packaging of filler metals. The filler metal available from manufacturers can be classified into three major categories: general use wire, filler metal that conforms to specific fabrication specifications, and filler metal that has undergone a detailed chemical analysis.

Filler metals are easily contaminated by oil, moisture, grease, smoke, soot, and salt. Contamination by foreign materials can cause weld defects. Filler metals should be stored in clean, dry areas and handled as little as possible.

Review Questions

Write your answers on a separate sheet of paper. Do not write in this book.

1. Shielding gases protect the electrode and the molten weld metal from atmospheric _____.
2. List two inert gases used as GTAW shielding gases.
3. What is the most commonly used shielding gas for GTAW?
4. Why does helium gas require higher flow rates than argon?
5. Which gas is used as an inexpensive initial purge gas for manifolds, vessels, and pipes?
6. Which argon-helium gas mixture is commonly used in GTAW for ac welding of aluminum?
7. What is the most popular argon-hydrogen mixture used for GTAW?
8. Which two metals have extremely low tolerances for contaminants in inert gases?
9. Transportable liquefied gas cylinders are commonly called _____.
10. Gas cylinders and containers should be stored in a(n) _____ position.
11. Gases distributed through a manifold require high-pressure _____ to reduce the pressure from the cylinder to the desired manifold pressure level.
12. What is the usual pressure range for manifolds?
13. What is the purpose of reverse flow check valves?
14. What should you do if a cylinder is not clearly labeled?
15. What is a Certificate of Conformance?
16. List four different forms of GTAW filler metals.
17. The _____ and _____ of welding wire used for automatic welding must be within specifications or the wire will cause excessive wear in the wire feeder liner and guide tip.
18. How are straight length rods identified?
19. List five ways in which filler wires may become contaminated.
20. Name two degreasers used to clean filler wires before use.

These tanks are refilled and ready for bulk shipment. Note the framework and roller cart used to contain and move the tanks. (Mark Prosser)

Chapter 5

Weld Joints and Weld Types

Objectives

After completing this chapter, you will be able to:

- ❏ Recognize the various types of weld joints.
- ❏ Recall common weldment configurations.
- ❏ Understand AWS standard terminology.
- ❏ Interpret welding symbols.
- ❏ Compare the advantages and disadvantages of different weld and joint designs.
- ❏ Understand the effect of weld joint design on the properties of the final weldment.
- ❏ Recall the AWS welding position classifications.

Key Terms

arrow	interference fit
autogenous weld	interpass
backing weld	joggle-type joints
basic weld symbols	joint
butt joint	lap joint
buttering	mismatch
chill bars	pi tapes
cladding	postheating
corner joint	preheating
dilution	purge blocks
edge joint	reference line
grain structures	surfacing
hardfacing	T-joint

Introduction

The American Welding Society defines a *joint* as *the manner in which materials fit together*. The five basic types of joints are butt, T-joint, lap, corner, and edge, **Figure 5-1**.

Weld joints can be initially prepared in a number of ways. This preparation can include mill scale removal, beveling, root face preparation, fitup, and tacking. These include the following processes:

- Shearing
- Casting
- Forging
- Machining
- Stamping
- Filing
- Etching
- Grinding
- Routing
- Oxyfuel cutting
- Plasma arc cutting

Final preparation of the weld joint includes making sure the material is clean and correctly prepared, fits together with the proper clearances, and is adequately tack welded. The final preparation is done just before the weld is made.

Weld Types

Two types of welds, fillet welds and groove welds, are used for many different types of weld joints. Fillet welds are used to weld T-joints, lap joints, and some corner joints. Groove welds are used mostly for butt joints, a joint type in which the two pieces of metal are butted together. Different types of butt joint groove welds are prepared with different edge profiles for different purposes.

Butt Joints

A *butt joint* is created when two pieces of metal are aligned edge-to-edge in the same plane. Butt joints are useful when a flat, smooth transition is needed between two pieces. This type of weld requires the material to be flat and even at the edges, all the way across the joint. Different types of groove welds are used to weld butt joints, depending mainly on the thickness of the base material to be welded.

The specific type of groove weld used on a butt joint depends on the way the edges of the metal are prepared. Some common types of groove welds used with butt joints are shown in **Figure 5-2**:

- Square-groove
- Bevel-groove
- V-groove
- J-groove
- U-groove
- Flare-bevel-groove

T-Joints

A *T-joint* is created by placing the edge of one piece of material against the surface of another piece of material at a 90° angle. The joint looks like a "T". T-joints are completed with the types of welds shown in **Figure 5-3**:

- Fillet
- Plug
- Slot
- Bevel-groove
- J-groove
- Flare-bevel-groove

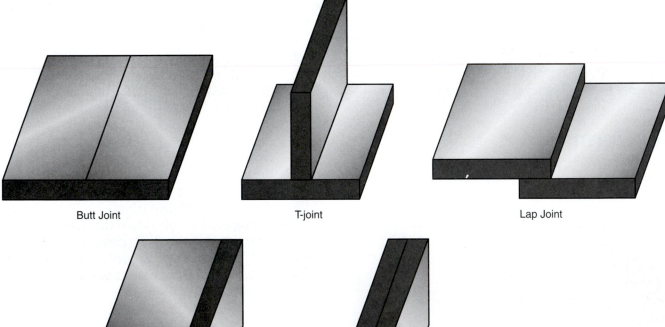

Butt Joint T-joint Lap Joint

Corner Joint Edge Joint

Figure 5-1. Basic types of joints.

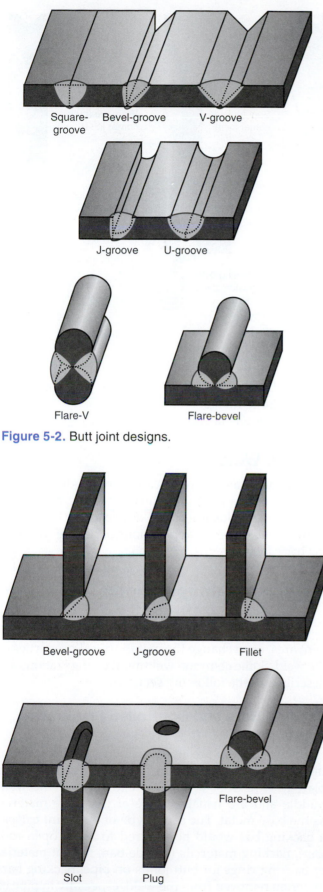

Figure 5-2. Butt joint designs.

Figure 5-3. T-joint configurations.

Lap Joints

A *lap joint* is created when the flat surface of one piece of material is placed against the flat surface of another piece of material, and at least one edge does not line up with an edge on the other piece of metal. This creates a small 90° angle weld joint. Lap joints are completed with the types of welds shown in **Figure 5-4**:

- Fillet
- Plug
- Slot
- Spot
- Bevel-groove
- J-groove

Corner Joints

A *corner joint* is created when two pieces of material are matched up, edge-to-edge, at a 90° angle or nearly 90° angle, creating a corner. If welded on the inside of the 90° angle, the joint is referred to as an inside corner joint. If welded on the outside of the corner, the joint is referred to as an outside corner joint.

Corner joints are completed with the types of welds shown in **Figure 5-5**:

- Fillet
- Spot
- Square-groove
- V-groove
- Bevel-groove
- U-groove
- J-groove

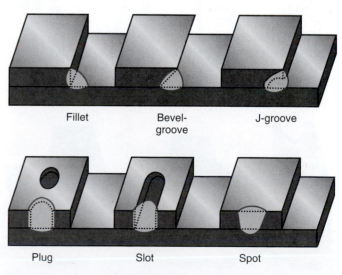

Figure 5-4. Lap joint configurations.

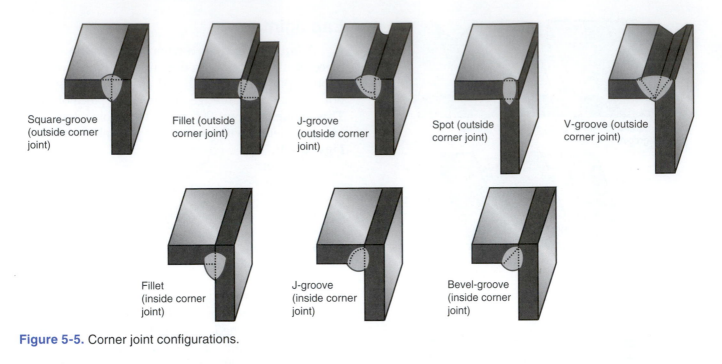

Figure 5-5. Corner joint configurations.

Edge Joints

An *edge joint* is created when two pieces of material are fit face-to-face with two or more edges aligned. If one piece of material is stacked perfectly on top of another piece of material so that the edges align, an edge joint is created. Edge joints can be completed with the types of welds shown in **Figure 5-6**:
- Square-groove or butt
- Bevel-groove
- V-groove
- J-groove
- U-groove
- Edge-flange
- Corner-flange

Double Welds

In some cases, a weld cannot be made from just one side of the joint, but must be made from both sides. These welds are called *double welds*. **Figure 5-7** shows some common applications of double welds on the basic joint designs.

Weldment Configurations

A basic joint is often modified to assist in assembly of the components, to gain access to the weld joint area, or to change the metallurgical properties of the weld. Some common weldment configurations are described in the following sections.

Backing Bars and Controlled Penetration Joints

Backing bars are used on butt joints to control the amount of penetration. These backing bars are placed at the root of the weld to support the weld puddle and are usually made of the same material as the base metal. The same type of butt joint *without* a backing bar would be referred to as an open root joint. Backing materials include bars for flat material or backing rings for butt joints on pipe. Backing bars are often removed from the weld, especially in pipe or tubing welds where the flow of a liquid or gas may be

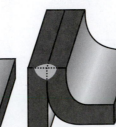

Square-groove   Bevel-groove   J-groove   V-groove

U-groove   Edge flange   Corner flange

Figure 5-6. Basic edge joint configurations.

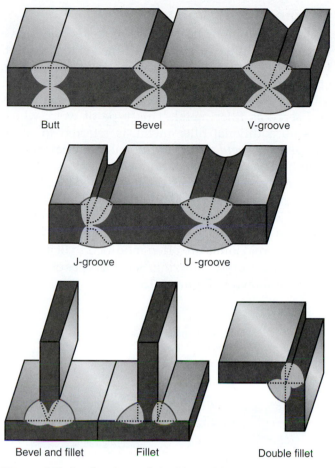

Butt   Bevel   V-groove

J-groove   U -groove

Bevel and fillet   Fillet   Double fillet

Figure 5-7. Applications of double welds.

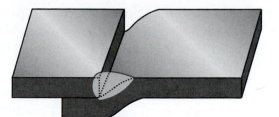

Figure 5-8. Joggle-type joint.

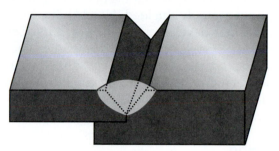

Figure 5-9. Built-in backing bar joint.

restricted. If the backing bar is to be removed from the weld, the letter *R* will appear in the weld symbol on the blueprint. Backing bars are sometimes left in position and become part of the weld. A *backing weld* is a weld made from the back side of the weldment to fill in the root of the weld before the front side of the joint is welded. A backing weld serves the same purpose as a backing bar or a backing ring.

In *joggle-type joints*, one piece of the metal is flanged, or stepped. The mating piece of metal rests on the flange or step, ensuring that the top surfaces are flush, **Figure 5-8**. The flange or step acts as a backing to control excessive penetration. The flange also supports the top piece until it can be tack welded in place. Joggle-type joints are used in cylinder and head assemblies where backing bars or tooling cannot be used. In the sheet metal industry, joggle-type joints are referred to as "step" joints.

Where sufficient material is available for machining the required back-up on pipe or tubing, a backing bar can be machined into the edge of the thicker tubing. The thinner piece will slide over the flange. The flange will hold the thinner piece in place and control penetration like a backing ring. See **Figure 5-9**.

In groove welds with a fabricated backing bar, the backing bars are fitted very tightly to the joint, **Figure 5-10**. Loose-fitting bars cause problems with heat flow and penetration. A loose-fitting backing bar is the same as an improper joint fitup. A loose backing bar can allow metal to flow where it does not belong and can cause overheating problems. A well-designed weld joint can control penetration, which is especially important where excessive penetration would cause a problem with assembly or liquid flow.

Purge Blocks

Purge blocks are small boxes of different sizes and shapes designed to provide shielding gas to the back side of weld joints. They have a fitting for connecting shielding gas and many small holes in the face of the box through which the shielding gas can flow. The face of the box is placed against the back

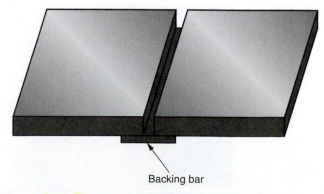

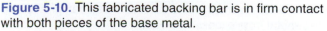

Backing bar

Figure 5-10. This fabricated backing bar is in firm contact with both pieces of the base metal.

side of a weld joint, shielding gas is connected to the block and turned on. Then, as the weld is made, the shielding gas from the purge block protects the root of the weld from the atmosphere.

One of the most common uses of purge boxes is to shield the back side of butt welds on stainless steel to prevent "sugaring." Sugaring refers to a condition in which the surface of the weld and surrounding metal is black and grainy, as if covered in burnt sugar. It is the result of oxidation that occurs if a stainless steel weld is not properly shielded. A weld with sugaring is weak and may develop cracks if it is put into service.

A purge block never becomes part of the weld. Purge blocks are slightly concave, so there is a small gap between the weldment and the block along the joint. The blocks do not bond to the weld because they are always made from copper.

Surfacing

The term *surfacing* refers to overlaying a series of weld beads on the surface of metal. The term *buttering* refers to a type of surfacing commonly used to prepare dissimilar metals to be welded together, **Figure 5-11**. In a buttered weld, each of the two dissimilar metals is surfaced with a compatible alloy. Then, a filler alloy that is compatible with the buttering alloys is used to weld the parts together. Other common uses for surfacing include *hardfacing* and *cladding*. In these surfacing procedures, the base metal is overlaid with weld beads intended to protect the base metal from wear or corrosion. See **Figure 5-12**.

Pipe and Tubing Joints

Tube-to-tube or tube-to-flange welds often have the required filler material machined into the joint. **Figure 5-13** illustrates how the mating tube sections are machined to create a flange at the end of each tube. The flange will melt during the welding process and serve as filler metal. This type of weld, which uses no additional filler metal, is referred to as an ***autogenous weld***. The additional material simply melts and becomes part of the weld.

Tubing weld joints are often modified by adding a sleeve either on the inside or outside of the joint. This type of joint, as shown in **Figure 5-14**, allows tubes to be joined without fitting them up end-to-end.

Pipe weld joints are often modified to fit a special type of insert, or backing ring. Several types of backing rings are shown in **Figure 5-15**.

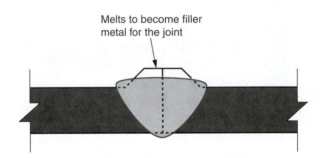

Figure 5-13. Tube-to-tube or tube-to-flange weld.

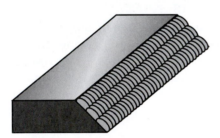

Figure 5-11. A weld joint face can be buttered.

Figure 5-12. Overlaid welds such as these can protect the base metal from excessive wear.

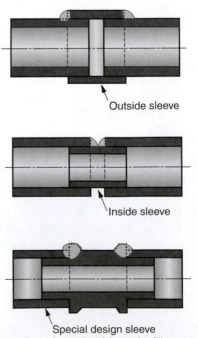

Figure 5-14. Sleeve-type joints are commonly used where the tubing is restrained and cannot shrink during the welding.

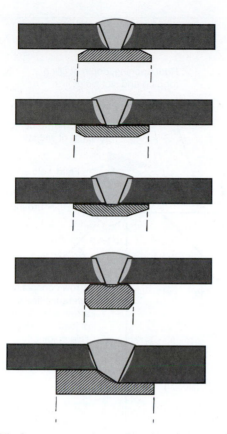

Figure 5-15. Common profiles of backing rings used on pipe joints.

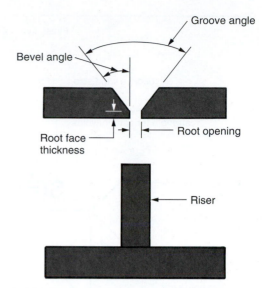

Figure 5-16. The terms used to describe different parts of a weld joint.

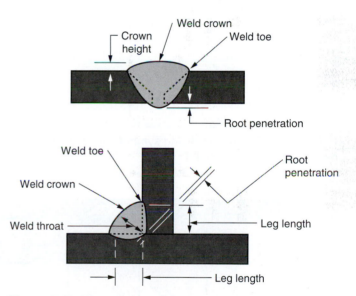

Figure 5-17. The terms used to describe different parts of groove and fillet welds.

Welding Terms and Symbols

Welding terms and symbols are critically important in fabrication or manufacturing settings in order to relay all needed information. Welding terminology and symbols are a standardized language that enables designers and engineers to accurately communicate their ideas to welders. The AWS document A3.0 *Welding Terms and Definitions* is the most commonly used document for consistency in the welding trade. A welder must be able to understand these standardized terms and symbols in order to correctly read a blueprint and accurately fabricate a given project.

Some of the terms used to describe a weld joint are shown in **Figure 5-16**. Common terms used to describe the weld and the weld area are given in **Figure 5-17**.

The AWS welding symbol shown in **Figure 5-18** indicates where the weld is to be made. The symbol contains all of the information needed to prepare the joint, as well as information describing the finished weld, such as finish and length of the completed weld.

Figure 5-19 contains the *basic weld symbols* that direct the welder to select the proper type of weld joint. By combining the basic weld symbols and the welding symbol, a designer or drafter can communicate any welding operation to a welder.

It is important to study and understand each part of the welding symbol. The *arrow* indicates and marks the point at which the weld is to be made. The line to the arrow is always at an angle to the *reference line* (long horizontal line in the welding symbol). Whenever the

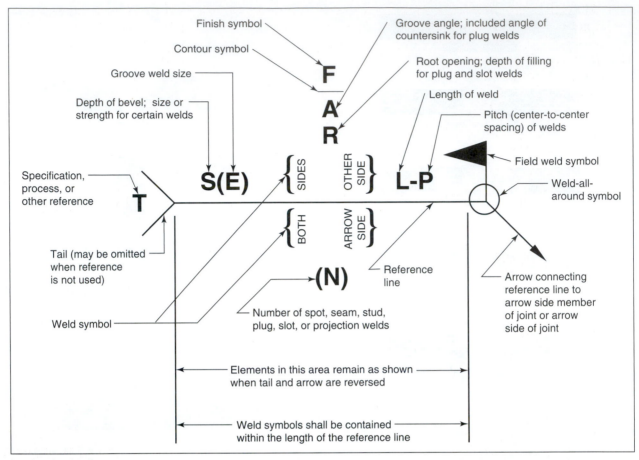

Figure 5-18. The AWS welding symbol conveys specific information to the welder. (Printed with permission of the American Welding Society)

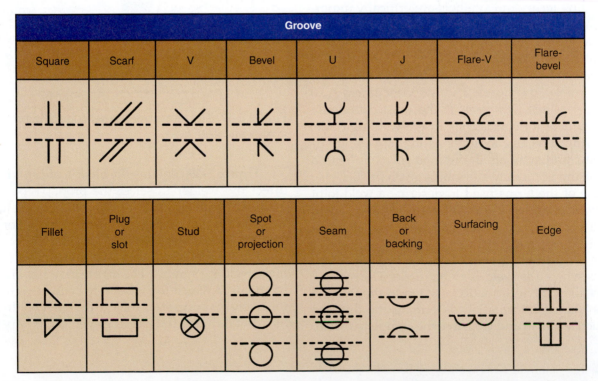

Figure 5-19. AWS basic weld symbols. (Printed with permission of the American Welding Society)

basic weld symbol is placed *below* the reference line, the weld is to be made at the point indicated by the arrow, as shown in **Figure 5-20**. Whenever the basic symbol is placed *above* the reference line, the weld is to be made on the other side of the joint, as shown in **Figure 5-21**. Dimensions on the symbol and drawing indicate the exact size of the weld. Study the examples of weld symbols and weldments shown in **Figure 5-22**.

Classes are offered that provide advanced study in print reading for welders. A welder can improve his or her ability to read and interpret welding drawings by taking such a class or by studying print reading texts.

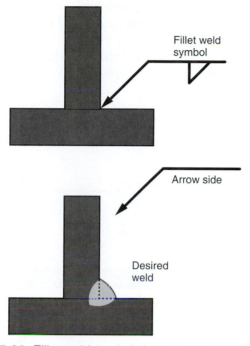

Figure 5-20. Fillet weld symbol shown on the arrow side. Note the resulting weld.

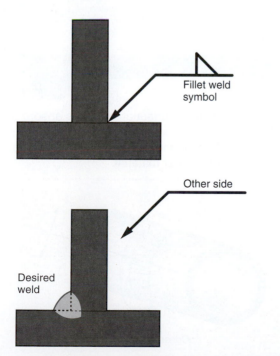

Figure 5-21. Fillet weld symbol on the other side, away from the arrow. Note the resulting weld.

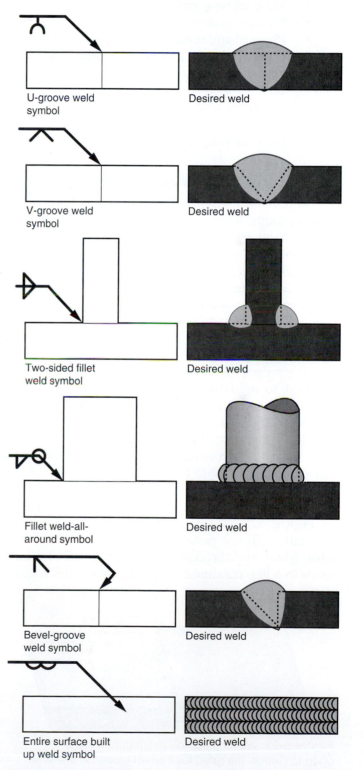

Figure 5-22. Typical weld symbols and their weldments.

Design Considerations

The design of the weld and weld joint must be carefully considered so that the resulting weldment does its intended job. The weld should be made with reasonable cost. Several weld design factors must be considered:

- Material type and condition
- Service conditions
- Physical and mechanical properties of the completed weld and heat affected zone
- Preparation requirements
- Cost
- Assembly configurations
- Accessibility of the joint
- Necessary equipment and tooling

Butt Joints and Welds

Butt joints are the strongest of the five basic types of weld joints. These joints are reliable and are able to withstand stress better than any other weld type. To achieve full stress value, the weld must have 100% penetration through the joint. This can be achieved by welding from one side. Otherwise, welding can be done from both sides, with the welds joining in the center.

Thin gauge metals are more difficult to fit up for welding and require more costly tooling to maintain the proper joint configuration. Tack welding can be used to hold the components during assembly. However, tack welds present many problems compared to the use of good tooling. Tack welds can conflict with the final joint penetration into the weld joint, add to the crown dimension (height), and crack during welding due to the heat and expansion of the joint. The expansion of the base metal during welding will often cause a condition called *mismatch*, shown in **Figure 5-23**. When mismatch occurs, the weld generally will not penetrate completely through the joint. Many specifications limit highly stressed butt joints to a 10% maximum mismatch of the thickness.

Whenever possible, butt joints should mate at the bottom of the joint, as shown in **Figure 5-24**. Taper joints of unequal thicknesses in the weld area, as shown in **Figure 5-25**, to prevent incomplete or inadequate fusion.

A very important factor affecting butt welds is the amount of shrinkage of the components at the weld joint. Butt welds always shrink across the joint (transverse shrinkage) during welding. For this reason, a shrinkage allowance must be made if the part dimensions of the finished parts have a small tolerance. Butt welds in pipe, tubing, and cylinders also shrink on the diameter of the material. This shrinkage is shown in **Figure 5-26**.

In areas where dimensions must be maintained, a shrinkage test should be made to determine the amount of shrinkage. **Figure 5-27** shows how the test is done.

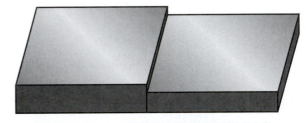

Figure 5-24. Mating the joint at the bottom equalizes stress loads.

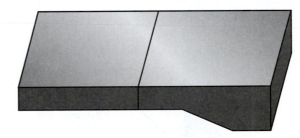

Figure 5-25. Joints of unequal thickness absorb different amounts of heat and expand at different ratios. Equalize the heat flow by tapering the heavier material to the thickness of the thinner material.

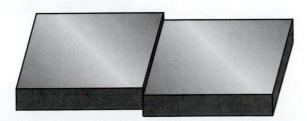

Figure 5-23. Welds made on mismatched joints will often fail below the rated load when placed in stressed conditions.

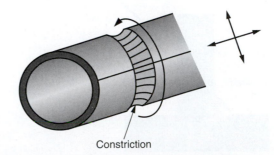

Constriction

Figure 5-26. Butt welds shrink during welding in both transverse and longitudinal directions.

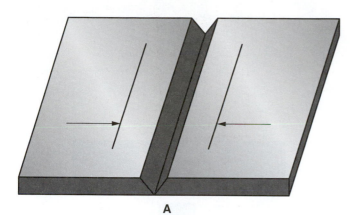

A

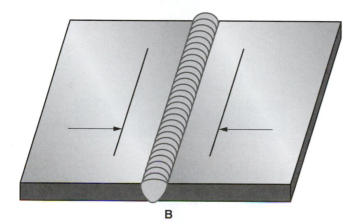

B

Figure 5-27. A simple test can be made to determine the amount of shrinkage across a specific joint. A—A straight line, parallel to the weld axis, is marked on each workpiece in the weldment. The distance between the lines is carefully measured. B—The weldment is welded and allowed to cool. Then, the distance between the two lines is measured again. The difference between the first and second measurement equals the material shrinkage. Future weldments can be resized to compensate for the shrinkage.

The amount of shrinkage is determined by making lines or marks on each component prior to welding. After the distance between the lines is measured, the weld is completed and the distance between the lines is measured again. The difference between the preweld and postweld distances equals the amount of shrinkage. This information is very important if weldments must be built to exact dimensions.

Thick materials shrink more than thin materials. Single-groove welds shrink less than double-groove welds, since less welding is involved and less filler metal is used.

Lap Joints and Welds

Lap joints are either single-fillet, double-fillet, or spot-welded. They require very little joint preparation and are generally used in static-load designs.

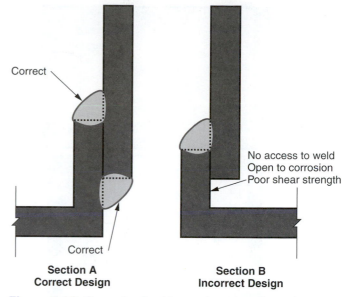

Section A
Correct Design

Section B
Incorrect Design

Figure 5-28. Corrosive liquids must not be allowed to enter the penetration side of the joint.

Both edges of a lap joint must be welded if the part will be exposed to corrosive liquids. See **Figure 5-28**. A major problem with lap joints is *bridging*, shown in **Figure 5-29**. Where the component parts are not in close contact, a bridging effect occurs that must be avoided. T-joint and lap joint fillet welds are prone to bridging, which occurs when there is not enough heat generated in the joint to wash the molten puddle down into the root of the joint. Bridging can also occur when too much filler metal is added without adequate heat to melt the filler metal down into the root of the joint. The result is a lack of penetration in the weld, which can make the weld rejectable. Proper fitup is a critical factor in making quality fillet welds. Use adequate clamps or tooling to maintain contact of the material at the weld joint.

Cylindrical components can be assembled with an *interference fit*, as shown in **Figure 5-30**, to prevent bridging. The outer part's inside diameter (ID) is made several thousands of an inch smaller than the inner part's outside diameter (OD). *Pi tapes* are used to determine the diameters of the inner and outer cylinder components. Pi tapes are flexible measuring

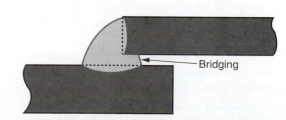

Bridging

Figure 5-29. The bridging effect in this lap joint creates a space for corrosion. The weld is weak due to lack of penetration.

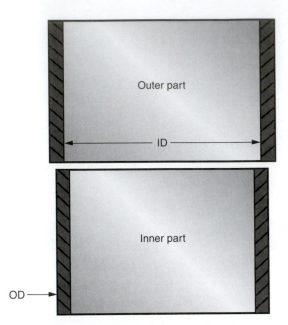

Figure 5-30. The diameters of the component parts to be assembled can be determined by using a pi tape around the inner and outer cylinder components. The tape measures in thousandths of an inch and full inches.

tapes that provide quick and accurate measurements on round and out-of-round forms. They convert circumferential distances into diameter measurements. When ready for assembly, the outer part is heated enough that the inner part will fit inside it. The outer part is then assembled and allowed to cool in place. When cool, the outer part will be locked tightly into place on the inner part.

T-Joints and Welds

T-joints are used for joining components at angles to each other. Depending on the use of the joint, T-joints can be made with a single-fillet weld or a double-fillet weld. The butting edges may be beveled for improved penetration. See **Figure 5-31**.

Fillet welds are made to specific sizes, which are determined by the desired design load. The sizes are measured as shown in **Figure 5-32**. If the weldment is not being constructed to handle a specific load, a simple rule of thumb can be used to determine the fillet size. In these cases, the fillet leg lengths must equal the thickness of the thinnest workpiece in the joint.

The main problems encountered in T-joint fillet welds are insufficient throat length and lack of penetration at the joint intersection, both of which result in weak welds. **Figure 5-33** shows the relationship between weld size, leg size, and effective throat length. In a fillet weld with a convex crown, the amount of

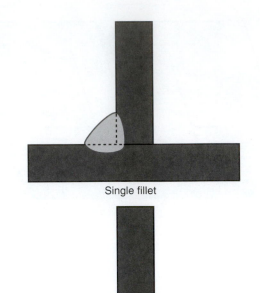

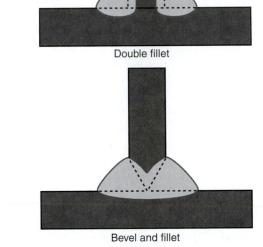

Single fillet

Double fillet

Bevel and fillet

Figure 5-31. Various types of T-joints and welds.

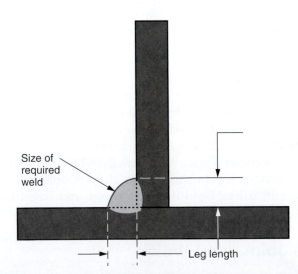

Size of required weld

Leg length

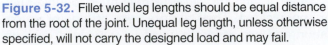

Figure 5-32. Fillet weld leg lengths should be equal distance from the root of the joint. Unequal leg length, unless otherwise specified, will not carry the designed load and may fail.

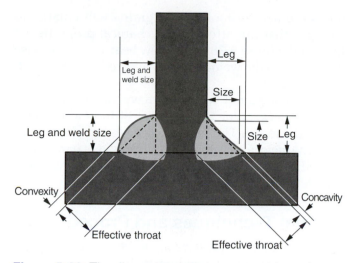

Figure 5-33. The dimensions of a convex weld are shown on the left, and the dimensions of a concave weld are shown on the right. Although the legs of each weld are the same length, the weld size and effective throat of the concave weld are smaller.

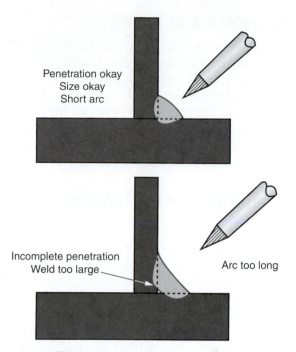

Figure 5-34. Fillet welds must penetrate into the intersection between the two pieces of metal, as shown in the top drawing. Lack of penetration into the root of the joint, as shown in the bottom drawing, can cause weld failure under load.

convexity does not affect effective throat length, leg size, or weld size. However, in a concave weld, weld size and effective throat length are both reduced in proportion to the amount of concavity in the weld. As a result, a concave weld will have a shorter effective throat and weld size than a convex weld with legs of the same length. Keep in mind that GTAW fillet welds are usually concave rather than convex. Lack of penetration is caused by insufficient heat at the root of the joint. The adjoining materials absorb the heat before it gets into the corner. With the addition of filler material, the problem gets worse as more arc energy is directed at melting the filler material. The weld must penetrate beyond the intersecting point, as shown in **Figure 5-34**, to be effective. To eliminate poor penetration, use a small diameter filler material and avoid large welds.

Corner Joints and Welds

Corner welds are very similar to T-joints because they consist of sheets or plates mating at an angle to one another. Corner welds are usually used in conjunction with groove welds and fillet welds. Some of the many different corner weld joint designs are shown in **Figure 5-35**. When thin gauge materials are used, it can be difficult to assemble the component parts without the proper tooling. Tack welding and welding heat will cause distortion of thin material. For the most part, corner joints should be used only with materials that are at least 1/8″ (3.2 mm) thick.

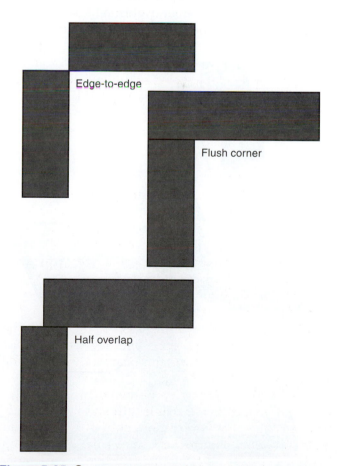

Figure 5-35. Common corner weld joint designs used in fabrication of component parts.

Edge Joints and Welds

Edge welds are used where the edges of two sheets or plates are adjacent, and the sheets or plates are in approximately parallel planes at the point of welding. **Figure 5-36** shows three edge weld designs. These designs are commonly used to hold metal in a fixed position. Since the weld does not penetrate completely through the joint thickness, it should not be used in stress or pressure applications.

Weld and Heat-Affected Zone Grain Structures

The design of the weld joint and the amount of welding done has a significant effect on the properties of the final weldment. The amount of heat used to weld the joint, the type of filler metal, and the number of passes or layers used in the joint determine the dilution of the base metal and filler metal. (*Dilution*, in welding terminology, describes the change from the original chemical composition to another composition with different physical or mechanical properties.) **Figure 5-37** shows various weld joint designs and the level of dilution associated with those designs. Multipass groove or fillet welds add considerable heat input to the weld area. Where this heat flow is

not controlled by tooling, the grains will enlarge to a degree that can affect the mechanical properties of the metal. **Figure 5-38** shows how heat affects the base metal next to the weld, which is referred to as the heat-affected zone.

The completed weld will have various types of *grain structures*, or arrangement of individual crystals in the metal. The grain structures are affected by all of the factors mentioned in the previous paragraph, as well as the final cooling and heat treatment of the completed weld.

Special Techniques and Procedures

Special techniques are often used in the fabrication of a weldment. Depending on the material's carbon content, alloying agents, and thickness, specialized procedures such as preheating, postheating, or both may be required to retain the desired properties of the base metal. Different welding techniques control overall heat input to help control distortion and other negative effects on particular material types.

Weld Reinforcement

Some materials are weakened by the welding process and cannot be heat treated after welding

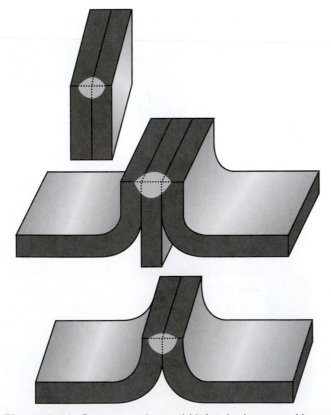

Figure 5-36. Common edge weld joint designs used in fabrication of component parts.

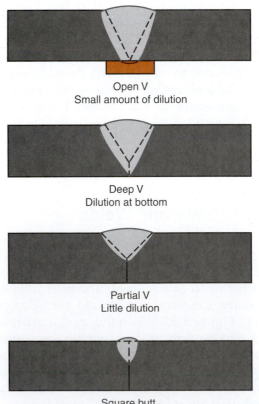

Open V
Small amount of dilution

Deep V
Dilution at bottom

Partial V
Little dilution

Square butt
High dilution

Figure 5-37. The amount of dilution created in various joints.

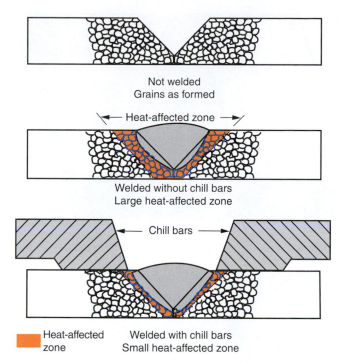

Figure 5-38. Chill bars and tooling are used to remove heat from the weld area and to restrict heat flow into the metal. A—Prepared but unwelded joint. B—If the joint is welded without chill bars, the heat-affected zone is relatively large, and there is significant grain enlargement within the zone. C—Chill bars reduce the size of the heat-affected zone, minimizing the effect of grain enlargement.

because of the weldment's configuration or size. In such cases, a joint with built-in reinforcement, **Figure 5-39**, can be used to achieve full strength in the weld. Because of the additional material in the joint, such a weld joint can be just as strong as the original, unwelded material as long as the proper welding temperatures, alloying agents or filler metal, and welding techniques are used.

Buttering

As mentioned earlier, buttering is the process of laying stringer beads to the faces of a weld joint before actually welding the joint. Buttering is commonly used to join dissimilar metals. Because each metal transfers heat at a different rate and has a different melting point, it may be impossible to weld dissimilar metals and achieve equal penetration and fusion on each side of the weld. Dissimilar metals can be successfully joined by buttering the face of each piece with an alloy that is compatible with both base metals. Buttering produces surfaces with identical physical properties on both pieces, which can then be properly welded together. See **Figure 5-40**.

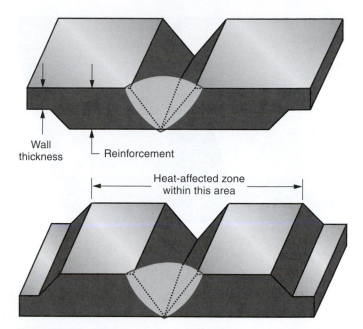

Figure 5-39. Because of the joint thickness and the filler metal's tensile strength, the strength of the weld is equivalent to the strength of the base material in this design.

In addition to its usefulness for joining dissimilar metals, buttering can also be used to control the mechanical properties of a weld. For example, a nickel-based alloy might be used to weld the stringer beads on the faces of mating parts before making the actual weld. The addition of nickel results in a better-quality weld that might otherwise be degraded by the amount of heat needed to complete the weld.

Buttering can also be used to help with fitup. An oversized gap can be closed by buttering the edges.

Preheating, Interpass, and Postheating

Preheating, interpass, and postheating operations can be used to control grain size. *Preheating* brings the metal up to temperature before the welding process begins. The temperature and duration of preheating depends on the material being welded and the thickness of the material. *Interpass* is the period of time on a multipass weld after a weld pass is completed and before a new weld pass is started. *Postheating* is the process of cooling the weld at a controlled rate to prevent cracking. Thicker sections require more postheating, and different materials require different postheating procedures. Postheating is the method most commonly used for stress relief.

Tooling and Chill Bars

Special procedures, tooling, and *chill bars* are used to localize and remove welding heat during the welding application. **Figure 5-41** shows tooling being used to remove heat from a part.

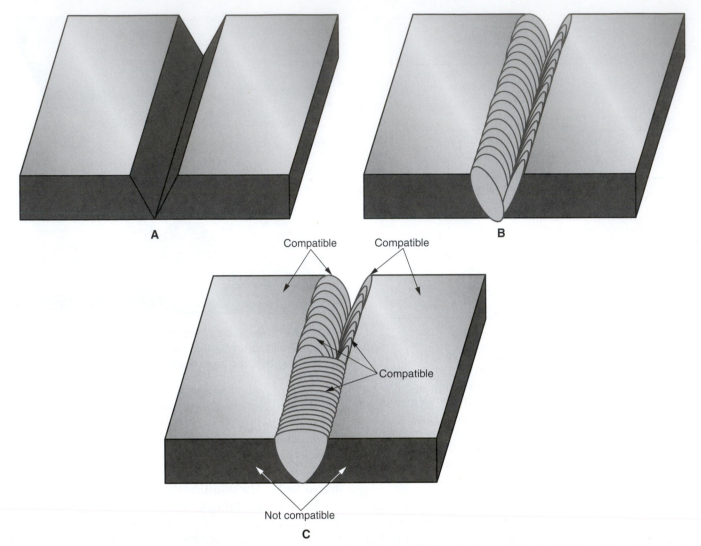

A

B

Compatible   Compatible

Compatible

Not compatible

C

Figure 5-40. Buttering techniques are commonly used to adapt dissimilar metals for welding. A—This weldment is made from dissimilar metals. B—The face of each workpiece is covered with an alloy that is compatible with the base metal and the filler that will be used for the final weld. C—A final weld is made to join the buttered surfaces.

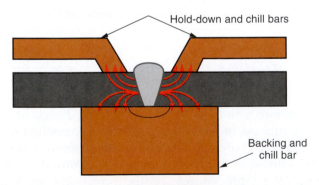

Hold-down and chill bars

Backing and
chill bar

Figure 5-41 Tooling and bars that are used to remove heat from the weld and resist heat flow into the base material are called *chill bars*.

Welding Positions

It is often necessary to make welds in positions other than the flat position. The American Welding Society welding position classifications include *flat*, *horizontal*, *vertical*, and *overhead*. **Figure 5-42** shows the four positions for fillet welds, grooved butt welds, and pipe welds.

With any weld, a number represents the position the weld is to be made in, and a letter represents the type of weld it is. The position of the weld is designated by 1, 2, 3, or 4. The 1 represents the flat position, 2 is the horizontal position, 3 is the vertical position, and 4 is the overhead position. Letters are used to designate the type of weld. Welds are classified into two categories: fillet welds, which are represented by the letter F, and groove welds, which are represented by the letter G. For example, a groove weld that is to be welded in the vertical position would be designated with *3G*.

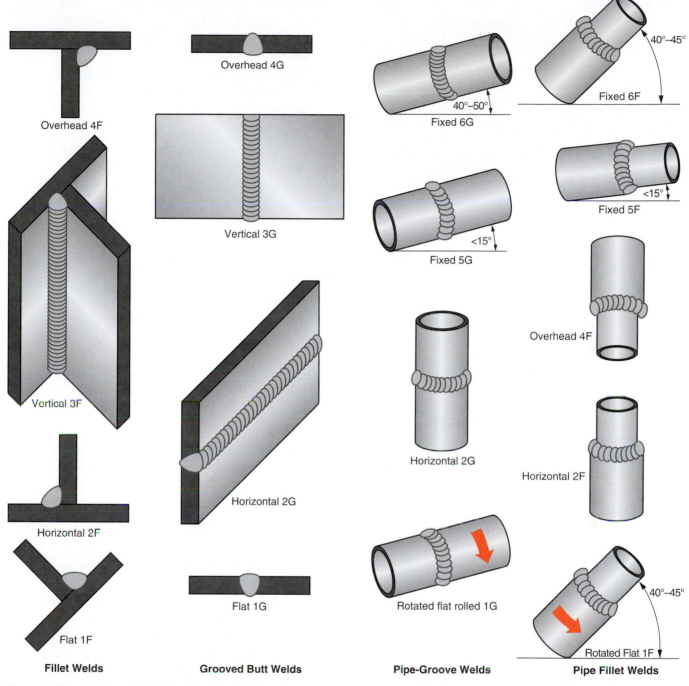

Figure 5-42. American Welding Society welding position designations.

There are also position designators for pipe welds. These are 1G, 1F, 2G, 2F, 4F, 5G, 5F, 6G, and 6F welding positions. In the 1G welding position, the pipe is horizontal, and is rotated so the entire joint can be welded in the flat position. The 1F position is similar, but since it is a fillet weld, the pipe is at a 45° angle and rotated so the weld can be made in the flat position. In the 2G and 2F positions, the pipe is in the vertical position and the weld is in the horizontal position. In the 4F position, the pipe is oriented vertically, with the larger diameter section above the smaller diameter section so the fillet weld is in the overhead position. In the 5G and 5F positions, the pipe is in a fixed horizontal position, requiring the welder to switch between multiple welding positions as the weld progresses around the pipe. In the 6G and 6F positions, the pipe is fixed at a 45° angle, again requiring the welder to switch between multiple welding positions.

Summary

Two types of welds, fillet welds and groove welds, are used for a variety of weld joints. Basic joint types include butt joints, T-joints, lap joints, corner joints, and edge joints.

Backing bars and joggle-type joints control excessive penetration. Purge blocks cover the root or the backside of a weld with shielding gas to prevent oxidation. Surfacing is the act of laying rows of weld beads on the surface of a part to form a weld joint or to protect the base material. Buttering refers to surfacing the faces of a weld joint before welding the joint. In many tube-to-tube or tube-to-flange welds, the mating tube sections are machined to create a flange at the end of each tube. The flange melts during welding, serving as filler metal.

Standard terminology and symbols are used in the welding trade. Basic weld symbols direct the welder to create the proper type of weld joint. The AWS welding symbol indicates where a weld is to be made and contains all necessary information for preparation of a joint.

The design of the weld joint is of prime importance. Butt joints are used where high strength is required. Lap joints require little joint preparation and are generally used in static-load designs. T-joints and corner joints are used for joining components at angles to each other. Edge joints, in which the edges of two sheets or plates are adjacent, are commonly used to hold metal in a fixed position, but cannot handle a large load.

The weld joint design and the amount of welding done affects the properties of the final weldment. There are various amounts of dilution with different weld joint designs.

Special designs and procedures include weld reinforcement; buttering; using preheating, interpass, and postheating operations to control grain size; and using tooling and chill bars to remove heat.

The American Welding Society classifies welding positions as flat, horizontal, vertical, and overhead. The welding position is represented by a number (1, 2, 3, or 4), and the type of weld is represented by a letter (F or G). There are additional position designators for pipe welds.

Review Questions

Write your answers on a separate sheet of paper. Do not write in this book.

1. Based of the joint configuration, most welds can be classified as either groove welds or ____ welds.
2. What are the five basic types of joints?
3. ____ joints are used to join two flat pieces of metal together edge-to-edge in the same plane.
4. What is the purpose of using backing bars on butt joints?
5. What are two advantages of a joggle-type joint compared to a butt joint?
6. The term ____ refers to laying weld beads on the surface of metal.
7. An operation known as ____ is often used to build up a weld joint for joining dissimilar metals.
8. What is an *autogenous* weld?
9. Whenever the basic weld symbol is placed above the reference line, where is the weld to be made?
10. Of the five basic types of weld joints, which is the strongest?
11. List three problems commonly encountered when using tack welds during assembly.
12. If the parts of a lap joint are not in close contact, a problem known as ____ can occur.
13. Whenever possible, butt joints should mate at the ____ of the joint.
14. Do double-groove welds shrink more or less than single-groove welds?
15. Are GTAW fillet welds usually concave or convex?
16. If a weldment is not being built to handle a specific load, what rule of thumb should be used to determine the leg length of fillet welds in the weldment?
17. The area in the base metal next to the weld where grain structure changes occur is called the ____.
18. In welding terminology, what does *dilution* mean?
19. What is the purpose of chill bars?
20. Which welding position is represented by the number 4?

Chapter

Tooling

Objectives

After completing this chapter, you will be able to:

- ☐ Recall tool design considerations.
- ☐ Distinguish between basic tool designs.
- ☐ Explain application of tools in producing the weldment or in aligning component parts.
- ☐ Describe various backing bar designs and their advantages.
- ☐ Describe various hold-down bar designs and their advantages.
- ☐ Describe effective maintenance practices for welding tools.
- ☐ Explain the purpose of test tooling.
- ☐ Recall various types of auxiliary tooling and their purposes.

Key Terms

atmosphere chambers
backing rings
hard tooling
lathes
manipulators
planishers
positioners
relief step design
seam trackers
seamers
segmented finger bar
shop aids
solid hold-down bar
test tooling

Introduction

Many tools are used in GTAW to help prepare the part for welding. These tools include clamping and aligning devices, machines to position and manipulate the weld joint, and a variety of other tooling used to ensure that a high quality weld is made. We classify all of these shop aids as tooling.

Tooling is a very important part of the manufacturing cycle of the weldment. If tooling is not designed and constructed properly, serious problems can result that affect both quality and production cost of the weldment.

The following factors are considered when tooling is designed:

- alignment of the components to be welded
- heat control of the weld zone
- positioning of the joint for welding
- assembly (loading) and disassembly (unloading) of the components
- providing atmosphere to prevent contamination of the weld
- accessibility for the weld torch and filler metal
- dimensional tolerances
- type of material to be welded
- complexity of the weld
- tool cost
- number of parts to be made
- quality requirements for the weld
- welding conditions

Design Concepts

Major tools designed for assembly of component parts and welding are considered *hard tooling*. They are generally used where large numbers of items are welded. Where a small number of parts are to be made, *shop aids* are generally used. These tools may use the weldment as part of the tooling, or they may be a complete tool. Simple shop aids and tools can be used to help align a butt joint for tacking and welding. A common piece of angle iron can be used to center the two pieces of material. The part is rotated by hand as each segment of the weld is completed.

Location and holding tools are used to position parts, maintain tolerances, and hold parts during tack welding. The tack welded assembly is then removed from the tool for the final welding operation.

Types of Tools

There are several different types of tooling used in a shop. The type of tooling used depends on the task being performed:

- **Internal tooling.** The weld is made from the outside of the joint. A typical tool is shown in **Figure 6-1**.
- **External tooling.** The weld is made from the inside of the joint. A typical tool is shown in **Figure 6-2**.
- **Combination assembly and welding tooling.** The weld may be made from inside or outside or the part may become part of the tool. Often the tooling is removed after tack welding. A typical tool is shown in **Figure 6-3**.

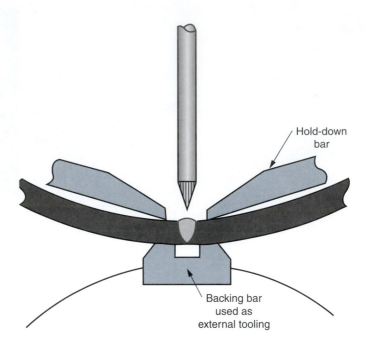

Figure 6-2. The inside diameter of the cylinder must be sufficient to accept the welding torch and component parts of the fixture in external tooling.

- **Heat sink tooling.** The tool is used only to remove heat from the part during the welding operation. This heat removal reduces excessive penetration and distortion. A typical heat sink tool is shown in **Figure 6-4**.

Since the weld and the weld areas are hot and expand during welding, the tool must not lock the parts in place. The tool must allow for this expansion, as well as shrinkage during cooling. If parts cannot shrink as they cool, the weld joint stresses can exceed the strength of the weld. This can result in weld centerline cracking, either in the form of visual cracks

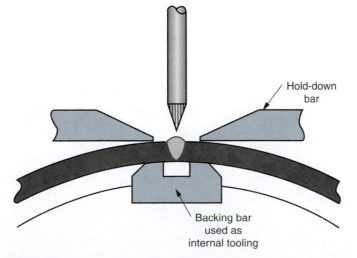

Figure 6-1. Air pressure is applied to the hold-down bars to firmly hold the cylinder long seam joint to the backing bar.

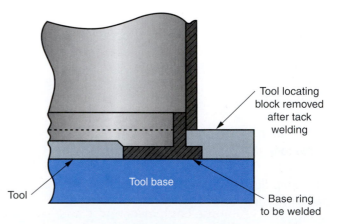

Figure 6-3. This circumferential tool accurately locates and holds the base ring. Locating blocks space the outer ring during assembly. The blocks are removed after tack welding the assembly for welding the inner or outer weld.

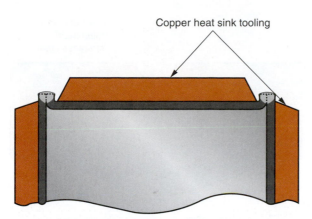

Copper heat sink tooling

Figure 6-4. A copper heat sink is placed on both sides of the weld joint to remove the heat generated during the welding cycle.

on the surfaces of the weld or micro-cracking in the weld grain structure.

Application and Operation of Tooling

The application of the tool in producing the weldment or in aligning the component parts usually consists of two parts. A fixed member is used for location of the part, and a movable member is used for clamping purposes. When using simple tools, a movable member may not be required. In some cases, the part is clamped directly to the fixed tool, as shown in **Figure 6-5**.

Mechanical tooling can use a number of methods for holding or clamping operations. An example of mechanical clamping is shown in **Figure 6-6**. The use of mechanical clamping devices often requires excessive amounts of time for loading and unloading of the tool. Other types of clamping forces, such as air pressure, hydraulic pressure, or a combination of the two, require less time. Pneumatic (air pressure operated) and hydraulic systems operate very rapidly. They are reliable and far more efficient than mechanical systems for applying clamping pressure.

Tooling Materials

The materials used in the tool's main structure or base must be capable of securely holding the component parts. Clamping equipment and the actual weld zone tooling components must operate consistently after repeated use. Square tubing, pipe, and box-type frames offer considerable rigidity and are used extensively in the manufacture of welding tooling. Gussets

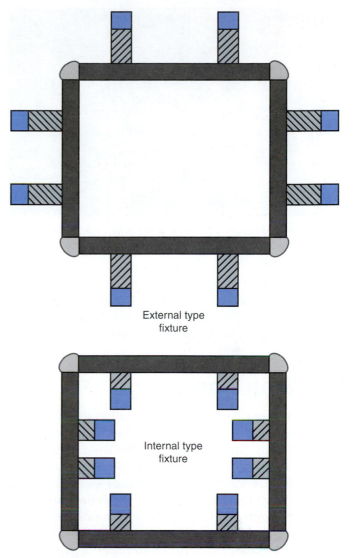

External type fixture

Internal type fixture

Figure 6-5. In an external type fixture, location points are on the outside of the frame. Since the welds will shrink, the part inside the location points will not become wedged in the fixture. In the internal type fixture, location points are located on the inside of the frame. Shim blocks are then used with the location points for assembly. They are removed before welding to allow for any possible shrinkage.

and other reinforcement can be built into the tooling to strengthen it.

Materials used in the weld joint area are a major tooling design consideration. Ferrous materials, such as iron and steel, are magnetic and may cause arc blow (magnetic influence on the arc direction) when DCEN is used. Nonferrous materials, such as aluminum and copper, are nonmagnetic; however, they are soft. Tools made from nonferrous materials can be easily deformed by welding heat and the pressures used for clamping purposes. Stainless steels in the 300 series are often used for tooling in the weld area. These stainless steels are nonmagnetic and not as soft as aluminum or copper. Combinations of

Figure 6-6. Clamping force may be provided by hydraulic, pneumatic, or even hand-adjusted systems. (Miller Electric Mfg. Co.)

materials are often used in tooling to diminish problems. See **Figure 6-7.** The use of material combinations retains tool rigidity and lowers maintenance cost.

Backing Bar Design

A backing bar for full penetration groove welds must be rigid enough to support the component parts and still allow for penetration of the weld metal at the root of the weld. The grooves cut into backing bars are of various designs. These grooves shape the penetrating metal for a specific need. If the backing bar is flat against the weld, it acts as a heat sink, removing necessary heat to ensure fusion of the root. Different groove styles can create different types of penetration and reinforcement. The welding industry often refers to this type of forming, or molding, of the penetration into a grooved backing bar as *casting the penetration*. Normally, a shallow radius groove is used for casting the penetration on the root of an aluminum weld. A deeper radius or square groove is used when an inert

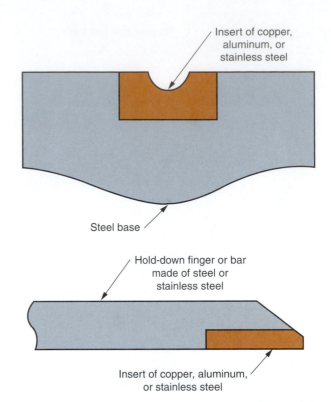

Figure 6-7. The major part of the tool is made of steel for rigidity. The area in contact with the weld area is made of stainless steel, aluminum, or copper.

gas backing is required. **Figure 6-8** shows common types of grooves.

Figure 6-9 shows recommended dimensions for backing bar grooves. Larger radius and square groove designs rely on surface tension of the weld penetration, rather than casting or molding, to control the amount of penetration.

Where inert gas backing is used, the gas must be admitted into the groove with very little pressure. This is done by drilling holes into a gas manifold in the base of the backing bar. The holes may be placed on or off center and chamfered at the top to prevent jetting of

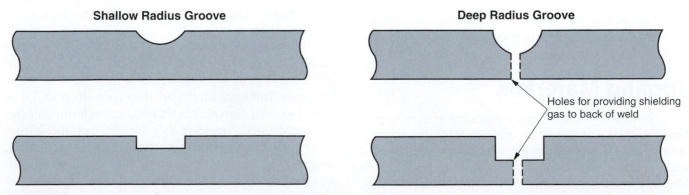

Figure 6-8. Types of weld backing grooves for full penetration groove welds. The backing bars on the right have holes through them to provide shielding gas to the back of the weld.

Metal Thickness	Weld Type	Groove Dimensions for Casting Penetration	Groove Dimensions for Gas Backing Penetration
.005–.012	Fusion	040W010D	040W125D
	Filler	—	
.013–.020	Fusion	063W010D	125W100D
	Filler	—	
.021–.032	Fusion	093W010D	187W100D
	Filler	125W020D	
.033–.040	Fusion	125W020D	
	Filler	187W025D	
.041–.050	Fusion	125W020D	
	Filler	187W025D	
.051–.062	Fusion	187W020D	
	Filler	250W040D	
.063–.072	Fusion	187W020D	250W100D
	Filler	250W040D	
.073–.125	Fusion	250W020D	
	Filler	312W040D	
.126–.250	Fusion	312W020D	312W100D
	Filler	375W050D	
.251–.375	Fusion	—	
	Filler	—	

Number reflects width and depth
e.g. 040W010D is .040 wide and .010 deep

Figure 6-9. Backing bar groove dimensions for full penetration welds.

the gas onto the weld penetration. **Figure 6-10** shows a backing bar groove with holes in place.

Manifolds for distribution of gas can be placed in or below the grooved bar, as shown in **Figure 6-11**. The manifold should always be as large as possible to

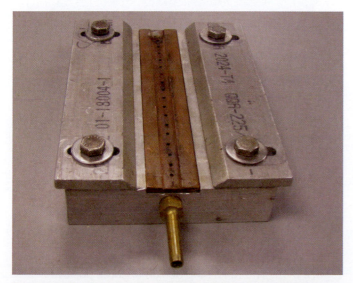

Figure 6-10. Common backing bar design for admitting gas to the penetration side of the weld joint. (Mark Prosser)

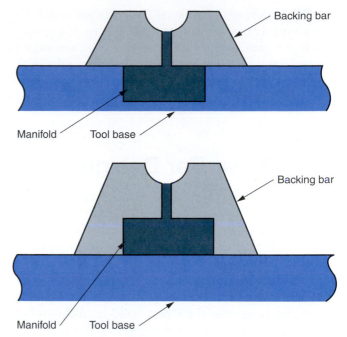

Figure 6-11. Backing bars designed with manifolds are costly to make and maintain. The smaller bar shown is less costly to replace.

provide inert gas to the weld area in sufficient volume with low pressure.

Some welding operations require preheat and interpass temperature control of the weldment prior to and during the welding operations. Backing bars can be designed with built-in heating elements to control these temperatures. Such backing bars are equipped with thermostats that precisely control their temperature. Heat from the backing bar is transferred into the weld zone of the part to be welded. The major benefit of this type of heating is very close control of the weld area temperatures throughout the welding operation. **Figure 6-12** shows a cross section of a typical backing bar equipped with an electric heating element.

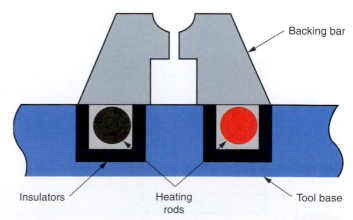

Figure 6-12. Backing bars that use strip heaters must be of sufficient thickness to accept the strip heater and insulators.

Backing rings are inserts that provide quick and easy alignment for pipe and tubing by providing small tabs that establish the root opening. They are easy to use and create a consistent gap with tight tolerances. The welder can snap off the alignment tabs once the joint is tack welded, or the tabs can be consumed in the weld.

Hold-Down Bar Design

There are two types of hold-down bars—solid and segmented finger. The function of the *solid hold-down bar* or *segmented finger bar* is to hold components in place and to control the heat flow from the weld joint. Solid bars are generally used for small or short welds. Segmented finger bars are used for longer welds. The type of basic tool design may dictate the use of one type over the other. Solid hold-down bars distort more easily and have higher replacement costs than segmented finger bars. Segmented finger bars are cheaper to make and replace. **Figure 6-13** shows the two types of designs.

Figure 6-14 shows some of the common hold-down bar designs in use. The sharp nose bar is fragile, regardless of the material used in its construction. For this reason, a sharp nose bar may require frequent maintenance to restore its shape so it will remain in close contact with the workpiece. The flat nose design provides greater rigidity than the sharp nose design. Flat nose bars are placed further back from the joint to give the operator more visibility. This also allows space for the torch and the wire feeder manipulator if used in the operation.

The *relief step design* with a flat nose, shown in **Figure 6-15**, has several advantages over the other

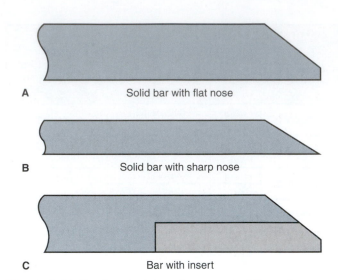

A   Solid bar with flat nose

B   Solid bar with sharp nose

C   Bar with insert

Figure 6-14. Design and use of segmented finger bar contact area. A—A flat nose bar is used where thick welds are made and good heat transfer is required from the weld joint. B—A sharp nose bar is used for thin gauge material with minimum hold-down pressure. C—This design is used where considerable pressure is required to hold the part in place.

hold-down bar designs. However, the overall cost is higher due to the added machining required for the relief step. The main advantage of the relief step design is that the bar can be used on slightly warped parts. The main pressure-bearing area is small compared to the pressure-bearing area of a solid finger or bar in complete contact with the part. The small pressure-bearing area ensures close contact with the part in the weld area. **Figure 6-16** shows a relief step hold-down bar in use.

Relief Step with Flat Nose

Figure 6-15. A relief step design hold-down bar requires a thicker cross section.

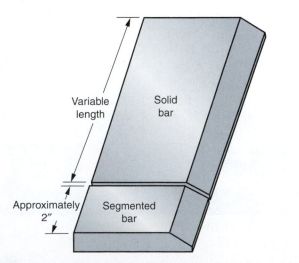

Variable length

Solid bar

Approximately 2″

Segmented bar

Figure 6-13. Solid and segmented finger hold-down bar designs.

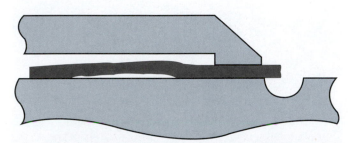

Figure 6-16. The relief step design ensures positive clamping of material and good heat transfer characteristics at the weld joint.

Tooling Maintenance

Welding tooling undergoes tremendous stresses and strains during welding operations due to the pressure and heat of operating conditions. The tools must be closely inspected and properly maintained to ensure that they continue to produce consistent welds. Even the most rigid tools will lose alignment after constant use. A tool that is designed to hold a close tolerance should be inspected often for proper alignment.

Backing bars and backing rings should be checked for burrs and nicks, which prevent close contact with the component part. Without the proper chill of the weld zone, a consistent weld cannot be made, and weld defects (low welds, no penetration) will occur. **Figure 6-17** shows how this condition can be detected.

Inert gas leaks can cause contamination of the weld penetration with oxygen, hydrogen, or nitrogen. This contamination would cause rejection in many welds. In some materials, lack of gas coverage prevents weld root formation, and the penetration is uneven and does not blend into the base metal. Gas lines and hoses to tooling need to be inspected regularly for leaks and cracks, and any problems found should be corrected.

Tool cleanliness is a must. A tool that provides an inert backing gas atmosphere must be clean and dry. Any moisture and grease on the tool will contaminate the inert gas.

Store backing bars and backing rings in a clean, dry area, and clean the tool again immediately before using it. Steel fixtures absorb moisture when cool. If a steel tool is used near the weld, any moisture picked up by the tool will affect the gas purity.

Any portion of a tool's inert gas area that can be contaminated during storage should be disassembled and thoroughly cleaned before the tool is used. This should also be done before the tool is initially used, since oil and grease may have accumulated during manufacture.

Check mechanical/manual areas of the tool often for proper operation. Check for loose bolts, nuts, screws, toggles, cams, and other potential problems. All of these parts affect tooling operation and should be replaced or repaired whenever necessary.

Test Tooling

Test tooling is made to prove tooling design concepts, make test welds, and establish weld joint shrinkage dimensions. In some cases, welder training and certification is done on test tooling prior to the production task. Test tooling can be made rather quickly and modified as required at a much lower cost than changing the production tool. An example of tool modification is the addition of heater strips or rods for preheating and interpass temperature control.

Auxiliary Tooling

Auxiliary tooling includes seamers, manipulators, positioners and lathes, planishers, gas atmosphere welding chambers, and seam trackers.

Seamers

Seamers are used for making long seam internal or external welds on flat or cylindrical stock. Seamers hold the material in place, provide the proper type of backing bar and hold-down segmented finger bar design, and operate consistently. Seamers vary in size from a bench model to a large unit capable of welding cylinders several feet in diameter.

Side beam tracks and carriages that contain the welding torch and equipment for movement of the torch can be added to the seamer. These units can be aligned to a very close tolerance for tracking the weld seam. Weld repeatability with a seamer is very good. **Figure 6-18** shows a seamer with a mechanized carriage and the welding equipment mounted over the welding area.

Manipulators

Manipulators are often used to position the welding head at various locations and heights. Manipulators are available in many different designs. In some cases, the main base of the manipulator is mounted on rails to enable movement throughout the welding area. If the manipulator is used for long

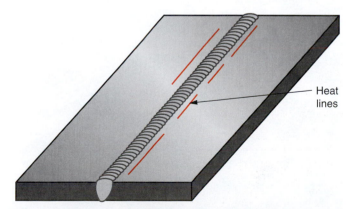

Figure 6-17. Inspection of the weld area after the weld is completed indicates whether or not the segmented finger bar was in proper contact. A heat line along the contact point indicates poor contact. Where this line varies or becomes larger, the bar did not transfer heat properly.

Heat lines

Figure 6-18. A commercial seamer used to make long seam welds in tubing, pipes, cylinders, and flat materials. The machine can be set up to perform almost any type of longitudinal weld. (Jetline Engineering, Inc.)

seam welding, accuracy and speed are very critical concerns. In other cases, the manipulator is used only to position and hold the head at a stationary spot, and the weld seam is then moved under the welding head, as shown in **Figure 6-19**. In these cases, the accuracy and speed of the manipulator are not critical.

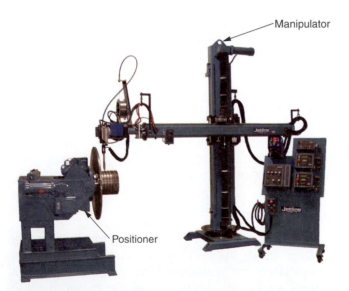

Manipulator

Positioner

Figure 6-19. An adjustable manipulator set in a fixed position. The weld joint is moved by the rotating the positioner. (Jetline Engineering, Inc.)

Positioners and Lathes

Positioners and *lathes* are machines that rotate the weldment so welds can be made along circular seams. Modern positioners and lathes are equipped with electronic components, such as SCR controls, that convert ac power to dc power with adjustable voltage to control the machine's turning direction and speed. These tools also have dynamic braking systems, which use the electricity to perform a smooth and even braking action. Positioners and lathes can be adapted to perform many different tasks with a very high degree of repeatability of the operation.

A *positioner* is a motor-driven turntable on which a weldment is mounted so it can be easily repositioned. Positioners range in size from very small, table-top positioners to those capable of handling weldments of several hundred tons. **Figure 6-20** shows a small positioner. The pipe or weldment is secured to the positioner's table. The table can then be rotated left or right and tilted up to 90°, as needed. Positioners are used specifically to move the part into the position where it is easiest for the welder to weld it.

Welding lathes are used to rotate a cylindrical or semicylindrical part so it can be welded all the way

Figure 6-20. This type of small welding positioner uses a dc motor for very accurate rotational drive. (Mark Prosser)

around its circumference with very good accuracy. Engine crankshafts are welded this way. Building up a weld that will later be machined off is also easily done this way. The lathe controls the speed of the rotating part and can create very accurate welds with tight tolerances if used correctly. A welding lathe is shown in **Figure 6-21**.

When selecting a positioner or lathe, thoroughly investigate the capacity of the equipment. For example, one positioner may turn from .5–10 rpm, while another positioner may turn from .1–3 rpm. Consider the diameter of the weldment and the welding speed when choosing a positioner or lathe. Another example is a 200 lb positioner that, when welding with the table flat, will handle a 200 lb load. However, when the table is vertical and the load is in the horizontal position with the center of gravity 4″ from the table, the maximum load is 130 lbs. With this in mind, consider the added weight of the part tooling and the possibility of overloading. See **Figure 6-22**.

Planishers

A *planisher* is a tool used to flatten and smooth a welded seam and prepare the part for final finishing. **Figure 6-23** shows one type of planisher. For some materials, cold working by planishing will improve the mechanical properties of the weld. If a part is to be formed or spun after welding, the weld cross section must be similar in size to the base metal. A planisher flattening wheel may be specially designed for specific contours. An "as welded" weld grain structure and a planished weld grain structure are compared in **Figure 6-24**.

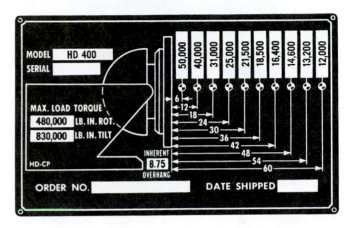

Figure 6-22. Welding positioner load capacity levels. Manufacturers establish safe operating limits for each type and model. (Aronson Machine Co.)

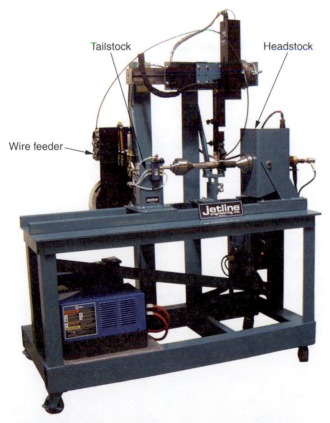

Figure 6-21. The tailstock on this welding lathe can be moved for adapting to long or short parts. (Jetline Engineering, Inc.)

Figure 6-23. Planishers operate on both longitudinal and circumferential welds to flatten and smooth welds. (Jetline Engineering, Inc)

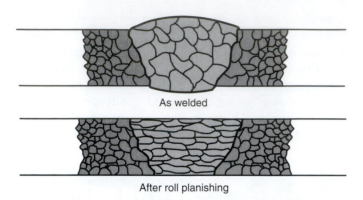

Figure 6-24. A comparison of weld structure after welding and after planishing.

Gas Atmosphere Welding Chambers

Atmosphere chambers are enclosures that provide shielding for the weld by surrounding the weld with a protective atmosphere. They are available in different sizes for different applications. The whole welded part is set inside the chamber, and the chamber is filled with shielding gas to protect the weld from contamination. These chambers are used when the highest quality welds are necessary. Atmosphere chambers are used where the weld must be fully protected from atmospheric contamination. Reactive materials such as titanium, zirconium, and tantalum readily absorb atmospheric contamination when heated during the welding operation. In some cases, tooling, trailing shields, and special torch nozzles may be used to provide the inert gas shield in the weld area. However, in many cases, this type of equipment cannot be used and gas atmosphere welding chambers must be used.

Atmosphere chambers are made in many sizes and configurations and from various materials including steel, stainless steel, and plastic. **Figure 6-25** shows a professional inert gas welding chamber. The air is removed, and then the chamber is flooded with argon. The parts are loaded through a loading port, which is essentially an air lock, so the atmosphere in the main chamber is unaffected.

Some chambers, like the one shown in **Figure 6-26**, are equipped with vacuum pumps to remove the air from the chamber. After the air is evacuated, the inert gas is admitted into the empty chamber. This type of chamber reduces the time required to achieve a welding atmosphere. However, initial costs are quite high due to the heavier design of the chamber and the pumping equipment.

Loading port

Figure 6-25. A high level atmosphere chamber used for welding of specialty materials. Note the loading port, which allows loading and unloading parts without affecting the atmosphere in the main chamber. (Jetline Engineering, Inc.)

Seam Trackers

Seam trackers are used in semiautomatic and automatic welding operations to align the welding torch along the weld joint. These machines are holding devices that move along a track and use sensors to locate the seam on a weldment. A sensor or probe tracks the joint, and controls a motorized carriage that moves on two axes to keep the torch positioned directly over the seam. The torch then moves through the use of motorized cross slides.

Different types of joints may be tracked by changing the type of probe. Where extensive tooling is not available and joint accuracy is required, seam trackers can produce excellent uniformity and consistency.

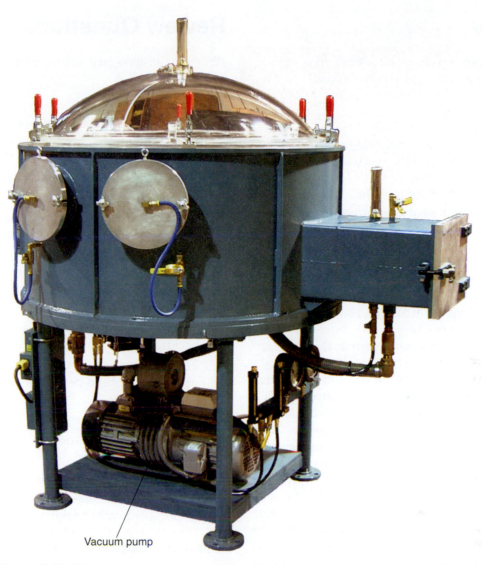

Vacuum pump

Figure 6-26. The vacuum pump mounted on this inert gas chamber reduces the time required to obtain a welding atmosphere. (Jetline Engineering, Inc.)

Summary

Tooling can affect every phase of fabrication, from welder training to selection of the process used. If tooling is not designed and constructed properly, the quality and production cost of the weldment are negatively affected.

Tooling can be classified as *hard tooling* or *shop aids*. Depending on the task, internal tooling, external tooling, combination assembly and welding tooling, or heat sink tooling is used.

Mechanical tooling can use a number of methods for holding or clamping. These methods include mechanical clamping forces, pneumatic pressure, hydraulic pressure, or a combination of pneumatic and hydraulic pressure.

The materials used in welding tooling are a major consideration. The strength and rigidity of the materials are important factors, as well as the magnetism of the materials.

Backing bars have grooves of various designs. The grooves shape the penetrating metal for a specific need. Inert gas backing must enter the groove with very little pressure. A backing bar can be used to control preheat and interpass temperatures. Backing rings provide quick and easy alignment for pipe and tubing.

Two types of hold-down bars—solid and segmented finger—are used to hold components in place and to control the heat flow from the weld joint. A relief step design hold-down bar can be used on slightly warped parts.

Tooling must be closely inspected and properly maintained so that consistent welds are produced. Tool cleanliness is essential. Tools that provide an inert backing gas atmosphere must be clean and dry. Mechanical/manual areas tools should be checked often for proper operation.

Test tooling can be made quickly and modified as necessary. Test tooling is used to prove tooling design concepts, make test welds, and establish weld joint shrinkage dimensions. Test tooling is also sometimes used in welder training and certification.

Auxiliary tooling includes seamers, manipulators, positioners and lathes, planishers, gas atmosphere welding chambers, and seam trackers. Seamers are used for making long seam internal or external welds on flat or cylindrical stock. Manipulators position the welding head at various locations and heights. Positioners and lathes rotate the weldment so welds can be made along circular seams. Planishers flatten and smooth a welded seam and prepare the part for final finishing. Atmosphere chambers are enclosures that surround the weld with a protective atmosphere. These chambers are used when the highest quality welds are necessary. Seam trackers are used in semiautomatic and automatic welding operations to align the welding torch along the weld joint.

Review Questions

Write your answers on a separate sheet of paper. Do not write in this book.

1. What is *hard tooling*?
2. What are the four basic types of tooling?
3. Why must tooling be designed to allow for shrinkage during welding?
4. List four methods of applying clamping pressure.
5. If the backing bar is flat against the weld, it acts as a(n) _____, removing heat to ensure fusion of the root.
6. What tooling materials are not used in the immediate weld area when using direct current for welding? Why?
7. What is a backing bar with a shallow radius backup groove normally used for?
8. The welding industry refers to the forming, or molding, of the penetration into a grooved backing bar as _____ the penetration.
9. What are the two basic types of hold-down bars?
10. Which type of hold-down bar distorts more easily?
11. Hold-down bars are made with a(n) _____ design for use on parts that are warped or distorted.
12. Backing bars and backing rings should be checked for _____ and _____, which prevent close contact with the component part.
13. Test tooling can be used to prove tooling design concepts, make test welds, and establish weld joint _____ dimensions.
14. What is the function of a seamer?
15. What are the two critical concerns if manipulators are used to make long seam welds?
16. _____ are motor-driven turntables that move the part into the position where it is easiest for the welder to weld it.
17. Lathes rotate the weldment so welds can be made along _____ seams.
18. What is a *planisher*?
19. Some gas atmosphere welding chambers use a(n) _____ to remove the air from the chamber.
20. _____ are used in semiautomatic and automatic welding operations to align the welding torch along the weld joint.

Chapter

Weld Preparation and Equipment Setup

Objectives

After completing this chapter, you will be able to:

❐ Understand the purpose of welding procedures.
❐ Summarize the techniques used to clean various metals prior to welding.
❐ Explain how the proper setup of the equipment and power source are determined.
❐ Set up welding machines and auxiliary equipment for DCEN.
❐ Set up welding machines and auxiliary equipment for DCEP.
❐ Set up welding machines and auxiliary equipment for alternating current.

Key Terms

abrasive pads
bright metal
welding procedures

Introduction

Some of the most important aspects of creating a quality weld take place before the arc is ever struck. Material preparation, joint design, and proper cleaning of the material are critical to ensure a quality weld. The best welding technique and highest skill levels cannot compensate for improperly cleaned materials or an improperly set up machine.

Different materials require different preparation. It is important to understand correct preparation techniques in order to avoid inconsistencies when making a weld. Some materials, such as titanium and aluminum, require more cleaning and protection than regular mild steel.

Equipment setup is also a critical aspect of making quality welds. The setup of a power source and accessory equipment requires careful consideration. Without proper setup, the equipment may be damaged, the components or parts may be ruined, or the operator may be injured. Careful selection of the welding process can also greatly affect efficiency in a high-production situation.

Welding Procedure Specifications

In situations where the weld must be qualified by testing prior to use, *welding procedures* are followed. A welding procedure is recorded in a welding procedure specification (WPS), which lists all specific variables and parameters that produce acceptable welds for a specific application. These values are obtained during the test welding and must be used for the production job. A blank welding procedure specification form for a manual GTAW operation is shown in **Figure 7-1**.

Some code books have a section devoted to the development of pre-qualified welding procedures. This section provides data for essential variables that are acceptable to create a pre-qualified procedure. The procedure produced by these guidelines then has to be proven through different types of testing, which might include both destructive and non-destructive testing procedures. Once a procedure has passed all the required testing, it is then considered a qualified procedure. Each procedure is specific to the welding process, position, joint design, and essential variables.

Since different jobs require different settings and variable values, and a particular job may be repeated in the future, a detailed welding procedure or setup should be written for each job. Record the variables used during successful test welding for future reference. During future operations, refer to these values to be sure the job is being performed properly.

Job and Equipment Considerations

Prior to making first time welds or welds without established procedures, select the equipment and adjust the settings based on the following variables:

Power source
- Type of current to be used.
- Amperage required for the job.
- Duty cycle required for the job.

Welding torch
- Duty cycle required for the job.

Electrode type and size
- Select an electrode based on the material being welded and the welding current polarity that will be used.
  - Pure tungsten for ac welding of aluminum and magnesium.
  - Zirconiated tungsten for ac welding of aluminum and magnesium where tungsten spitting cannot be tolerated.
  - Thoriated tungsten (1% or 2%) for welding all other metals using direct current straight polarity.
- Select an electrode size for the amperage range that will be used. (Use a chart like the one shown in **Figure 7-2**).

MANUAL GAS TUNGSTEN ARC WELDING PROCEDURE SPECIFICATION

PROJECT_____NAME_____

WELDING SPEC._____

BASE MATERIAL_____THICKNESS_____BASE METAL PREP_____

TYPE OF JOINT_____WELDING POSITION_____

FILLER MATERIAL_____SPEC._____DIAMETER_____

TORCH GAS_____CFH_____BACKUP GAS_____CFH_____

ELECTRODE DIAMETER_____TUNGSTEN TYPE AND CONFIGURATION_____

TYPE OF CURRENT_____AMPERAGE_____POWER SUPPLY_____

PREHEAT TEMP._____INTERPASS TEMP._____POST HEAT TEMP._____

NOTES:

Figure 7-1. Welding procedure specifications are used to record the proper machine settings for different jobs.

Electrode Diameter (inches)	Current–Amperes		Approximate Argon Gas Flow cfh at 20 psi	
	Pure Tungsten	Thoriated Tungsten	Aluminum	Stainless Steel
.010	0–8	0–8	3–8	3–8
.015	5–12	5–12	5–10	5–10
.020	8–20	8–20	5–10	5–10
.040	20–50	20–50	5–10	5–10
1/16	40–120	50–150	13–17	9–13
3/32	100–160	140–250	15–19	11–15
1/8	150–210	220–350	19–23	11–15
5/32	190–270	300–450	21–25	13–17
3/16	250–350	400–550	23–27	18–22
1/4	300–490	500–800	28–32	23–27

Figure 7-2. Select the correct electrode size for job, and then set the amperage range on the machine for the amount of current desired.

Electrode preparation
- Pure and zirconiated tungsten—Grind a radius on the end to be used. Form a ball of the desired diameter on the end of the electrode if it going to be used with ac.
- Thoriated tungsten—Grind a taper on one end (do not grind the painted end) for direct current electrode negative.

Shielding gas
- Select the appropriate shielding gas.
  - Argon—Arc is easy to start. Good blanket coverage.
  - Argon and helium mix—Arc is harder to start. Welds have deeper penetration. Requires higher flow rates as helium rises up from weld, and welds can be made at faster speeds.
- Shielding gas flow must be maintained at the torch nozzle for weld protection. If the electrode is extended beyond 1/4″ (6.5 mm), use a gas lens for a more stable flow.
- Nozzle size should be as large as possible to cover the weld area.

Preweld Cleaning

Both practice welding and production welding must be done on clean material. Improperly prepared material causes many problems during the welding operation and will result in poor weld quality. Oxides on the material will cause porosity and other discontinuities in the weld.

Cold-Rolled Steel

Cold-rolled steel usually has a bright metallic finish with a light coating of oil on the surface. Remove the oil by washing with a solvent such as alcohol or acetone.

The weld joint, the immediate area, and the filler metal must be clean in order to produce a weld with satisfactory properties. The cleaning operation must completely remove rust and mill scale (scale that formed when the material was cooled in air after forming or rolling) from the weld area.

Any rust on the material must be removed down to *bright metal* (shiny bare metal). Rust is formed by oxidation of iron. Oxides cause porosity in the weld. The weld pool does not burn rust away; if you weld over rust, it will become part of the weld.

Several methods can be used to clean the joint. These cleaning methods include grinding, chipping, sanding, filing, chemical etching, and media blasting. Weld joints can be sanded with a sanding wheel, **Figure 7-3**, to remove scale. After removal of the scale, brush the area to remove oxide particles and residue left over from the sanding wheel. After media blasting, the surface must be wire brushed to remove any remaining particles.

Immediately prior to welding, wash the joint area with alcohol or acetone. Follow this cleaning with a short drying period or a blast of clean, dry air to evaporate any residual liquid. Do *not* attempt to weld on a wet surface.

All tooling in the weld area must also be cleaned prior to welding. Backing bars, chill bars, clamps, and other tooling retain moisture. When this tooling is heated by the welding operation, hydrogen may be

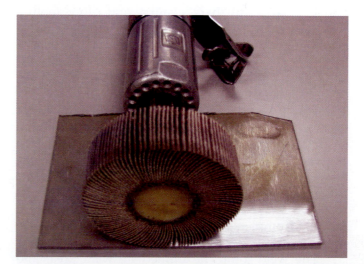

Figure 7-3. The edge has been sanded to bright metal to prepare the part for welding. This type of sander is often called a PG wheel. (Mark Prosser)

released. Cleaning the tooling is just as important as cleaning the base material and welding wire.

Hot-Rolled Steel

The surface of hot-rolled steel often has a flaky blue scale called *mill scale*. Mill scale must be removed by media blasting, grinding, or sanding to bright metal. Media blasting, if used, must be followed by grinding, sanding, or wire brushing to remove embedded grit from the surface. Use the same preparation for rusted hot-rolled steel as for cold-rolled steel.

Stainless Steel

The weld area must be clean in order to create a satisfactory weld. Remove oil and grease from all types of stainless steels prior to welding. Use acetone, alcohol, or any commercial degreaser.

Austenitic stainless steels, which are the most commonly used stainless steels, are usually bright and shiny. They require only a washing with solvent, alcohol, or acetone to remove oils and grease. Remove any heavy oxidation on the metal prior to welding by chemical etching, media blasting, grinding, or sanding. Follow media blasting with sanding, grinding, or wire brushing to remove embedded grit from the surface.

Martensitic, ferritic, and various precipitation-hardening stainless materials have a hard oxide film on the surface that must be removed down to bright metal prior to welding. Remove heavy oxides by grinding or media blasting. After blasting is completed, rotary sand to bright metal. Wire brushing will not sufficiently remove grit particles.

Prior to welding, clean all tooling in the weld area. Backing bars, chill bars, clamps, and other tooling retain moisture and release hydrogen when heated by the welding operation.

Aluminum

All aluminum materials have an oxide film on the surface. This film results from exposure of the material to air after forming or rolling operations. The oxide film melts at approximately 3500°F (1927°C), while the aluminum melts at approximately 1200°F (649°C). The difference in the melting points of aluminum and its oxide can make the metal difficult to weld. It can also lead to incomplete fusion and weaken the weld due to inclusions of oxides in the weld.

The weld joint, the immediate weld area, and the filler metal must be clean if the weld is to have satisfactory properties. After the metal is clean, begin welding as soon as possible. An oxide film quickly forms on freshly cleaned aluminum. Such a film can prevent fusion between the filler metal and the base metal. An oxide film can also cause the formation of flakes of oxide. Incomplete cleaning can result in dross becoming trapped within the weld metal.

Cleaning can be done using chemical or mechanical means. Oxide film is often removed with a stainless steel wire brush, either by hand or with a power tool. Any tools used to clean aluminum must be dedicated to aluminum only. If a tool is used on other materials and then used to clean aluminum, it can impregnate impurities into the aluminum, causing problem with the weld.

Chemical cleaning includes the following methods:

- Commercial degreasers.
- Commercial aluminum cleaners, as shown in **Figure 7-4**.
- Chemical etching. Usually done with a commercial etch and followed by a deionized water rinse and oven drying.

Mechanical cleaning methods include the following:

- Filing with a coarse file, **Figure 7-5**.

Figure 7-4. A commercial aluminum cleaner. (Mark Prosser)

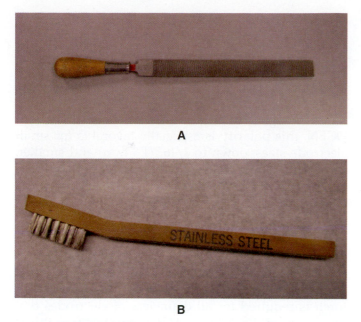

A

B

Figure 7-5. Mechanical cleaning can be done with a coarse file or a stainless steel wire brush. A—File. B—Stainless steel wire brush. (Mark Prosser)

- Using stainless steel wire brushes. Avoid a heavy wire-type brush because it can tear the material.
- Chipping or routing. This method is limited to heavier materials such as plate, castings, or forgings.
- Abrasive pads. *Abrasive pads* are mineral-impregnated fiber pads. Typical pads are approximately 6″ × 9″ and readily form to various contours.

Immediately prior to welding, wipe the joint with alcohol or acetone. Follow this operation with a short drying period. A blast of clean, dry air can be used to evaporate any residual liquid. Do *not* attempt to weld on a wet surface.

Prior to welding, clean all tooling in the weld area. Backing bars, chill bars, and clamps retain moisture and release hydrogen when heated by the welding operation.

Magnesium

Magnesium is generally received from the mill with a lightly oiled or chemically coated surface. Mechanical cleaning with stainless steel wool, a power wire brush, or even an abrasive scratch pad is generally adequate for removing surface contaminants. Cleaning can also be done with a chemical solution of chromic acid, ferric nitrate, potassium fluoride, and water. However, mechanical cleaning is acceptable in most situations.

Copper and Copper Alloys

Remove oil, grease, and other contaminants with any standard degreaser, such as acetone or alcohol, prior to welding. If corrosion or scale is present, remove it prior to welding. Commercial copper cleaners or acid baths can be used for this purpose. Copper can also be mechanically cleaned with abrasives to remove corrosion or pitting. Abrasives gum up quickly due to the softness of the copper. Abrasives can also contaminate the base material. The chemical method is often the best way of preparing copper materials.

Titanium

GTAW can produce excellent welds in titanium, but proper preparation is crucial. Prior to welding, oil, grease, and other contaminants must be removed. Any standard degreaser, such as acetone or alcohol, can be used for this purpose.

Thermally Cut Edges

Metal edges cut by oxyacetylene cutting, plasma arc cutting, or carbon arc gouging must be cleaned to bright metal before being welded. These joint edges usually have an oxide film on the surface that is heavier than the oxide film of hot-rolled material. Remove the oxide film by sanding, grinding, or media blasting until only bright metal is visible all around the joint area. Removing oxide film by media blasting should be followed up with grinding, sanding, or wire brushing to remove residual grit. Any indentations into the side of the weld joint must be removed by grinding, as they contain heavy oxides. Refer to **Figure 7-6**.

Butted Edges

All butted edges of joints have oxides on the surface regardless of the preparation process. This oxide film must be removed to obtain high quality welds. (This is especially important on aluminum, as the oxide film can prevent fusion in the root pass.) The surface can easily be cleaned by drawing a file across the edge. The operation needs to remove only a small amount of material.

Rough edge caused by thermal cutting

Figure 7-6. Thermal cutting gouges and oxide films must be removed to bright metal. Note the indentations on the edge, which can contain oxides that contaminate the weld. The entire edge must be ground smooth. (Mark Prosser)

Equipment and Welding Machine Setup

The actual setup of the equipment and the power source will vary considerably depending on the type and make of the equipment. Equipment from different manufacturers varies in design. However, all will have the same basic controls. For manual all-purpose work in a typical shop, an industrial-rated 300 ampere, 60% duty cycle, ac/dc power source is a good choice.

Selecting a Welding Current

Before setting up a welding machine for a job, the welder must determine which welding current polarity will be used. Detailed information about the different polarities (and when they should be used) was presented in Chapter 2, *Power Sources*. The welding machine settings, electrode selection and preparation, and shielding gas selection vary depending on the welding current polarity selected.

DCEN Polarity

DCEN is the preferred current type for welding steels. DCEN has excellent penetration characteristics because it concentrates two-thirds of the arc heat onto the base metal and one-third of the heat onto the electrode.

DCEN can be used with a liquid-cooled torch or an air-cooled torch, depending on the amperage

range and the thickness of the material. The amount of heat generated in the torch depends on the amount of amperage being used. Higher amperages require a liquid-cooled torch. The amount of heat in the torch is monitored by the welder and is determined by the temperature of the torch handle.

Although any size electrode can be used with DCEN, this polarity is especially favorable for small electrode sizes, down to 1/16″ (1.6 mm) and smaller. Small electrode sizes can be used because two-thirds of the arc heat is concentrated on the material not the electrode.

DCEP Polarity

With DCEP polarity, all of the welding heat is generated in the electrode. Because of this, a large-diameter electrode must be used. A water-cooled torch is required for all diameters of electrodes that may be used. DCEP current concentrates two-thirds of the heat onto the electrode and one-third onto the base metal. For this reason, DCEP current is only used in situations where a wide but shallow bead is desired. The penetration characteristics of DCEP are very limited compared to those of DCEN current.

Since GTAW torches are not rated in the DCEP mode, the duty-cycle rating cannot be used. The torch temperature must be monitored by the welder. A warm torch indicates insufficient cooling. If the torch is warm, the welder must lower the amperage or reduce the amount of time the torch is used.

A 2% thoriated tungsten electrode should be used to carry the high heat. The end of the electrode should not be tapered. A slight radius on the end will help form a ball when positive polarity is used.

AC Polarity

Alternating current switches back and forth from DCEN to DCEP 60 times a second (or up to 200 times a second with the newer inverter-type machines). The current switches back and forth between electrode positive current, which breaks up the oxide film on the metal, and electrode negative current, which provides good penetration characteristics. Another difference between ac and DCEN or DCEP current is that ac current can make use of high-frequency voltage.

Setting Up the Welding Machine

After choosing a polarity, the welder must set up the correct shielding gas and prepare and install the proper electrode in the torch. These tasks are covered in detail in Chapter 3, *Auxiliary Equipment and Systems*. After choosing the polarity and preparing the auxiliary equipment, the welding machine can be set up for the job at hand.

Each manufacturer provides a booklet or instruction manual that explains the machine's operation in detail. Before the equipment and machine are set up, the instructions should be read until they are fully understood.

The major controls used in manual GTAW are shown on an industrial-rated welding power source in **Figure 7-7**. Refer to this figure as you read through the following setup procedure:

1. Connect the primary power.
2. Connect the argon gas supply to the welding machine.
   A gas line is typically a rubber hose that connects the outlet of the shielding gas regulator to the back of the machine. If the gas flow is adjusted manually, it is controlled by a solenoid that is operated by a foot pedal or a control on the torch. Gas lines must be handled with care to avoid punctures and tears. The gas lines and the fittings on the ends of the lines need to be checked periodically to ensure they are tight and not leaking. Keep gas lines off the floor and do not run the wheels of the machine over the lines.
3. Connect the water supply (if the torch is water-cooled) to the welding machine.
   Water-cooled torches are used primarily for welding at higher amperage ranges. The water supply line is connected to the back of the welding machine. The water cooling system in a welding machine works much like the cooling system of an automobile. The machine has a pump that circulates water through the torch

Figure 7-7. Industrial-rated (60% duty cycle) power source. (Photo used with the permission of Lincoln Electric, Cleveland, OH)

line to the head of the torch, where it absorbs heat, cooling the torch. The warm water returns to a radiator in the welding machine, where it is air cooled and then recirculated.

4. Set the *Current Type* switch to *DCEN*, *DCEP*, or *AC*, depending on the desired polarity.
5. Connect the torch power cable to the electrode terminal and connect the ground cable to the ground terminal. Always check the cable connections to ensure they are tight. Loose connections can cause overheating and result in damage to the machine.
6. Set the *Current Range* switch (if available) to the range required.
   Current Range switches are found on older machines. On newer machines, the current (amperage) is controlled by a rheostat dial. Amperage settings are based on the type and thickness of the material to be welded.
7. Set *Current* control for maximum current desired. (Note: This control is 100% of set range.) When a scratch start or lift arc system is used, the amount of amperage set on the machine begins flowing as soon as the arc is struck. When a foot pedal control is used, the amperage can be increased or decreased by pushing the pedal further down or releasing the pedal.
8. Set the *Panel/Remote* control switch to the desired position.
   If the switch is in the *Panel* position, the current is held steady at the amperage set by the maximum current setting on the welding machine. Any adjustments to welding current must be made by adjusting the controls on the machine itself. If the switch is in the *Remote* position, the welder can adjust the amperage using the torch control or foot pedal. With either setting, current will not exceed the panel rheostat current control setting.
9. Set the *High Frequency* control.
   DCEN and DCEP Polarities
   • Set the *High Frequency* control to *Start Only*.
   • High frequency is set to *Start Only* for welding steels. High-frequency voltage is an arc-starting device that initiates the arc without the electrode touching the work. This is the system used with a foot pedal control. Machines without high-frequency voltage capability require the electrode to be touched to the base metal.
   AC Polarity
   • Set the *High Frequency* control to *Continuous*.
   • When welding is being performed with ac current, high-frequency voltage provides

the initial starting arc without the need to touch the electrode to the base metal. It also provides arc stability when the current switches back and forth between different polarities. When the current switches from electrode negative to electrode positive, the arc normally extinguishes. Adding high-frequency voltage stabilizes and maintains the arc, which would otherwise die out with each switch of polarity.

- Set *Wave Balance* control to 50% (or *Normal*). A typical ac sine wave is divided evenly between electrode negative and electrode positive polarity. Using this control, the percentage of time spent in electrode negative polarity can be adjusted. Increasing the percentage of time spent in electrode negative polarity increases penetration. Decreasing the percentage of time spent in electrode negative polarity increases the percentage of time spent in electrode positive polarity, which improves the cleaning action. When working with new materials, it is best to start with a "balanced" wave.
- Set the *Soft Start* control.
- The *Soft Start* control is used for starting low currents for welding thin materials only.

10. Set the shielding gas *Postflow Time* (if it is controlled by the welding machine).
    Postflow of gas is required to prevent electrode contamination during the postweld cooling period. The postflow time is based on the size of electrode and the type of material being welded. The larger the electrode diameter, the longer the postflow time should be. To prevent contamination, be sure the postflow timer setting provides enough time for the electrode to cool to a silver color before the gas shuts off. Different base metal materials, such as stainless steels, can require longer postflow times to ensure that the weld cools in an inert atmosphere.

11. Set the position of the *Mode* switch (if available) to GTAW. The *Mode* switch can be used to configure the welding machine for either SMAW or GTAW capabilities. Some machines, such as the one shown in **Figure 7-7**, describe the settings as TIG and Stick.
    When this switch is in the GTAW or TIG position, it actuates the machine's control circuits for starting and stopping the cover gas flow and the coolant flow on water-cooled torches and energizes the contactor circuit. *Mode* switches are more common on newer machines. If the GTAW machine does not have a *Mode* switch, the cover gas and the coolant flow must be started and stopped manually.

12. Turn on the power source.

13. Energize the contactor.
    Once the contactor is energized, current is controlled through a foot pedal, a hand control that is located on the torch handle, or a panel control on the welding machine. The gas flow rate must also be set. The flow rate is usually set to 20 to 30 cfh (cubic feet per hour), depending on the type of material and the joint design. The system must be initiated in order to set the flow of gas. When a pedal is being used for current control, the power is turned on and the pedal is pushed to start the process. The pressure is adjusted by slowly turning the regulator valve until the desired pressure is displayed on the regulator pressure gauge. If the system is equipped with a thumb control rather than a foot pedal, the thumb control must be turned on in order to set the regulator.

Summary

Improper material preparation, cleaning, and machine setup can cause weld discontinuities. A welding procedure specification (WPS) lists all specific variables and parameters that produce acceptable welds for a specific application. When making first time welds or welds without established procedures, a welder must consider the current type to be used, amperage required, power source duty cycle; welding torch duty cycle, electrode type and size, electrode preparation, and shielding gas selection.

Welding on improperly prepared material causes many problems during the operation and affects the final weld quality. Oxides will cause porosity in the weld. Oil, grease, rust, and oxide film, must be removed prior to welding. Acetone, alcohol, or commercial degreasers can be used to remove oil and grease.

Material can be cleaned using chemical or mechanical means. Cleaning methods include grinding, chipping, routing, sanding, filing, abrasive pads, commercial cleaners, chemical etching, and media blasting. Removing oxide film by media blasting should be followed up with grinding, sanding, or wire brushing to remove residual grit. Any tools used to clean aluminum must be dedicated to aluminum only.

The welding machine must be set up correctly for the particular job. If the machine is not operated correctly, malfunctions can occur that can ruin the machine or the components being welded. The manufacturer's instructions should be read and understood before the equipment and machine are set up.

Review Questions

Write your answers on a separate sheet of paper. Do not write in this book.

1. A(n) _____ lists the required welding variables for a specific application.

2. Oxides on material to be welded will cause _____ in the weld.

3. How does the finish of cold-rolled steel differ from the finish of hot-rolled steel?

4. What does the term *bright metal* refer to?

5. Rust is formed by _____ of iron.

6. A commercial degreaser, _____, or _____ can be used to remove oil or grease from material.

7. Which stainless steels have a hard oxide film on the surface?

8. At what temperature does the oxide film on aluminum melt?

9. A(n) _____ wire brush is often used to remove oxide film from aluminum.

10. Why should welding begin as soon as possible after cleaning aluminum?

11. List three methods that can be used to remove grit after media blasting.

12. Tooling in the weld area retains moisture and can release _____ when heated by the welding operation.

13. Oxides can be removed from butted edges of joints by drawing a(n) _____ across the edge.

14. What should be used to remove corrosion or scale from copper?

15. DCEN is the preferred current type for welding steels because of its excellent _____ characteristics.

16. What is the function of high-frequency voltage when welding aluminum?

17. Amperage settings are based on the _____ of the material to be welded.

18. What is the function of wave balance control when using ac current?

19. To prevent contamination, the postflow timer setting should provide enough time for the electrode to cool to a(n) _____ color before the gas shuts off.

20. What is the *Mode* switch used for?

In order to produce a satisfactory weld, any rust on the material must be removed down to bright metal. Several methods, including grinding, can be used to clean the joint. (Dainis Derics/Shutterstock)

Chapter

Manual Welding Techniques

8

Objectives

After completing this chapter, you will be able to:

- ❑ Identify types and grades of steel and steel alloys.
- ❑ Identify shapes and forms of steel and steel alloys.
- ❑ Select the appropriate steel filler based on the steel to be welded.
- ❑ Explain joint preparation, weld backing, and preheating for steels.
- ❑ Recall welding procedures and techniques for welding steel using DCEN.
- ❑ Recall correct torch positioning for various types of welds.
- ❑ Differentiate between stringer beads and weave beads.
- ❑ Recognize groove and fillet weld defects.
- ❑ Recall factors that influence postweld treatment.

Key Terms

annealing
casting
chromium-molybdenum steels
forging
heat treating
martensite
quenching
stringer beads
tempering
travel angle
weave beads
wetted
work angle

Introduction

The GTAW technique required for a given job depends on many factors, including the weld joint design. Welders must constantly monitor and adjust many critical variables in order to produce a high-quality weld. The effects of these variables can be seen in the behavior of the weld pool as the weld is being made. Welding skill is acquired by practice, practice, and more practice. Changing one variable at a time is the best way to determine what the weld needs.

The following questions should be considered when practicing welding:

- Is the weld the right size?
- Is the weld too high or too low?
- Is the proper travel speed being used?
- Is the torch angle correct?
- Is the welding rod being held at the proper angle?
- Is the weld pool flowing properly?
- Is the current (amperage) correct?
- Is the voltage (arc length) correct?
- Is the weld in the proper location?
- Is the fillet weld leg the proper size?

Also, it is important to remember to fill the crater at the end of the weld and hold the torch over the end of the weld until it cools. This allows the postflow of shielding gas to protect the hot metal.

Base Materials (Steel)

Many types and grades of steel are included in the steel family. These materials are typically magnetic and melt at approximately 2500°F (1371°C).

Carbon Steels

Carbon steels are identified as a group that contains the following materials:

- Carbon—1.70% maximum
- Manganese—1.65% maximum
- Silicon—0.60%

Carbon steels are further classified as low, mild, medium, and high carbon:

- Low-carbon steel—less than 20% carbon
- Medium-carbon steel—0.20%–0.50% carbon
- High-carbon steel—over 50% carbon

Low-Alloy Steels

Low-alloy steels contain varying amounts of carbon and a variety of alloying elements. These elements include chromium, molybdenum, nickel, vanadium, and manganese. These elements increase the strength and toughness of the material and, in some cases, increase resistance to corrosion.

Heat-Treated Materials

Heat-treated steels are used for many different applications. *Heat treating* is a process in which the material is heated and cooled to specific temperatures for specific amounts of time. A material is heat treated to obtain desirable qualities or to reduce undesirable qualities. Many materials and alloys, including aluminum, copper, titanium, nickel, and chromium-molybdenum, can be heat treated to make the material perform in different ways. Heat treating of a material is most commonly done as part of the manufacturing process.

Quenching is the process of rapidly cooling a material to obtain certain material properties that increase the material's toughness. The quenching process changes the crystalline structure of the material.

Tempering is a heat treatment that can improve the mechanical properties of a material. *Tempering* involves heating the metal to a certain temperature below the material's melting temperature for a certain length of time. This process allows trapped carbon to produce a different crystalline structure. The material is then cooled at a controlled rate to a low temperature. The temperature and the duration of the tempering process determine which mechanical properties will be enhanced.

Annealing is a heat treatment that involves heating the material to a temperature that rearranges the crystalline structure and reduces internal stress of the material, making the material softer and more ductile. Annealing is typically done to improve the cold working properties of a material.

Tool Steels

Tool steels are steels with a combination of high carbon and alloy content. Tool steels are used for making dies, cutting bits, and many other types of tools. Dies and other tools are formed from annealed tool steel. Later in the manufacturing process, after the tools are formed, the metal is treated to harden or temper the metal as needed.

The various types of tool steel include W1, W2, SI, S5, S7, 01, 06, A2, A4, and D2. Also included are H11, H12, H13, and M1, M2, M10. The numbers for the various tool steels identify the various compositions. The letters usually indicate the type of quenching required to achieve full mechanical values.

Forms and Shapes of Steel

Steel is supplied in many forms and shapes for welding. Some of the more common shapes are hot-rolled steel, cold-rolled steel, castings, and forgings.

Hot-Rolled Steel

Hot-rolled steel includes plate and structural forms which are allowed to cool in air after rolling. After cooling, the oxide film on the surface is light blue in color.

Cold-Rolled Steel

This material is final-rolled to the required dimensions in the cold condition and does not have an oxide film on the surface. A light coating of oil is placed on the material to prevent rusting.

Castings

Casting refers to pouring molten metal into a mold that has the desired shape. The metal is kept in the mold until it solidifies, at which point the mold is removed. The cast metal inside has the same shape as the inside of the mold. Castings are usually made in a sand mold and have a rough surface. New castings are sandblasted, leaving a dull silver surface.

Forgings

Forging refers to pressing a shaped die into hot metal under high pressure. The metal is heated to a plastic state before it is forged so it will take on the shape of the die that is pressed into it. Forgings are made from billets, bars, round stock, or square stock. Forgings can be identified by the remnants of the flashing usually extending around the center of the part. Since the forging operation is done while the part is hot, a scale forms on the surface, similar to the scale on hot-rolled steel.

Steel Classifications

Steels are manufactured to specifications developed by various organizations, including ASTM International. One grouping of steels under this classification is ASTM A335. The assigned designation identifies the material as high-ferritic steel alloy pipe for high-temperature applications. Other steel specifications include SAE International's Aerospace Material Specifications (AMS), the American Iron and Steel Institute (AISI) classifications, and corporation specifications.

Filler Metals

Filler metals used to join carbon steels, low-alloy steels, and tool steels must be selected to produce the desired mechanical properties of the weldment after any required welding or heat treatment. Carbon and low-alloy steel filler metals are selected from specifications, such as AWS A5.18, AWS A5.28, MIL-E-23765, MIL-S-6758, and AMS6370.

When selecting a steel filler metal, the welder should consider the basic type of steel to be welded and the possibility of excessive porosity within the completed weld. **Figure 8-1** lists three classifications of filler metal, each with a 70,000 psi tensile strength that can be used to weld porosity-prone material. **Figure 8-2** lists several types of filler metals that can be used to weld HSLA (high-strength, low-alloy) steels.

Chromium-molybdenum steels (chrome-moly steels) are a class of extremely strong and hard steels. If the chrome-moly weldment is to be hardened after being welded, one of the following filler metals should be used:

- 1-1/4% chrome—1/2% moly filler metal
- 2-1/4% chrome—1% moly filler metal
- 4%–6% chrome (AISI 502)

Solid Wires for Carbon Steel[a] Per AWS A5.18						
Classification	Composition					
	C	Mn	Si	Ti	Zr	Al
ER70S-2	0.07	0.90–1.40	0.40–0.70	0.05–0.15	0.02–0.12	0.05–0.15
ER70S-3	0.06–0.15	0.90–1.40	0.45–0.70	–	–	–
ER70S-5	0.07–0.19	0.90–1.40	0.30–0.60	–	–	0.50–0.90

[a] P, 0.025; S, 0.035; Cu 0.50.

Figure 8-1. The letters *E* and *R* in the classification code identify these filler metals as solid electrode wires that can be used in the GTAW process. The letter *E* indicates the wire is designed to carry current for the GMAW process. Some electrode wires can also be used (without current) as filler metal for the GTAW process. If the electrode wire can also be used for GTAW, the letter *E* is followed by the letter *R* (which stands for "rod").

If the chrome-moly steel weldment will not be hardened after welding, the following stainless steel filler metals can be used:

- 25% Cr—20% Ni stainless steel filler metal
- 25% Cr—12% Ni stainless steel filler metal

Do not use stainless steel filler metals for welds in service over 1000°F (538°C). Welds subjected to temperatures above 1000°F (538°C) will have carbon migration from the steel to the stainless steel, which can cause weld failure.

When selecting filler metals for tool steels, refer to the filler metal manufacturer for the recommended type. These filler metals have been specially developed for specific applications and are not made to precisely conform to specifications issued by organizations such as ASTM or AISI.

Filler Metal Quality

Steel filler metals are supplied by the manufacturer with a bright finish, oiled finish, or copper flash finish. Since these materials are susceptible to rusting, store them in a dry, heated area until they are needed. Always clean the filler metal with acetone or alcohol just before use. Return all unused material to the storage area.

Solid Wires for HSLA Steel Per AWS 15.28											
Classification	**Composition**[a,b]										**Characteristics, Applications**
	C	Mn	Si	P	S	Ni	Cr	Mo	V	Cu	
Cr-Mo ER80S-B2	0.07–0.12						1.20–1.50	0.40–0.65			Use to weld 1/2 Cr-1/2 Mo, 1 Cr-1/2 Mo, 1-1/4 Cr-1/2 Mo steels for high-temperature and corrosive service and for Cr-Mo to carbon-steel welds.
ER80S-B2L	0.05	0.40–0.70	0.40–0.70	0.025	0.025	0.20	1.20–1.50	0.40–0.65	–	0.35	Lower carbon content improves resistance to cracking. Use where postweld heat treatment is not performed.
ER90S-B3	0.07–0.12						2.30–2.70	0.90–1.20			Use for 2-1/4 Cr-1 Mo steel for high-temperature high-pressure piping and vessels. For all Cr-Mo-to-carbon-steel joints.
ER90S-B3L	0.05						2.30–2.70	0.90–1.20			Lower carbon content improves resistance to cracking. Use where postweld heat treatment is not performed.
Ni steel ER80S-Ni3	0.12	1.25	0.40–0.80	0.25	0.25	3.00–3.75	–	–	–	–	Use to weld HSLA steels and nickel steels up to 3-1/2%. Gives high strength and good toughness down to -100°F (-73°C).
Mn-Mo ER80S-D2	0.07–0.12	1.60–2.10	0.50–0.80	0.025	0.025	0.15	–	0.40–0.60	–	0.50	High Mn and Si contents reduce porosity. For single and multipass welding of carbon and low-alloy steels. Use on S-bearing (free machining) steel. Farm implements, auto parts, light-gauge steel, AISI 4130, T-1.
Other[d] ER100S-1	0.08	1.25–1.80	0.20–0.50	–	–	1.40–2.10	0.30	0.25–0.55	0.05	0.25	High-strength, tough weld metal for critical welds. HY80, HY100 and higher-strength steels, structural applications for service as low as -60°F (-51°C).
ER100S-2	0.12	1.25–1.80	0.20–0.60	0.010	0.010	0.80–1.25	0.30	0.20–0.55	0.05	0.35–0.65	
ER110S-1	0.09	1.40–1.80	0.20–0.55			1.90–2.60	0.50	0.25–0.55	0.04	0.25	
ER120S-1 ERXXS-G[c]	0.10	1.40–1.80	0.25–0.60			2.00–2.80	0.60	0.30–0.65	0.03	0.25	

a. As manufactured.
b. Total other elements, 0.50.
c. Minimum 0.50 nickel, 0.30 chromium, or 0.20 molybdenum.
d. Ti, Zr, Al, 0.10 each, except –G wires.

Figure 8-2. The letters *HSLA* identify the steels as high-strength, low-alloy materials.

Joint Preparation

Joint edges prepared by thermal cutting processes have a heavy oxide film on the surface, as shown in **Figure 8-3**. This oxide should be completely removed prior to welding in order to prevent porosity in the weld metal.

Joint edges prepared by shearing should be sharp without tearing or ridges. Otherwise, dirt or oil can become trapped in the joint and result in faulty welds. Rough edges can be removed prior to welding with a light grinding.

Weld Backing for Steel

Groove welds that are welded from one side of the joint, where 100% penetration is required, can be backed with argon gas or a solid bar. Solid bars can be made of copper or stainless steel.

Excluding air from the backside or penetration side of the joint will assist in wetting of the penetration. A *wetted* penetration has a smooth junction and even flow of the penetration onto the base metal. The exclusion of air also prevents the formation of scale and oxide.

Preheating Steels

Carbon steels less than 1" (25.4 mm) thick and with less than .30% carbon generally do not require preheat. Welding on highly restrained joints is an exception. These joints should be preheated to 50°–100°F (10°–38°C) to minimize shrinkage cracks in the base metal and the weld deposit.

Low-alloy steels, such as the chrome-moly steels, have hard heat-affected zones after welding if the preheat temperature is too low. The hard heat-affected zones are caused by the rapid cooling rate of the base material and the formation of martensitic grain structures. A 200°–400°F (93°–204°C) preheat temperature

slows down the cooling rate and prevents the formation of a martensitic structure.

Martensite is a metallurgical term that defines a type of grain structure obtained by heating and quenching. The martensitic grain structure makes the metal hard. Parts that are welded are, in effect, heated and quenched. If the carbon content of the material is sufficient, and the preheat temperature is too low, the material in the heat affected zone will harden during welding. This results in high tensile properties with very low ductility. The formation of hard martensite-type grains increases the possibility of cracking as the weld metal cools and shrinks.

Tool steels have a very high carbon content and are prone to cracking in the heat-affected zone without sufficient preheat. **Figure 8-4** lists some recommended preheat temperature ranges for welding tool steels.

Quenched and tempered steel requires preheat and interpass temperature control to retain the original mechanical properties of the metal. The manufacturer's recommendations for these temperatures should be strictly followed.

Torch Angles and Bead Placement

Proper manipulation of the welding torch is very important in making a good weld. The torch is usually held like a pencil, as shown in **Figure 8-5**. When welding is done in the flat position, the hand should be placed lightly on a surface, so the hand can

Figure 8-3. The oxide scale and small gouge indentations were formed on this piece of metal by the oxyacetylene cutting process. (Mark Prosser)

Material Type	Annealed Base Material Preheat and Postheat Temperature	Hardened Base Material Preheat and Postheat Temperature
W1, W2	250–450°F (121–232°C)	250–450°F (121–232°C)
S1	300–500°F (149–260°C)	300–500°F (149–260°C)
S5	300–500°F (149–260°C)	300–500°F (149–260°C)
S7	300–500°F (149–260°C)	300–500°F (149–260°C)
01	300–400°F (149–204°C)	300–400°F (149–204°C)
06	300–400°F (149–204°C)	300–400°F (149–204°C)
A2	300–500°F (149–260°C)	300–400°F (149–204°C)
A4	300–500°F (149–260°C)	300–400°F (149–204°C)
D2	700–900°F (371–482°C)	700–900°F (371–482°C)
H11, H12, H13	900–1200°F (482–649°C)	700–1000°F (371–538°C)
M1, M2, M10	950–1100°F (510–593°C)	950–1050°F (510–566°C)

Figure 8-4. Thoroughly preheat the part to be welded to the required temperature. Do not allow the part to cool below the minimum temperature until postheat is complete.

Figure 8-5. Holding the torch in the manner shown allows the welder to move the torch easily to change torch angles. (Mark Prosser)

move across the joint evenly. Movement of the torch by the fingers alone usually results in incorrect torch angles and a poor weld.

Torch angles are very important to maintain and understand. The torch angles control the heat of the arc, and the weld pool will move where it is directed by the torch angles. Two different torch angles, the

work angle and the travel angle, greatly affect how the weld bead is placed into the weld joint.

The *work angle* is the angle of the torch in a plane that is perpendicular to the weld axis and is measured from the perpendicular position to the torch's actual position. See **Figure 8-6A**. Work angles usually split the weld pool evenly between the two pieces of base metal. When the work angle points to one side of the joint more than the other, the weld will move in that direction. For example, on a flat butt joint, the work angle is 90°, straight up and down into the joint. Work angles are generally straight into the middle of the joint. The weld pool will show if the weld is favoring one side or the other. When this happens, the welder must adjust the work angle to correct the uneven weld.

The *travel angle* is the angle of the torch in the plane parallel to the weld axis and is measured from the perpendicular position to the torch's actual position. See **Figure 8-6B**. The torch should always be pushed in the direction of the weld. Torch travel angles are usually 15° to 20°. The main reason to maintain a good travel angle is so that the welder can see the weld pool. The welder must be able to see the weld pool when making any weld. The actual weld pool is much smaller than the ceramic cup on the end of the torch, so the torch must be laid over so the weld pool can be seen. However, travel angles that are too severe can suck air into the weld from the back side, causing porosity or other discontinuities due to atmospheric contamination.

For welding butt joints in any position, the suggested work angle is generally 90°. Because the work angle is perpendicular to the workpiece, it provides maximum penetration. The suggested travel angle for welding butt joints is typically a push angle

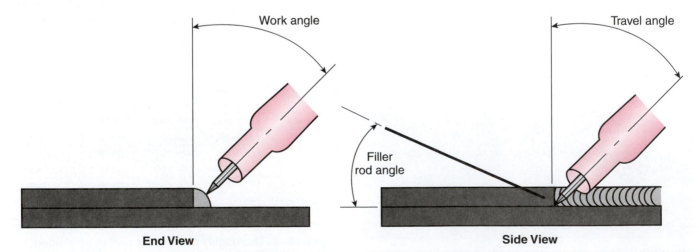

End View   **Side View**

Figure 8-6. Welders must be mindful of the travel angle, work angle, and filler rod angle. A—This end view of a fillet weld in progress shows how the work angle is measured. B—This side view of a fillet weld in progress shows how travel angle and filler rod angle are measured.

of 15°–20°. The term *push angle* refers to the torch being pointed in the direction the weld progresses. This angle allows the welder to see the weld pool, but still provides good penetration. A larger travel angle would create an elongated weld pool and decrease the penetration into the joint. The welding rod is typically held 10°–20° from the surface of the base metal. This angle makes it easy for the welder to see the weld pool and to feed welding rod into the weld pool as needed. See **Figure 8-7.**

When making fillet welds, the torch is generally held at a 45° work angle and a 15°–20° travel angle. If an inside corner joint is being welded, the tip of the electrode is pointed directly at the joint between the two workpieces. If a lap joint is being welded, the tip of the electrode should be pointed slightly more toward the surface piece than the edge piece. The edges of a metal plate are not able to dissipate heat as quickly as the large surfaces of the plate. As a result, if the torch is pointed directly at the joint between the two plates, the edge of one piece heats up faster than the surface of the other, which can lead to undercutting at the edges of the joint. Pointing the torch slightly more toward the surface than the edge provides even heating through the entire joint. See **Figure 8-8.**

When adding welding rod, the welder should grip the rod in the fingers as shown in **Figure 8-9.** The hand should be kept as close as possible to the arc to hold the rod steady. The rod should be moved in conjunction with the torch movement. When additional rod is required, it can be moved forward through the fingers using a forward movement of the thumb. Too much extension of the welding rod from the fingers results in a wobbly rod end, making addition to the weld pool very uneven. Adding rod to the pool requires steadiness and concentration in order to place the correct

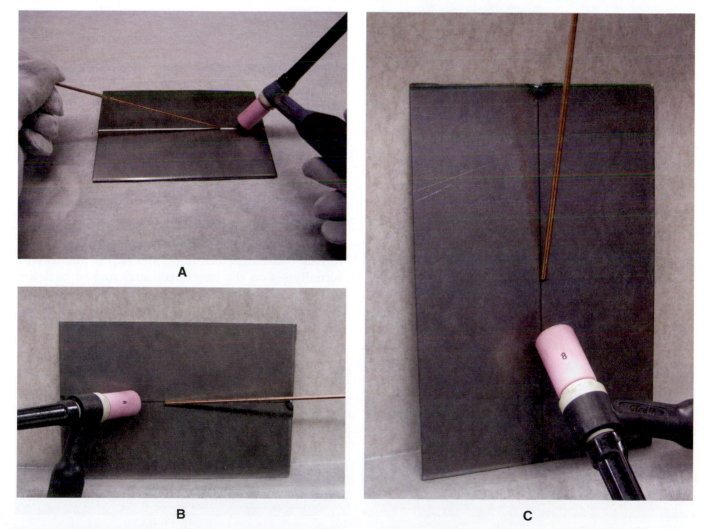

Figure 8-7. Typical welding angles for welding a butt joint in any position. A—This butt weld is in the flat position. Note that the torch is at a 90° work angle. The torch is tilted so the electrode is pointing in the direction the weld is progressing. This is referred to as the *travel angle*. B—A butt weld in the horizontal position. C—A butt weld in the vertical position. (Mark Prosser)

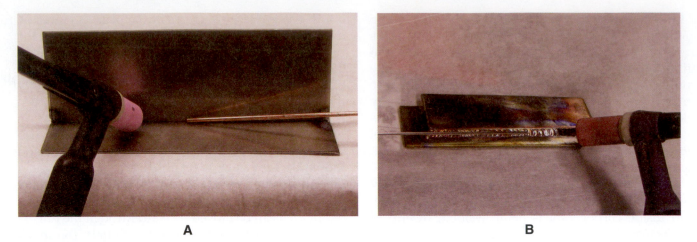

Figure 8-8. The proper work and travel angles for making fillet welds. A—Proper angles for T-joints and inside corner joints. B—Proper angles for lap joints. (Mark Prosser)

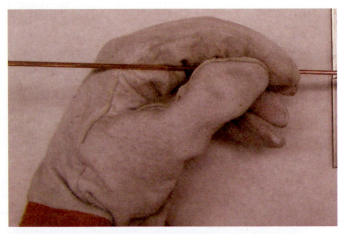

Figure 8-9. Welding rod held in this manner can be added to the weld pool as needed. (Mark Prosser)

amount of material at the right place at exactly the right time. A 10°–20° entrance angle of the rod into the weld pool should always be maintained.

Types of Beads/Passes

Welds made without any side-to-side movement (oscillation) of the torch are called *stringer beads*, or *stringer passes*. Welds made with side-to-side movement of the torch are called *weave beads*. When a weave bead is being made, filler metal should be added at the edges of the weld pool to prevent under-cutting. **Figure 8-10** compares a stringer bead and a weave bead. A technique known as *walking*, or *rocking*, the cup is commonly used to create a weave pattern in V-groove butt joints in thicker sections of material. In this oscillation technique, the cup of the torch is actually set on the base metal and rocked back and forth, much like moving a refrigerator across the floor. Walking or rocking the cup to make a weave pattern takes a great deal of practice and determination.

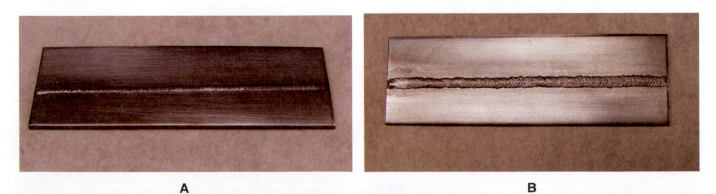

Figure 8-10. A stringer bead (A) is narrower than a weave bead (B). (Mark Prosser)

Welding Steels Using DCEN (Straight) Polarity

DCEN polarity is used for welding steels because the higher concentration of heat input into the base material results in deeper penetration. DCEN polarity is the best option for obtaining full penetration welds and controllable weld bead profiles. With DCEN polarity, two-thirds of the arc's heat is concentrated on the workpiece and one-third is concentrated on the electrode. Since the majority of the heat is concentrated on the work and not on the electrode, smaller electrodes can be used without worrying about electrode deterioration.

Electrodes Used on Steels

For a long time, 1% or 2% thoriated electrodes were recommended for welding steel. When thorium is added to tungsten in small amounts, it increases the electrode's current-carrying capabilities. Thoriated electrodes do not work well with ac current and are not recommended for use with the newer inverter-type machines. If thoriated electrodes are overheated, the end of the electrode can begin to vaporize and spit small amounts of the electrode into the weld. **Grinding thoriated tungsten electrodes results in radioactive dust that can be hazardous.** These electrodes usually have a yellow or red color band near the top of the electrode, depending on the percentage of thorium.

Today, electrodes with more desirable characteristics are available. Lanthanated tungsten electrodes are relatively new to the US, but are becoming more popular. Lanthanum is a "rare earth" material that is not radioactive. Lanthanated electrodes work well for ac or dc currents, have good current-carrying capacities, and can be used universally, eliminating the confusion of selecting an electrode. Lanthanated electrodes usually have a black color band near the top of the electrode.

Other types of electrodes, such as ceriated, zirconiated, and pure tungsten, all have some desirable traits. Ceriated tungstens usually have an orange color band, zinconiated tungstens usually have a brown color band, and pure tungstens usually have a green color band. Research the base materials to be welded to determine which electrode is best suited for your application. More detailed information about different electrode types and when they should be used was presented in Chapter 3, *Auxiliary Equipment and Systems*.

Amperage ranges must be considered when selecting the size of the electrode. Higher welding amperages require larger electrodes.

When preparing an electrode for welding steel, keep in mind that different tip configurations will result in different bead profiles. For welding steel, it is generally recommended to sharpen the electrode to a point and make a slight flattened tip. Pointed tips direct the arc in a more concentrated area. Flattening the very tip of the electrode helps concentrate the arc even more.

Shielding Gases for GTAW on Steels

Pure argon gas can be used when welding metal up to 1/8" (3.2 mm) thick. When thicker materials are being welded, helium should be added to the shielding gas to improve weld pool control and penetration. Helium increases the temperature of the arc, resulting in deeper penetration. The following standard shielding gas mixes can be used:

- 75% argon–25% helium
- 50% argon–50% helium
- 25% argon–75% helium

Pure helium gas can be used for automatic welding of steel or steel alloys. Gas purity must be sufficient to prevent hydrogen pickup. Hydrogen will cause porosity and cracking.

Techniques for Welding Steels and Steel Alloys

1. Clean the oxide scale from the weld joint immediately prior to welding. Clean the entire weld area to bright metal.
2. Select a welding rod that has the required properties. When welding porosity-prone material, always use a welding rod that contains deoxidizers to prevent porosity.
3. Preheat low-alloy steels and tool steels to prevent cracking in the weld and heat-affected zones.
4. Weld porosity-prone materials by maintaining the arc on the molten metal. Do not hold the arc on the base metal. On crack-sensitive materials, do not make concave welds. These welds are prone to cracking through the centerline.
5. On multipass welds, always remove any scale on the surface of each pass to reduce the possibility of oxide entrapment.
6. Always fill craters on crack-sensitive materials.
7. When welding tool steels, make small stringer beads, (**Figure 8-10A**), to reduce the expansion of the base metal and prevent cracking. Immediately after welding, the weld metal can be peened to reduce shrinkage as the metal cools.

Welding Procedure Specifications (WPS)

The following eleven procedures have been developed for GTAW practice and production. Any scrap material can be used for practice. The weld joint should fit properly for good results. Clean all of the materials, as previously discussed, before tack welding and welding.

When steel is being welded, a small white dot may form on the weld pool surface. This is a normal condition and does not require any corrective action. The material in the dot is silicon, which separates from the base material and the welding rod. It floats on the surface because it is lighter than the metal.

Gas nozzle sizes are not specified, since many variables are involved. Always use the largest possible size. However, do not use a size that will obstruct your view of the weld pool.

Tooling can be used when welding practice plates. Tools assist in holding the material in the proper plane and prevent warpage and misalignment during the welding sequence. The tooling may also have a gas manifold to permit use of a backing gas, if desired. See **Figure 8-11**.

Welding Procedure Number 8-1

Weld joint type:	Flat plate autogenous weld
Position:	Flat
Material type:	Cold-rolled steel
Thickness:	1/16″ (1.6 mm)
Machine setup:	DCEN high-frequency start
Shielding gas:	Argon
CFH:	15–25
Tungsten type:	1%, 1.5%, or 2% lanthanated, ceriated, or thoriated
Diameter:	1/16″ (1.6 mm) (tapered)

Procedure:

1. Prepare and clean the materials.
2. Raise the part to be welded 1/8″ (3.2 mm) above the table with metal blocks.
3. Align the torch to the angles shown in **Figure 8-12** (90° work angle and 15°–25° travel angle) and lower the torch until the tip of the electrode is approximately 1/8″ (3.2 mm) from the top surface.
4. Start the arc at low current; lower the torch until the electrode tip is approximately 1/16″ (1.6 mm) from the surface.

B

A

Figure 8-11. Tooling can support the weld and provide backing gas. A—Hand-built copper backing plates for providing cover gas to the back side of a butt weld. The weld is positioned on top of the plate and welded. B—This fixture includes a manifold for backing gas. The weldment is clamped and held in place by the top bars. (Mark Prosser)

Figure 8-12. Torch position for an autogenous weld in the flat position. (Mark Prosser)

5. Increase the amperage and form a weld pool approximately 3/16″ (4.8 mm) in diameter.
6. Move the torch forward while:
   A. Maintaining the weld pool size.
   B. Maintaining the torch height.
   C. Maintaining the torch angles.
7. Stop the weld at the end of the plate.
   A. Do not lift the torch away from the plate. Postflow gas will protect the hot metal during the cooling period.

Problem Areas and Corrections

1. Uneven top weld width and depression.
   Possible cause: Variation in welding speed.
   Solution: Maintain a consistent travel speed.
   Possible cause: Heat buildup in the weld pool.
   Solution: Maintain travel speed to reduce heat.
2. Uneven contour.
   Possible cause: Incorrect torch angle.
   Solution: Align the torch vertically over the weld.
3. Uneven penetration or lack of penetration.
   Possible cause: Insufficient amperage.
   Solution: Increase amperage.
   Possible cause: Variable torch height.
   Solution: Keep the torch the same height above the weld pool.
   Possible cause: Variable travel speed.
   Solution: Maintain a travel speed that keeps the weld pool a consistent size.

Welding Procedure Number 8-2

Weld joint type:	Bead on plate
Position:	Flat
Material type:	Cold-rolled steel
Thickness:	1/16″ (1.6 mm)
Filler metal:	ER70S-2 or 6
Diameter:	.045″ (1.1 mm)
Machine setup:	DCEN high-frequency start
Shielding gas:	Argon
CFH:	15–25
Tungsten type:	1%, 1.5%, or 2% lanthanated, ceriated, or thoriated
Diameter:	1/16″ (1.6 mm) (tapered)

Procedure:

1. Prepare and clean the materials.
2. Raise the part to be welded 1/8″ (3.2 mm) above the table with metal blocks.
3. Align the torch to the angles shown in **Figure 8-13** and lower the torch until the electrode tip is approximately 1/8″ (3.2 mm) from the top surface.

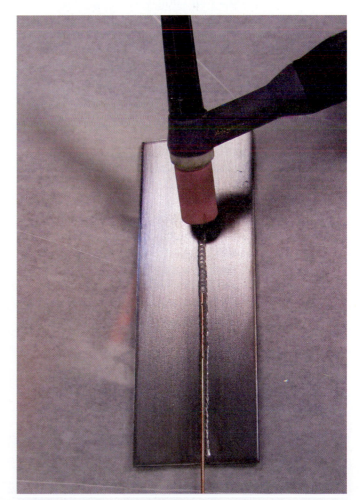

Figure 8-13. Torch and welding rod positions for running a bead on a plate in the flat position. (Mark Prosser)

4. Hold the welding rod as shown in **Figure 8-13** and move it so the end is approximately 1″ (25.4 mm) from the electrode.
5. Start the arc at low current. Lower the torch so the electrode tip is approximately 1/16″ (1.6 mm) from the surface.
6. Increase the amperage and form a weld pool approximately 1/4″ (6.5 mm) in diameter.
7. Bring the rod into the front edge of the pool and melt enough rod to form a slight crown on the surface.
8. Draw the rod approximately 1/2″ (12.7 mm) away from the pool.
9. Move the torch forward approximately 3/16″ (4.8 mm) and add rod again as in Step 7.
10. Continue moving the arc across the joint, adding rod as before.
11. Stop the weld at the end of the plate. Remember to hold the torch at the end until the weld cools. This protects the weld with shielding gas as it cools.

Problem Areas and Corrections:

1. Uneven top weld width and low crown.
   Possible cause: Variation in travel speed.
   Solution: Maintain a consistent travel speed.
   Possible cause: Heat buildup in the weld pool.
   Solution: Increase the travel speed as the plates get hotter.
   Possible cause: Variation in adding rod.
   Solution: Add rod to form a consistent crown.
2. Undercut.
   Possible cause: Torch not aligned vertically.
   Solution: Maintain a 90° work angle.
   Possible cause: Not enough filler metal added to the weld pool.
   Solution: Add rod to the center of the weld pool.
3. Penetration uneven or lack of penetration.
   Possible cause: Inconsistent travel speed.
   Solution: Maintain a consistent travel speed.
   Possible cause: Inconsistent rod addition.
   Solution: Add filler metal to create a consistent crown on the weld pool.
   Possible cause: Insufficient amperage.
   Solution: Increase amperage and add more welding rod.
   Possible cause: Inconsistent torch height.
   Solution: Hold the torch at a consistent height.

Welding Procedure Number 8-3

Weld joint type:	Square-groove butt
Position:	Flat
Material type:	Cold-rolled steel
Thickness:	1/16″ (1.6 mm)
Filler metal:	ER70S-2 or 6
Diameter:	.045″–.062″ (1.1 mm – 1.6 mm)
Machine setup:	DCEN high-frequency start
Shielding gas:	Argon
CFH:	15–25
Tungsten type:	1%, 1.5%, or 2% lanthanated, ceriated, or thoriated
Diameter:	1/16″ (1.6 mm) (tapered)

Procedure:

1. Prepare and clean the materials.
2. Tack weld the plates together.
3. Raise the part to be welded 1/8″ (3.2 mm) above the table with metal blocks.
4. Align the torch to the angles shown in **Figure 8-14** and lower the torch until the electrode tip is approximately 1/8″ (3.2 mm) from the surface.
5. Position the end of the welding rod approximately 1″ (25.4 mm) from the electrode.

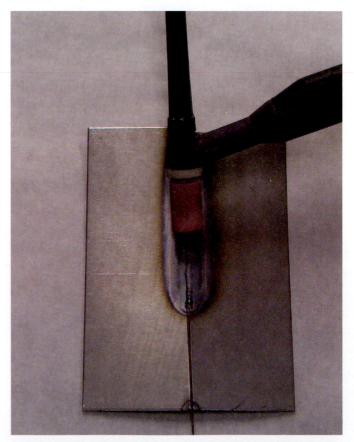

Figure 8-14. Torch position for a square-groove butt weld in the flat position. (Mark Prosser)

6. Start the arc at low current. Lower the torch until the electrode tip is approximately 1/16″ (1.6 mm) from the surface.
7. Increase the amperage and form a weld pool approximately 1/4″ (6.5 mm) in diameter.
8. Bring the rod into the front edge of the pool and melt enough rod to form a slight crown on the surface.
9. Draw the rod approximately 1/2″ (12.7 mm) away from the weld pool.
10. Move the torch forward approximately 3/32″ (2.4 mm) and add rod again as in Step 8.
11. Continue moving the arc across the joint, adding rod as before.
    A. Maintain the proper torch angles.
    B. Keep the electrode on the centerline of the joint.
12. Stop the weld at the end of the plate. Be sure to fill the crater completely. Remember to hold the torch at the end until the weld cools.

Problem Areas and Corrections:

1. Uneven top weld width and low crown.
   Possible cause: Variation in travel speed.
   Solution: Maintain a consistent travel speed.
   Possible cause: Heat buildup in the weld pool.
   Solution: Increase the travel speed as the plates get hotter.
   Possible cause: Variation in adding rod.
   Solution: Add rod to form a crown.
2. Undercut.
   Possible cause: Torch not aligned vertically.
   Solution: Maintain a 90° work angle.
   Possible cause: Not enough filler metal added to the weld pool.
   Solution: Add rod to the center of the weld pool.
3. Penetration uneven or lack of penetration.
   Possible cause: Inconsistent travel speed.
   Solution: Maintain a consistent travel speed.
   Possible cause: Inconsistent filler metal application.
   Solution: Add filler metal to form a crown.
   Possible cause: Insufficient amperage.
   Solution: Increase amperage and add more filler metal.
   Possible cause: Inconsistent torch height.
   Solution: Maintain a consistent torch height.
   Possible cause: Improper torch position or angle.
   Solution: Maintain the proper torch angles and position the torch on the centerline of the joint.

Welding Procedure Number 8-4

Weld joint type:	Square-groove butt
Position:	Horizontal
Material type:	Cold-rolled steel
Thickness:	1/16″ (1.6 mm)
Filler metal:	ER70S-2 or 6
Diameter:	.045″–.062″ (1.1 mm – 1.6 mm)
Machine setup:	DCEN
Shielding gas:	Argon
CFH:	15–25
Tungsten type:	1%, 1.5%, or 2% lanthanated, ceriated, or thoriated
Diameter:	1/16″ (1.6 mm) (tapered)

Procedure:

1. Prepare and clean the materials.
2. Tack weld the plates together.
3. Align the parts to be welded with the joint in the horizontal position.
4. Align the torch to the angles shown in **Figure 8-15** and move the torch toward the joint until the tip of the electrode is approximately 1/8″ (3.2 mm) from the surface.
5. Hold the welding rod as shown in **Figure 8-15** and position it so the end is approximately 1″ (25.4 mm) from the electrode.
6. Start the arc at low current. Move the torch toward the joint until the tip of the electrode is approximately 1/16″ (1.6 mm) from the surface.
7. Increase the amperage to form a weld pool approximately 3/16″ (4.8 mm) in diameter.
8. Bring the rod into the upper part of the weld pool and melt enough rod to form a slight crown.
9. Draw the rod approximately 1/2″ (12.7 mm) away from the pool.

Figure 8-15. Torch and welding rod positions for a square-groove butt weld in the horizontal position. (Mark Prosser)

10. Move the torch forward approximately 3/32″ (2.4 mm) and add welding rod again as in Step 8.
11. Continue moving the arc across the joint, adding welding rod as before.
    A. Maintain the proper torch angles.
    B. Maintain the proper filler rod angles.
12. Stop the weld at the end of the plate. Remember to hold the torch at the end until the weld cools.

Problem Areas and Corrections:

1. Undercut at top of weld crown.
   Possible cause: Torch angle flat.
   Solution: Maintain the proper torch angles.
   Possible cause: Not enough filler metal added to the weld pool.
   Solution: Add welding rod to the top of the weld pool.
2. Crown sags.
   Possible cause: Gravity.
   Solution: Add welding rod to the top of the weld pool.
   Solution: Do not add welding rod in large amounts.
   Solution: Move the torch with a slight circular motion to hold pool in position.
3. Penetration is uneven on the centerline of the weld.
   Possible cause: Torch not centered on the joint.
   Solution: Keep the electrode centered on the joint.

Welding Procedure Number 8-5

Weld joint type:	Square-groove butt
Position:	Vertical uphill
Material type:	Cold-rolled steel
Thickness:	1/16″ (1.6 mm)
Filler metal:	ER70S-2 or 6
Diameter:	.045″–.062″ (1.1 mm – 1.6 mm)
Machine setup:	DCEN high-frequency start
Shielding gas:	Argon
CFH:	15–25
Tungsten type:	1%, 1.5%, or 2% lanthanated, ceriated, or thoriated
Diameter:	1/16″ (1.6 mm) (tapered)

Procedure:

1. Prepare and clean the materials.
2. Tack weld the plates together.
3. Align the parts to be welded with the joint in the vertical position.
4. Align the torch to the angles shown in **Figure 8-16** and move the torch toward the joint until the tip of the electrode is approximately 1/8″ (3.2 mm) from the surface.

5. Hold the welding rod so the end is approximately 1″ (25.4 mm) from the electrode.
6. Start the arc at low current. Move the torch toward the joint until the tip of the electrode is approximately 1/16″ (1.6 mm) from the surface.
7. Increase the amperage to form a weld pool approximately 3/16″ (4.8 mm) in diameter.
8. Bring the welding rod into the upper part of the pool and melt enough filler metal to form a slight crown on the surface.
9. Draw the welding rod approximately 1/2″ (12.7 mm) away from the weld pool.
10. Move the torch upward approximately 3/32″ (2.4 mm) and add welding rod again as in Step 8.
11. Continue moving upward, adding welding rod as before.
12. Stop the weld at the end of the plate. Remember to hold the torch at the end until the metal cools.

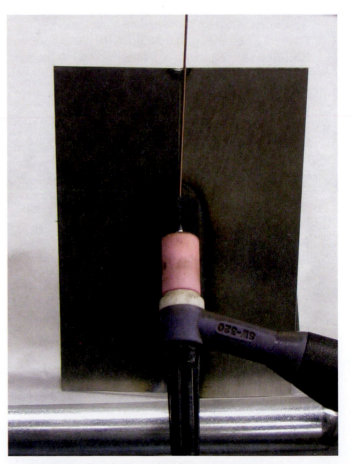

Figure 8-16. Torch and welding rod positions for a square-groove butt weld in the vertical position. (Mark Prosser)

Problem Areas and Corrections:

1. Undercut.
   Possible cause: Insufficient amount of filler metal added.
   Solution: Add rod in the top and center of the weld pool.
   Possible cause: Improper torch angles.
   Solution: Maintain proper torch angles.
2. High crown.
   Possible cause: Too much filler metal added to weld pool.
   Solution: Add filler metal in smaller amounts.
   Solution: Use smaller diameter welding rod.
3. Uneven penetration or lack of penetration.
   Possible cause: Inconsistent travel speed.
   Solution: Maintain a consistent travel speed.
   Possible cause: Inconsistent filler metal addition.
   Solution: Add filler metal to maintain a consistent crown on the weld pool.
   Possible cause: Incorrect torch angles.
   Solution: Maintain the proper torch angles.
   Possible cause: Heat buildup in the weld pool.
   Solution: Increase travel speed as the plates get hotter.

Welding Procedure Number 8-6

Weld joint type:	Square-groove butt
Position:	Flat
Material type:	Cold-rolled steel
Thickness:	11 gauge
Filler metal:	ER70S-2 or 6
Diameter:	3/32″ (2.4 mm) or 1/8″ (3.2 mm) (welder's preference)
Machine setup:	DCEN high-frequency start
Shielding gas:	Argon
CFH:	15–25
Tungsten type:	1%, 1.5%, or 2% lanthanated, ceriated, or thoriated
Diameter:	3/32″ (2.4 mm) (tapered)

Procedure:

1. Prepare and clean the materials.
2. Tack weld the plates with 3/32″ (2.4 mm) or 1/8″ (3.2 mm) spacing, depending on the filler being used. Mount the joint approximately 1/8″ (3.2 mm) above the table with metal blocks.
3. Weld the root pass to ensure full penetration.
   A. Maintain the torch on the weld pool with the proper angles. Use the same work and travel angles shown in **Figure 8-14**.

4. Wire brush the weld to remove the oxide film (both passes).
5. Realign the torch and weld the second pass.
   A. Add sufficient filler metal to create a slightly convex weld bead and widen the weld pool just beyond the edges of the joint.

Problem Areas and Corrections:

1. Undercut.
   Possible cause: Insufficient filler metal added to the weld pool.
   Solution: Add welding rod to the edge of the weld pool on the crown pass.
   Solution: Hold the torch at the edge of the weld pool when adding welding rod.
2. Uneven crown.
   Possible cause: Inconsistent travel speed.
   Solution: Maintain a consistent travel speed.
   Possible cause: Inconsistent filler metal addition.
   Solution: Add filler metal to create a consistently sized crown.
3. Uneven penetration.
   Possible cause: Inconsistent travel speed.
   Solution: Maintain a consistent travel speed.
   Possible cause: Inconsistent filler metal addition.
   Solution: Add filler metal to create a consistently sized crown.
   Possible cause: Electrode not centered on pool.
   Solution: Make sure torch is centered over the centerline of the joint.
   Possible cause: Inconsistent torch height.
   Solution: Keep the tip of the electrode a consistent distance from the weld pool during each pass.

Welding Procedure Number 8-7

Weld joint type:	Square-groove butt
Position:	Horizontal
Material type:	Cold-rolled steel
Thickness:	11 gauge
Filler metal:	ER70S-2 or 6
Diameter:	3/32″ (2.4 mm) or 1/8″ (3.2 mm) (welder's preference)
Machine setup:	DCEN
Shielding gas:	Argon
CFH:	10–15
Tungsten type:	2% thoriated
Diameter:	3/32″ (2.4 mm) (tapered)

Procedure:

1. Prepare and clean the materials.
2. Tack weld the plates with 3/32″ (2.4 mm) or 1/8″ (3.2 mm) spacing, depending on the filler metal used. Mount the joint in the horizontal position.
3. Weld the root pass with 3/32″ (2.4 mm) or 1/8″ (3.2 mm) diameter welding rod.
   A. Add the welding rod to the upper part of the weld pool. Use the work and travel angles shown in **Figure 8-15**.
4. Wire brush the weld to remove the oxide film (both passes).
5. Realign the torch and weld a second pass. Use a 3/32″ (2.4 mm) or 1/8″ (3.2 mm) diameter welding rod.
   A. Do not oscillate the torch. The weld should be approximately 1/16″ (1.6 mm) wider than the groove, and the weld crown should not be over 1/16″ (1.6 mm) high.

Problem Areas and Corrections:

1. Undercut.
   Possible cause: Sagging due to gravity.
   Solution: Add welding rod at the top of the weld pool.
   Possible cause: Weld bead is excessively large.
   Solution: Hold the torch and welding rod at the proper angles.
   Solution: Do not oscillate the torch. Keep the weld pool small.
   Possible cause: Excessive amperage.
   Solution: Use a lower amperage range and add filler metal more often.
   Possible cause: Heat buildup in the weld pool.
   Solution: Allow the plates to cool between passes.
2. Uneven bead width.
   Possible cause: Variation in the size and shape of the weld pool.
   Solution: Determine the size of each pass before starting.
   Solution: Determine the location of each pass before starting.
   Solution: Maintain the same pass dimensions for the full length of the joint.

Welding Procedure Number 8-8

Weld joint type:	Square-groove butt
Position:	Vertical uphill
Material type:	Cold-rolled steel
Thickness:	11 gauge
Filler metal:	ER70S-6
Diameter:	3/32″ (2.4 mm) or 1/8″ (3.2 mm) (welder's preference)
Machine setup:	DCEN high-frequency start
Shielding gas:	Argon
CFH:	15–25
Tungsten type:	1%, 1.5%, or 2% lanthanated, ceriated, or thoriated
Diameter:	3/32″ (2.4 mm) (tapered)

Procedure:

1. Prepare and clean the materials.
2. Tack weld the plates with 3/32″ (2.4 mm) or 1/8″ (3.2 mm) spacing depending on the filler metal used. Mount the joint in the vertical position.
3. Weld the root pass with 3/32″ (2.4 mm) or 1/8″ (3.2 mm) diameter welding rod. Use the same travel and work angles shown in **Figure 8-16**.
   A. Add the welding rod to the top of the weld pool.
   B. Add the welding rod directly on the pool centerline.
   C. Keep the electrode centered over the molten pool.
4. Wire brush the weld to remove the oxide film (all passes).
5. Realign the torch and weld the second pass using a 3/32″ (2.4 mm) or 1/8″ (3.2 mm) diameter welding rod.
   A. Use a slight oscillation for a wider bead.
   B. Add welding rod at the edge of the weld pool and always wait for a moment before dipping the filler into the weld pool. Keep the weld pool in a fluid state.

Problem Areas and Corrections:

1. Undercut.
   Possible cause: Sagging due to gravity.
   Solution: Add welding rod at the top of the weld pool.
   Possible cause: Weld bead is excessively large.
   Solution: Hold the torch and welding rod at the proper angles.
   Solution: Do not oscillate the torch. Keep the weld pool small.
   Possible cause: Excessive amperage.
   Solution: Use a lower amperage range and add filler metal more often.

Possible cause: Heat buildup in the weld pool.
Solution: Allow the plates to cool between passes.
2. Crooked or uneven beads.
   Possible cause: Variation in the size and shape of the weld pool.
   Solution: Determine the size of each pass before starting.
   Solution: Determine the location of each pass before starting.
   Solution: Maintain the same pass dimensions for the full length of the joint.

Welding Procedure Number 8-9

Weld joint type:	T-joint
Position:	Horizontal
Material type:	Cold-rolled steel
Thickness:	1/16″ (1.6 mm)
Filler metal:	ER70S-2 or 6 (welder's preference)
Diameter:	.045″–.062″ (1.1 mm – 1.6 mm)
Machine setup:	DCEN high-frequency start
Shielding gas:	Argon
CFH:	15–25
Tungsten type:	1%, 1.5%, or 2% lanthanated, ceriated, or thoriated
Diameter:	1/16″ (1.6 mm) (tapered)

Procedure:

1. Prepare and clean the materials.
2. Align the plates.
3. Tack weld the two plates at approximately right angles.
4. Weld the joint using the torch and welding rod angles shown in **Figure 8-17**.
   A. Add the welding rod often and in small amounts.
   B. The filler metal will flow evenly to both pieces and the crown should be flat to slightly convex.
   C. Feed the welding rod directly into the intersection of the joint.

Problem Areas and Corrections:

1. Concave weld crown.
   Possible cause: Insufficient filler.
   Solution: Add welding rod more often.
2. Irregular crown height or width.
   Possible cause: Improper feeding of welding rod.
   Solution: Feed welding rod into the weld pool often, and in small quantities.

Figure 8-17. Torch and welding rod positions for welding a T-joint in the horizontal position. (Mark Prosser)

Possible cause: Too large a welding rod size causing irregular heating of the weld pool.
Solution: Use a smaller welding rod.
3. Concave root surface or melt-through.
   Possible cause: Weld pool is too hot.
   Solution: Use a smaller diameter welding rod and add it often to control pool temperature.
   Solution: Lower the amperage setting.

Welding Procedure Number 8-10

Weld joint type:	T-joint
Position:	Horizontal
Material type:	Cold-rolled steel
Thickness:	11 gauge
Filler metal:	ER70S-2 or 6
Diameter:	3/32″ (2.4 mm) or 1/8″ (3.2 mm) (welder's preference)
Machine setup:	DCEN high-frequency start
Shielding gas:	Argon
CFH:	15–25
Tungsten type:	1%, 1.5%, or 2% lanthanated, ceriated, or thoriated
Diameter:	3/32″ (2.4 mm) (tapered)

Procedure:

1. Prepare and clean the materials.
2. Align the plates.
3. Tack weld the two plates at approximately right angles.
4. Weld the root pass using the torch and welding rod angles shown in **Figure 8-17**.
   A. Add filler metal in small amounts.
   B. Feed the welding rod directly into the intersection of the joint for the root pass.

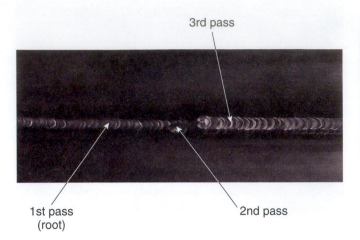

Figure 8-18. The three passes used to weld a T-joint in the horizontal position. (Mark Prosser)

C. Wire brush the weld to remove the oxide film (all passes).
D. Weld the second pass. Make sure the weld pool fuses into the bottom plate and overlaps the root pass by approximately 50%. See **Figure 8-18**.
E. Weld the third pass. Make sure the weld pool fuses into the top plate and the second pass, creating a flat weld in the joint.

Problem Areas and Corrections:

1. Concave weld crown or improper contour.
   Possible cause: Insufficient filler metal.
   Solution: Add welding rod more often.
   Possible cause: Incorrect torch angle and improperly placed welds.
2. Leg sizes are not equal.
   Possible cause: Incorrect torch angle.
   Solution: Reposition the torch to keep the weld pool even.
   Possible cause: Too much filler metal. Weld pool is too heavy.
   Solution: Add less filler metal.
3. Undercut on top weld.
   Possible cause: Weld pool is too hot.
   Solution: Add welding rod more often.
   Possible cause: Incorrect torch angle.
   Solution: Reposition the torch to keep the weld pool even.

Welding Procedure Number 8-11

Weld joint type:	T-joint
Position:	Vertical uphill
Material type:	Cold-rolled steel
Thickness:	11 gauge
Filler metal:	ER70S-2 or 6
Diameter:	3/32″ (2.4 mm) or 1/8″ (3.2 mm) (welder's preference)
Machine setup:	DCEN high-frequency start
Shielding gas:	Argon
CFH:	15–25
Tungsten type:	1%, 1.5%, or 2% lanthanated, ceriated, or thoriated
Diameter:	3/32″ (2.4 mm) (tapered)

Procedure:

1. Prepare and clean the materials.
2. Align the plates.
3. Tack weld the two plates at approximately right angles.
4. Weld the joint using the torch and welding rod angles shown in **Figure 8-19**.
   A. Add the filler metal in small amounts.
   B. Feed the welding rod directly into the intersection of the joint.
   C. Wire brush the joint to remove the oxide film.

Figure 8-19. Torch and welding rod positions for uphill welding a T-joint in the vertical position. (Mark Prosser)

Problem Areas and Corrections:

1. Convex weld crown.
   Possible cause: Too much filler metal or travel speed too slow.
   Solution: Add less filler metal.
   Solution: Increase the travel speed.
2. Concave weld crown.
   Possible cause: Not enough filler metal or travel speed too high.
   Solution: Add more filler metal.
   Solution: Decrease the travel speed.
3. Leg sizes are not equal.
   Possible cause: Incorrect torch angles.
   Solution: Reposition the torch to keep the weld pool even.
   Possible cause: Variation in the size of the weld pool.
   Solution: Maintain a consistent travel speed and feed the welding rod as needed to keep the weld pool size consistent.

4. Undercut.
   Possible cause: Improper torch work angle.
   Solution: Maintain the proper torch angles.
   Possible cause: Uneven weld pool.
   Solution: Add welding rod to the center of the pool.

Reading the Weld Pool

The ability to read a weld pool is a very important skill for welders, particularly in GTAW because the weld pool is relatively small. This skill develops with practice and an increased understanding of the variables that affect the weld. By reading the weld pool, a welder can tell when the travel speed is too fast or too slow and when torch angles are incorrect. The weld pool reveals discontinuities as they happen. Reading the weld pool enables the welder to make "on-the-fly" adjustments to ensure a quality weld.

Groove Weld Defects and Corrective Actions

Common groove weld defects may be noticed during the welding operation or later during the inspection of the completed weld. These defects are listed and illustrated with suggestions for corrective actions.

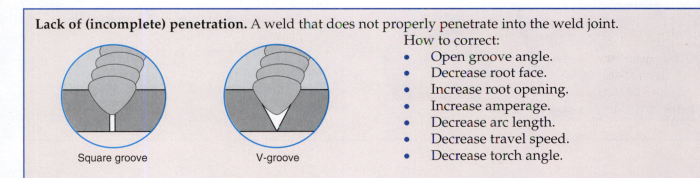

Lack of (incomplete) penetration. A weld that does not properly penetrate into the weld joint.
How to correct:
- Open groove angle.
- Decrease root face.
- Increase root opening.
- Increase amperage.
- Decrease arc length.
- Decrease travel speed.
- Decrease torch angle.

Square groove   V-groove

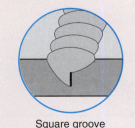

Lack of fusion. Fusion did not occur between the weld metal and fusion faces or adjoining weld beads.
How to correct:
- Clean weld joint before welding.
- Remove oxides from previous welds.
- Open groove angle.
- Decrease root face.
- Increase root opening.
- Increase amperage.
- Decrease arc length.
- Decrease travel speed.
- Decrease torch angle.

Square groove   V-groove

Overlap. Weld metal that has flowed over the edge of the joint and improperly fused with the base metal.

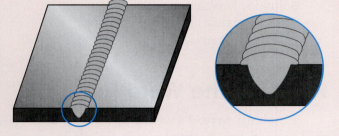

How to correct:
- Clean edge of weld joint.
- Remove oxides from previous welds.
- Reduce size of bead.
- Increase travel speed.

Undercut. Lack of filler metal at the toe of the weld metal.

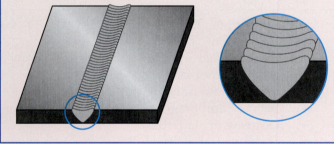

How to correct:
- Decrease travel speed.
- Increase dwell time at edge of joint on weave beads.
- Decrease arc length.
- Decrease amperage.
- Decrease torch angle.

Concave weld. A concave weld occurs when insufficient metal is added to fill the joint above the groove edges.

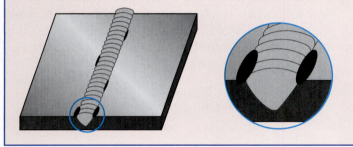

How to correct:
- Add additional filler metal on the fill passes.
- Use stringer beads on crown pass.

Convex crown. A weld that is peaked in the center.

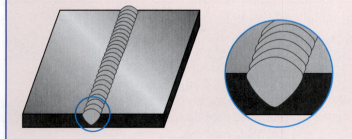

How to correct:
- Use less filler metal on the crown bead.

Craters. Formed at the end of a weld bead due to a lack of weld metal fill or weld shrinkage.

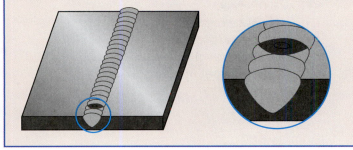

How to correct:
- Do not stop welding at end of joint (use tabs).
- Fill craters to the proper crown height and reduce amperage slowly.

Cracks. Caused by cooling stresses in the weld and/or base metal.

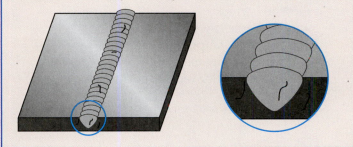

How to correct:
- Use wire with lower tensile strength or different chemistry.
- Increase joint preheat to slow the weld cooling rate.
- Allow joint to expand and contract during heating and cooling. Increase size of the weld.

Porosity. Caused by entrapped gas that did not have enough time to rise through the melt to the surface.

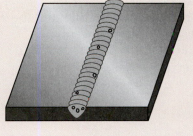

How to correct:
- Remove all heavy rust, paint, oil, or scale on joint before welding.
- Remove oxide film from previous passes or layers of weld.
- Protect welding area from wind.
- Remove all grease and oil from the filler metal. Handle filler metal with clean gloves.
- Check gas supply for flow rate, possible contamination, loose connections, dirty torch nozzle.

Linear porosity. Forms in a line along the root of the weld at the center of the joint where penetration is very shallow.

How to correct:
- Make sure root faces are clean.
- Increase current.
- Decrease arc length.
- Increase amperage.
- Decrease travel speed.

Melt-through. Occurs at a gap in the weld joint or place where metal is thin.

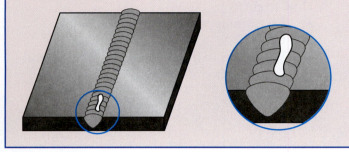

How to correct:
- Decrease current.
- Increase arc length.
- Increase travel speed.
- Decrease root opening.

Concave root surface. The root of the weld penetration does not extend below the lower edges of the joint.

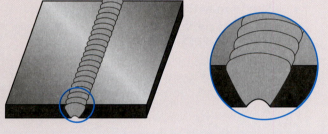

How to correct:
- Decrease travel speed.
- Decrease amperage.
- Decrease arc length.
- Increase filler wire deposition.

Icicles. Weld penetration beyond the joint root is uneven.

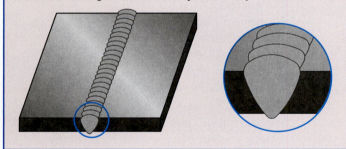

How to correct:
- Move the torch throughout the weld length in a consistent pattern.
- Add filler metal as required to maintain an even penetration.

Fillet Weld Defects and Corrective Actions

Common fillet weld defects may be noticed during the welding operation or later during the inspection of the completed weld. These defects are listed and illustrated with suggestions for corrective actions.

Lack of penetration. Insufficient weld metal penetration into the joint intersection.

How to correct:
- Maintain slight forward torch angle.
- Increase amperage.
- Decrease arc length.
- Decrease size of weld deposit.
- Use stringer beads
- Do not make weave beads on root passes.

Lack of fusion. Occurs in multiple-pass welds where layers do not fuse.

How to correct:
- Remove oxides and scale from previous weld passes.
- Increase amperage.
- Decrease arc length.
- Decrease travel speed.

Overlap. Occurs in horizontal, multiple-pass fillet welds when too much weld is placed on the bottom layer.

How to correct:
- Remove oxides and scale from weld area.
- Reduce the size of the weld pass.
- Increase travel speed.

Undercut. Occurs at the top of the weld bead in horizontal fillet welds.

How to correct:
- Make a smaller weld.
- Make a multiple-pass weld.
- Decrease amperage.
- Decrease arc length.

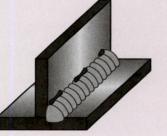

Concave weld. A concave weld is formed when insufficient metal is added to the molten pool.

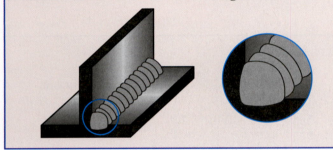

How to correct:
- Deposit more material into the pool with each application of the filler metal.

Convex weld. A weld that has a high crown.

How to correct:
- Deposit less metal into the pool with each application of the filler metal.

Craters. Formed when weld metal shrinks below the full cross section of the weld.

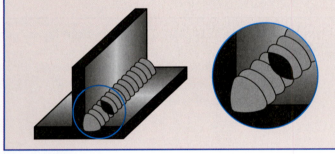

How to correct:
- Do not stop welding at the end of the joint (use tabs).
- Using the same travel torch angle, move the torch back on the full cross section before stopping.

Cracks. Occur in fillet welds just as they do in groove welds.

How to correct:
- Use suggestions for groove weld cracks.

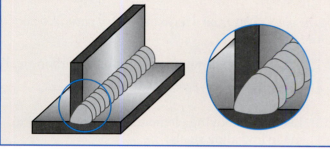

Excessive penetration. Formed when excessive heat is applied to the upper part of the joint.

How to correct:
- Change torch angle to lower plate.
- Decrease bead size.
- Increase welding speed.
- Use chill bar behind joint to remove excessive heat.

Melt-through. Occurs when the molten pool melts through the base material and creates a hole.

How to correct:
- Change torch angle to lower plate.
- Decrease bead size.
- Increase welding speed.
- Use chill bar behind joint to remove excessive heat.

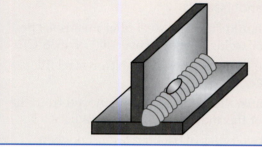

Concave root surface. The molten weld penetration has retracted from the root surface.

How to correct:
- Increase welding speed.
- Decrease amperage.
- Increase arc length.
- Increase filler metal feed.

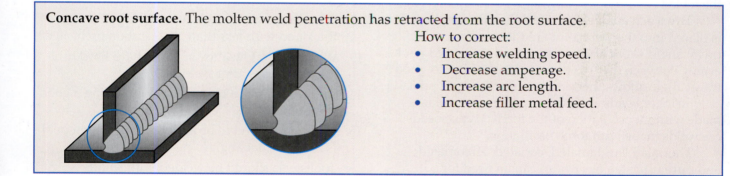

Postweld Treatment

Postweld treatment depends on the type of material welded, joint restraint, and the desired mechanical values. In postweld treatment, the cooling rate of the welded part is controlled, allowing the mechanical properties of the metal to return to a normal condition. Cooling some materials too rapidly severely affects their mechanical properties, such as making the part very hard and brittle. Preweld and postweld heat treatments are used on steels with higher carbon content, castings, and other materials that are prone to cracking.

A stress-relieving operation performed on carbon steels at approximately 1150°F (621°C) removes residual stresses caused by weld shrinkage. This can be done by local or furnace heating. The weld area is heated to 1150°F (621°C) and this temperature is maintained for one hour per inch of material thickness. The part can then be air-cooled. Low-alloy steels can be stress-relieved in the same manner; however, the cooling period should be lengthened by covering the part with heat-resistant materials.

Summary

Types of steel include carbon steels, low-alloy steels, heat-treated steels, and tool steels. Carbon steels are classified as low-, medium-, and high-carbon steels. Alloying elements in low-alloy steels include chromium, molybdenum, nickel, vanadium, and manganese. Heat treatments include quenching, tempering, and annealing. Tool steels are steels with a combination of high carbon and alloy content.

Common steel forms include hot-rolled steel, cold-rolled steel, castings, and forgings. Steels are manufactured to specifications developed by various organizations, including ASTM International.

Filler metals must be selected to produce the desired mechanical properties in the weldment after any required welding or heat treatment. When selecting a steel filler metal, the type of steel to be welded and the possibility of excessive porosity within the completed weld should be considered.

Joint preparation considerations include removing oxide film from edges prepared by thermal cutting processes, providing weld backing where needed, and preheating. Quenched and tempered steel requires preheat and interpass temperature control to retain the original mechanical properties of the metal.

Two torch angles that affect how the weld bead is placed into the weld joint are the work angle and travel angle. Work angles usually split the weld pool evenly between the two pieces of base metal. A good travel angle allows the welder to see the weld pool.

Stringer beads, or stringer passes, are welds made without oscillation of the torch. Weave beads are welds made with torch oscillation.

Thoriated, lanthanated, ceriated, zirconiated, and pure tungsten electrodes can be used for welding steel. Lanthanated electrodes work well for either ac or dc current and have good current-carrying capacities. Higher welding amperages require larger electrodes. An electrode that is sharpened to a point with a slight flattened tip is recommended for welding steel.

Pure argon gas can be used when welding metal up to 1/8" (3.2 mm) thick. For increasing thicknesses, helium should be added to the shielding gas.

Manual GTAW requires a great deal of skill to perform correctly. The ability to read a weld pool must be developed so that the welder can make the appropriate adjustments to ensure a quality weld.

Review Questions

Write your answers on a separate sheet of paper. Do not write in this book.

1. What are the three classifications of carbon steel?
2. Medium-carbon steel has a content of _____ carbon.
3. List three alloying elements commonly used to make low-alloy steels.
4. What is heat treating?
5. The oxide film on the surface of hot-rolled steel is _____ in color.
6. What color is the surface of castings that have been sandblasted?
7. Why should stainless steel filler metals *not* be used for welds in service over 1000°F (538°C)?
8. List the three finishes available on steel filler metals.
9. Why must the oxide scale be removed from the edges of thermally cut joints before use?
10. What two types of backing are used when welding 100% penetration steel welds?
11. Why is backing used for 100% penetration steel welds?
12. When are carbon steels less than 1" (25.4 mm) thick and with less than .30% carbon preheated?
13. What causes low-alloy steels to have hard heat-affected zones after welding?
14. What is the main reason for maintaining a good travel angle?
15. The welding rod to be added to the molten pool is held at what angle?
16. Tool steels should be welded with small _____ beads to reduce the amount of heat input and expansion of the base material.
17. At what approximate temperature can postweld stress relief of carbon steel weldments be done?
18. Welds made *without* any side-to-side movement of the torch are called _____ beads.
19. Welds made with side-to-side movement of the torch are called _____ beads.
20. When steel is being welded, a small white dot may form on the top of the molten metal. What is this material?

Chapter

Manual Welding of Aluminum

9

Objectives

After completing this chapter, you will be able to:

- ☐ Identify the characteristics of aluminum.
- ☐ Identify wrought material, castings, and tempers by their number systems or designations.
- ☐ Distinguish between heat-treatable or nonheat-treatable aluminum.
- ☐ Summarize the filler metal choices for various aluminum alloys.
- ☐ Recall joint preparation techniques, including preweld cleaning, weld backing, preheating, and tack welds.
- ☐ Select the correct power source, shielding gases, and electrodes for welding aluminum using ACHF, DCEN, and DCEP.
- ☐ Apply correct procedures for welding aluminum using ACHF.
- ☐ Apply correct procedures for welding aluminum using DCEN.
- ☐ Apply correct procedures for welding aluminum using DCEP.

Key Terms

artificially aged
cold work
crater
dross
ductility
hot short
keyhole method
naturally aged
solution heat-treated
stabilized
strain-hardened
swaging
tempers
wrought

Introduction

Aluminum is a widely used material in the manufacturing and fabrication industries because of its highly desirable characteristics. Aluminum is very light in comparison to steels, has a good strength-to-weight ratio, and is highly resistant to corrosion. There are several different types of aluminum, so identifying the particular type of material is important and sometimes difficult to do. Aluminum is an excellent material for many applications, but requires some slightly different welding techniques than those used on steel.

Base Materials

Aluminum is a nonferrous material readily welded with GTAW. The following are some important characteristics of aluminum:

- good thermal conductivity
- good electrical conductivity
- good ductility at subzero temperatures
- light weight
- high resistance to corrosion
- nonsparking
- nontoxic
- does not change color as it is heated

Pure aluminum melts at approximately 1200°F (649°C). The melting points of aluminum alloys range from approximately 900°F to 1200°F (482°C to 649°C).

The oxide film that forms on the surface of aluminum and aluminum alloys has a melting point of about 3500°F (1927°C). This oxide film must be removed prior to the welding operation or the resulting welds will be defective. If the oxide film is not removed, it will sink into the weld pool and contaminate the weld.

Wrought Material Identification

Aluminum and aluminum alloys are identified by their composition, thermal treatment, and work-hardening characteristics. The word *wrought* identifies material made by processes other than casting and is often used in specifications and codes. Plates and sheets are manufactured using a rolling process and pipe is extruded.

The numbering system established by the Aluminum Association to identify the various groups is shown in **Figure 9-1**. The four-digit number indicates alloy groups, modifications or impurity level, and aluminum alloy or aluminum purity as follows:

- **First number**—alloy groups
- **Second number**—modifications or impurity level
- **Third number**—aluminum alloy or aluminum purity
- **Fourth number**—also aluminum alloy or aluminum purity

For example, the numbers in alloy 4043 aluminum indicate the following:

- **4**—indicates the principle alloying element (silicon) and is used to describe the aluminum alloy series
- **0**—no modification to the specific alloy
- **43**—arbitrary numbers given to a specific alloy in the series of alloys

Aluminum Castings

Aluminum castings are grouped by major alloying elements. Most producers of primary casting alloys use the number system shown in **Figure 9-2** to identify the alloy used in their ingots. A decimal point zero (.0) following the designation identifies the material as a casting. A decimal point one (.1) or decimal point 2 (.2) identifies the material as an ingot.

Tempers

Aluminum materials are also identified by the condition in which the material is supplied. The Aluminum Association Temper Designation System applies to all forms of wrought and cast aluminum and aluminum alloys except ingot. The term *tempers* is used to describe materials with various mechanical properties that have been imparted to the metal by basic treatments. The basic temper designations are as follows:

—**F**   As fabricated.

—**O**   Annealed, recrystallized (softest temper). Annealed materials have undergone a series

Wrought Aluminum Alloy Groups	
Alloy Group	**Series Designation**
Aluminum, 99% minimum purity	1XXX
Aluminum-copper	2XXX
Aluminum-manganese	3XXX
Aluminum-silicon	4XXX
Aluminum-magnesium	5XXX
Aluminum-magnesium-silicon	6XXX
Aluminum-zinc	7XXX
Other	8XXX

Figure 9-1. Wrought material identification numbers.

Cast Aluminum Alloy Groups	
Alloy Series	**Principal Alloying Element**
1xx.x	99.000% minimum aluminum
2xx.x	Copper
3xx.x	Silicon plus copper and/or magnesium
4xx.x	Silicon
5xx.x	Magnesium
6xx.x	Unused series
7xx.x	Zinc
8xx.x	Tin
9xx.x	Other elements

Figure 9-2. Casting material identification numbers.

of heating cycles to allow the grains of the material to re-form (recrystallize) and cooling cycles to produce varying degrees of softness.

—H Strain-hardened (wrought products only). *Strain-hardened* metals have been strained by stretching, pulling, or forming to produce a grain structure with more desirable mechanical properties.

—H1 Strain-hardened only.

—H2 Strain-hardened and partially annealed.

—H3 Strain-hardened and stabilized.

—H4 Strain-hardened and lacquered or painted.

—W Solution heat-treated. *Solution heat-treated* materials have been heated to a predetermined temperature for a suitable length of time to allow a certain element in the material to enter into a "solid solution" with the other elements. A solid solution is a mixture of elements (often metals) that combine to form a single homogenous (uniform) crystalline structure. After the solid solution has been achieved, the alloy is quickly cooled to hold the element in this solution.

—T Thermally treated to produce stable tempers other than —*F*, —*0*, —*H*.

The —*T* is always followed by one or more digits. Specific sequences of basic treatments are as follows:

—T1 *Naturally aged* to a substantially stable condition from the as-cast condition. Aged materials are held at room temperature or at a predetermined temperature for a period of time for the purpose of increasing the hardness and strength of the material.

—T2 Naturally aged and then cold worked.

—T3 Solution heat-treated and then cold worked. *Cold work* identifies an operation of mechanically working a material without heat.

—T4 Solution heat-treated and naturally aged to a substantially stable condition.

—T5 *Artificially aged* from the as-cast condition.

—T6 Solution heat-treated and then artificially aged.

—T7 Solution heat-treated and then stabilized. A *stabilized* material has been strain-hardened and then heated to a predetermined low temperature to slightly lower the strength and to increase the ductility. *Ductility* is a property of a material that allows it to deform permanently or to exhibit plasticity without breaking while under tension (strain).

—T8 Solution heat-treated, cold worked, then artificially aged.

—T9 Solution heat-treated, artificially aged, then cold worked.

—T10 Artificially aged and then cold worked.

Heat-Treatable and Nonheat-Treatable Classifications

Aluminum is classified into two categories: nonheat-treatable and heat-treatable.

- **Nonheat-treatable.** These materials attain strength levels by the addition of alloying elements such as manganese, silicon, iron, and magnesium. They can also be cold worked (strain-hardened) by stretching, drawing, or swaging to increase their strength levels. *Swaging* is the process of changing the shape of a material with mechanical tools, such as hammers and dies.

- **Heat-treatable.** Initial strengths of heat-treatable alloys are produced by the addition of copper, magnesium, zinc, and silicon. Subjecting the material to various degrees of thermal treatment, quenching, and aging provides additional strengthening. Quenching is the process of rapid cooling of metal from a higher temperature for the purpose of hardening.

The grouping of the nonheat-treatable and heat-treatable wrought alloys is shown in **Figure 9-3**.

Nonheat-Treatable (alloys normally cold worked)	Heat-Treatable (alloys normally heat-treated)
1060	2011
1100	2014
3003	2017
3004	2018
4043	2024
5005	2025
5050	2117
5052	2218
5056	2618
5083	4032
5086	6053
5184	6061
5252	6063
5257	6066
5357	6101
5454	6151
5456	7039
5557	7075
5657	7079
	7178

Figure 9-3. Nonheat-treatable and heat-treatable aluminum alloys.

Filler Metals

It is important to choose the correct filler metal for welding different aluminum alloys. The metal in the weld pool is a combination of the base metal and the filler metal, and these materials must match each other as closely as possible. Weld metal must have the strength, ductility, and resistance to cracking and corrosion required by the application. The correct choice of filler metal improves ductility (plasticity under tension) in aluminum welds.

The table in **Figure 9-4** shows various recommendations for general purpose welding. In cases where maximum strength or maximum elongation is preferred, the filler metals listed in **Figure 9-5** should be used.

Maximum weld quality can only be obtained if the filler metal is clean and of high quality. Filler metal contaminants that most often cause weld porosity are hydrocarbons, such as oil, and hydrated oxides, which are a combination of an oxide and water. The heat of welding releases the hydrogen from these sources, causing porosity in the weld. Aluminum filler metals are manufactured under rigorous specifications and packaged to stay clean.

Handle the filler metal with the utmost care. This includes putting on clean gloves before touching the filler metal. This will help keep the material clean and ready to use.

Joint Preparation for Welding Aluminum

Joint edges prepared by the plasma arc cutting process or the carbon arc gouging process have a heavy oxide film on the surface. Prior to welding, this surface must be thoroughly cleaned to prevent dross inclusion and porosity in the final weld. *Dross* is oxidized metal or impurities produced during the cutting process. These impurities need to be removed from the base metal by grinding. If the dross is left on the base metal, it can become an impurity in the weld.

Joint edges prepared by the shearing process should be sharp without tears or ridges. Dirt and oil can become trapped in tears or ridges, resulting in faulty welds. Aluminum oxidizes easily, and

Base Metal	6070	6061, 6063 6101, 6151 6201, 6951	5456	5454	5154 5254a	5086	5083	5052 5652 a	5005 5050	3004 Alc. 3004	2219	2014 2024	1100 3003 Alc. 3003	1060 EC
1060, EC	ER4043h	ER4043h	ER5356c	ER4043e,h	ER4043e,h	ER5356c	ER5356c	ER4043i	ER1100c	ER4043	ER4145	ER4145	ER1100c	ER1100
1100, 3003 Alclad 3003	ER4043h	ER4043h	ER5356c	ER4043e,h	ER4043e,h	ER5356c	ER5356c	ER4043e,h	ER4043e	ER4043e	ER4145	ER4145	ER1100c	
2014, 2024	ER4145	ER4145									ER4145a	ER4145a		
2219	ER4043t,h	ER4043t,h	ER4043	ER4043h	ER4043h	ER4043i	ER4043	ER4043i	ER4043	ER4043	ER2319f,t,h			
3004 Alclad 3004	ER4043e	ER4043b	ER5356e	ER5654b	ER5654b	ER5356e	ER5356e	ER4043e,h	ER4043e	ER4043e				
5005, 5050	ER4043e	ER4043b	ER5356e	ER5654b	ER5654b	ER5356e	ER5356e	ER4043e,h	ER4043d					
5062, 5652 a	ER5356b,c	ER5356h,c	ER5356b	ER5654b	ER5654b	ER5356e	ER5356e	ER5654a,b,c						
5083	ER5356e	ER5356e	ER5183e	ER5356e	ER5356e	ER5356e	ER5183							
5086	ER5386e	ER5356e	ER5356e	ER5356e	ER5356b	ER5356e								
5154, 5254 a	ER5356b,c	ER5356h,c	ER5356b	ER5654b	ER5654a,b									
5454	ER5356b,c	ER5356h,c	ER5356b	ER5554a,b										
5456	ER5356e	ER5356e	ER5556e											
6061, 6063, 6101 6201, 6151, 6951	ER4043b,h	ER4043b,h												
6070	ER4043b,e,h													

Where no filler metal is listed, base metal combination is not recommended for welding.

a. Base metal alloys 5652 and 5254 are used for hydrogen peroxide service.
   ER5654 filler metal is used for welding both alloys for low-temperature service (150°F and below).
b. ER5183, ER5356, ER5554, ER5556, and ER5654 may be used.
c. ER4043 may be used.
d. Filler metal with same analysis as base metal is sometimes used.
e. ER5183, ER5356, or ER5556 may be used.
f. ER4145 may be used.
g. ER2319 may be used.
h. ER4047 may be used.
i. ER1100 may be used.

Figure 9-4. Wrought aluminum alloys—filler metal specification AWS A5.10.

Base Metal	Filler Alloys[1]		Base Metal	Filler Alloys[1]	
	Preferred for Maximum As-Welded Tensile Strength	Alternate Filler Alloys for Maximum Elongation		Preferred for Maximum As-Welded Tensile Strength	Alternate Filler Alloys for Maximum Elongation
EC	1100	EC/1260	5086	5183	5183
1100	1100/4043	1100/4043	5154	5356	5183/5356
2014	4145	4043/2319 [3]	5357	5554	5356
2024	4145	4043/2319 [3]	5454	5554	5356
2219	2319	—	5456	5556	5183
3003	5183	1100/4043	6061	4043/5183	5356 [2]
3004	5554	5183/4043	6063	4043/5183	5183 [2]
5005	5183/4043	5183/4043	7039	5039	5183
5050	5356	5183/4043	7075	5183	—
5052	5356/5183	5183/4043	7079	5183	—
5083	5183	5183	7178	5183	—

The above table shows recommended choices of filler alloys for welds requiring maximum mechanical properties. For all special services of welded aluminum, inquiry should be made of your supplier.
1. Data shown are for "0" temper.
2. When making welded joints in 6061 or 6063 electrical conductor in which maximum conductivity is desired, use 4043 filler metal.

However, if strength and conductivity both are required, 5356 filler may be used and the weld reinforcement increased in size to compensate for the lower conductivity of the 5356 filler metal.
3. Low ductility of weldment is not appreciably affected by filler used. Plate weldments in these base metal alloys generally have lower elongations than those of other alloys listed in this table.

Figure 9-5. Filler metals used to obtain specific properties of completed welds.

aluminum welds are sensitive to contamination, so the metal must be properly cleaned before welding to ensure a quality weld.

Weld Backing

Two types of weld backing can be used for welding aluminum groove welds. The first type is integral backing, or a backing ring. The second type is a removable backing bar. Backing bars or backing rings are used with open root butt joints to prevent the weld metal from falling through the back side of the weldment.

Integral Backing

Joints with integral backing are designed so that the backing is a fixed part of the weldment, as shown in **Figure 9-6**. Another method of weld backing is to place a backing ring on the back side of the weld, as shown in **Figure 9-7**. The backing ring becomes part of the weldment when the weld is complete. The main problem with this design is the difficulty in achieving penetration at the root. This difficulty can be overcome if the root face is not too large, or if a gap exists between the parts. See **Figure 9-8**.

Figure 9-6. Joints designed with integral backing reduce tooling requirements.

Figure 9-7. Backing rings are often used where tooling is difficult to install. The bar must fit tightly to the root side of the joint if a good weld is to be obtained.

Removable Backing Bar

Backing bars are found in stakes and seamers, which are holding devices that can be custom constructed to fit an application. Stakes and seamers typically use copper backing bars. Removable backing bars may also be a part of the assembly tooling. The groove in the bar can be designed for forming or

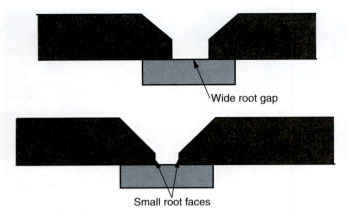

Figure 9-8. Large root faces or narrow gaps at the root of the weld obstruct penetration at the weld root. Whenever possible, use small root faces or wider root spacing to assist in obtaining penetration.

molding the penetration, as illustrated in **Figure 9-9**. Another design, shown in **Figure 9-10**, passes argon through the backing bar to prevent oxidation on the back side of the weld. The argon gas-type removable backing bar is used where very high quality welds are required.

If the groove in the backing bar is narrow, it will cause the weld to cool quickly, which can trap hydrogen gas in the weld bead, resulting in porosity. Preheating the backing bar causes the weld to solidify slowly. This allows the gas to escape before the weld solidifies, which helps to eliminate porosity.

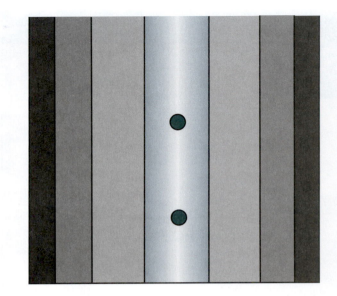

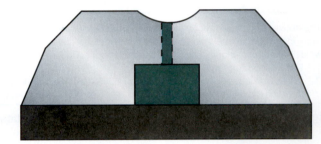

Figure 9-10. Argon gas shields the penetration during welding and prevents the formation of oxides. This also assists in the forming of the root bead and prevents oxide folds.

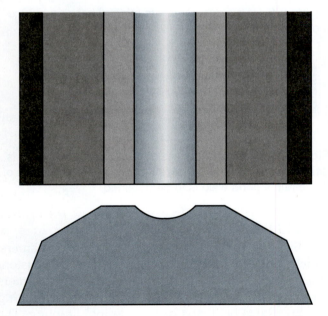

Figure 9-9. Always use generous grooves when casting the penetration of aluminum groove welds.

Preheating

Since aluminum spreads heat rapidly, it is often necessary to preheat thick pieces in order to achieve proper penetration and bead contour. Preheating also allows greater travel speeds, regardless of the type of joint. The entire joint area and tooling in the immediate area should be preheated to the same temperature.

Aluminum is generally preheated to temperatures of 350°–400°F (177°–204°C). Temperature-indicating paints, crayons, and pellets can be used to verify the actual temperature. Refer to **Figure 9-11**. Make sure the temperature-indicating material does not get into the weld area, where it can contact the molten metal in the weld pool.

Tack Welds

When tack welds must be made to maintain joint alignment for welding aluminum, the following considerations apply:

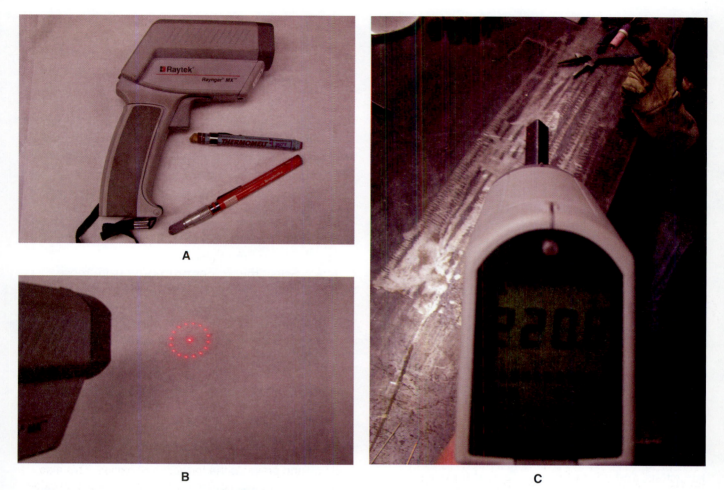

Figure 9-11. Temperature indicators. A—An electronic thermometer and crayons are the most accurate and easy way of monitoring temperatures. Temperature crayons are designed to melt at specific temperatures. B—An infrared thermometer casts a laser onto the part to be monitored and measures the temperature. C—The digital readout of an infrared thermometer checking the temperature of the weld. (Mark Prosser)

- If the joint welding procedure requires preheating, the preheating should be done before tack welding.
- Use the same filler metal for tack welding as will be used for the main welding procedure.
- If the joint requires full penetration, the tack weld requires full penetration.
- Keep the tack weld as small as possible. If possible, make it smaller than the main weld.
- If the tack weld cracks, leave it and make another close to the broken one.
- Do *not* grind out tack welds. Grinding grit in a joint makes it almost impossible to achieve a high-quality weld.
- Do *not* leave a crater in the tack weld. The term *crater* refers to a depression in a weld. The crater of a tack weld is its weakest point.

Welding Aluminum Using AC High-Frequency (ACHF) Polarity

When using a transformer-type machine or a combination ac/dc power source not designed for GTAW, do not exceed the derated capacity of the unit. If the power source has a wave balancer control, set it on *Normal* to start welding. Then, adjust the control to produce the desired penetration or cleaning action from that point. Be sure the machine is properly grounded to prevent high-frequency radiation.

If the oxide cleaning action is lost or insufficient during welding, check the high-frequency point gap. The high-frequency spark is needed to keep current flowing during the electrode positive (reverse polarity) portion of each cycle. Without the reverse polarity, the cleaning action is lost.

When using inverter-type machines, the switch between DCEN and DCEP current happens so fast that high-frequency voltage is not required for arc stabilization, and the sine wave can be controlled in many ways other than with the balance control.

Gases for Welding Aluminum Using ACHF

Argon gas is used as the shielding gas for most aluminum applications. When welding thicker sections of aluminum, the 75Ar-25He combination provides better penetration on all types of aluminum joints. Using a gas mixture with larger amounts of helium can create problems with arc starting. An argon-helium mixture produces a hotter arc, which results in deeper penetration, but the helium can also negatively affect the control and the stability of the arc. Only high-quality shielding gases should be used. There should be no leaks in the supply system.

Electrodes for Welding Aluminum Using ACHF

Pure tungsten has been used for a long time to weld aluminum. However, it was discovered that pure tungsten has the poorest heat resistance of all the tungstens, and it is not recommended for use with the highly popular inverter-type machines. Pure tungsten is used because of its ability to be balled on the end; however, a balled electrode is not a requirement for welding aluminum. It is a popular misconception that a ball on the end of the electrode affects the quality of the weld being made.

Pure tungsten is becoming obsolete because there are better electrode choices for any application. Zirconiated, ceriated, and lanthanated tungstens are being more frequently recommended for ac welding of aluminum. Thoriated tungstens cannot be used for ac welding because they "spit" after slight usage. *Spitting* refers to small particles of tungsten being separated from the electrode. The particles are carried by the arc stream into the weld pool, causing contamination.

Techniques for Welding Aluminum Using ACHF

To produce satisfactory welds on aluminum when using ACHF, follow these steps:
- Select the proper procedure from those shown in the chart in **Figure 9-12**. The values given for a groove weld can also be used to produce a fillet weld in the same position.
- Select and prepare the proper electrode. Taper large electrodes to the diameter required, as shown in **Figure 9-13**.
- Extend the electrode beyond the end of the cup a distance not greater than the inside diameter of the cup.
- Blunt the very tip of the electrode on the grinder.
- Start an arc on scrap material, and increase amperage until a radius is formed and the end of the electrode is rounded.
- Clean the metal to be welded. See **Figure 9-14**. Always use a dedicated stainless steel brush for removing oxides from the surface of the material or the weld.

Aluminum Thickness (inch)	Welding Position	Joint Type	Alternating Current (amperes)	Diameter of Tungsten Electrode (inch)	Argon Gas Flow	Filler Rod Diameter (inch)	Number of Passes
1/16	Flat	Square butt	70–100	1/16	20	3/32	1
	Horizontal and vertical	Square butt	70–100	1/16	20	3/32	1
	Overhead	Square butt	60–90	1/16	25	3/32	1
1/8	Flat	Square butt	125–160	3/32	20	1/8	1
	Horizontal and vertical	Square butt	115–150	3/32	20	1/8	1
	Overhead	Square butt	115–150	3/32	25	1/8	1
1/4	Flat	60° Single-bevel	225–275	5/32	30	3/16	2
	Horizontal and vertical	60° Single-bevel	200–240	5/32	30	3/16	2
	Overhead	100° Single-bevel	210–260	5/32	35	3/16	2

Figure 9-12. Recommended procedures for manual welding of aluminum with ac and argon shielding gas.

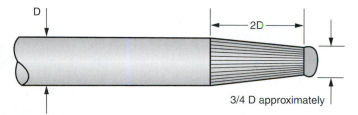

Figure 9-13. Electrodes used for alternating current should be ground as shown. The ball or rounded tip will form as the welding current is increased.

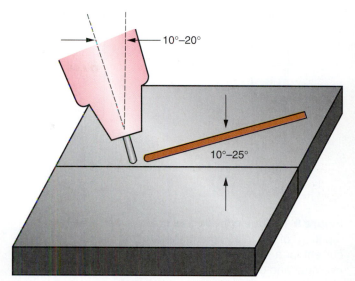

Figure 9-15. Torch and welding rod positions for welding groove welds in the flat position.

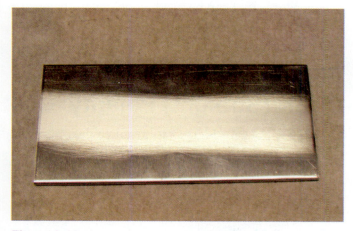

Figure 9-14. A dedicated stainless steel wire brush should be used to remove oxides from aluminum. (Mark Prosser)

- Set up the joint for welding. Aluminum is *hot short*, meaning it has low strength at high temperatures. There must be support near the weld area or the metal will collapse.
- Position the torch and welding rod as shown in **Figure 9-15** and start the arc.
- Move the torch across the joint using the step technique, shown in **Figure 9-16**.

- Watch the weld closely as it progresses. Look for any signs of incorrect settings or technique. **Figure 9-17** shows the problems encountered if the electrode is too large or if the current is excessive. Keep in mind that the weld pool does not change color as it changes temperature.
- Stop and replace or regrind the electrode if it contacts the molten filler metal or the weld pool. If this is not done, the aluminum on the electrode will burn away and contaminate the weld.
- Remember to fill the crater at the end of the weld. The completed weld should look similar to the one shown in **Figure 9-18**. The penetration side of the weld is shown in **Figure 9-19**. **Figure 9-19A** shows a weld with even penetration. **Figure 9-19B** shows penetration that did not fuse together.

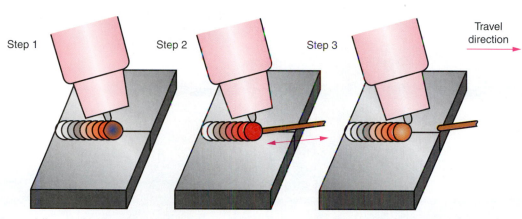

Figure 9-16. Step welding technique for welding aluminum groove welds. Step 1—Heat the metal until the pool drops. Step 2—Add welding rod until a crown is formed, then remove the rod. Step 3—Move the torch forward, stop, and repeat Steps 1–3. Continue this way until the weld is complete. Note: The color changes in the figure are used only to illustrate the heating of the metal. Aluminum does not change color as it is heated.

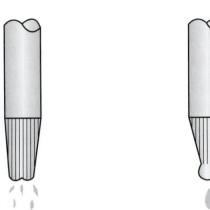

Figure 9-17. An electrode that is too large can cause "spitting" of particles across the arc gap. Excessive current for a small electrode diameter may cause the tip to overheat and drop off.

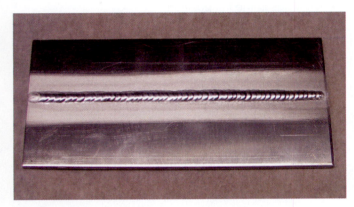

Figure 9-18. The step welding technique produces a weld with a well-defined ripple. (Mark Prosser)

- Thicker base metal requires a wider gap between the pieces. It is welded using the keyhole method. See **Figure 9-20**.

The *keyhole method* is used with all welding processes. This technique is used in GTAW to ensure full penetration on butt welds, especially on aluminum. A keyhole is produced in the root of the weld as the base metal fuses with the filler metal. The keyhole is produced with the right combination of travel speed and heat, and indicates full penetration in the joint.

Figure 9-21 lists some problem areas and corrective measures that can be used when gas tungsten arc welding aluminum.

Welding Aluminum Using DCEN (Straight) Polarity

DCEN polarity is used to produce deeply penetrating welds on aluminum. It is typically used only on thick metal. The joints require a butting surface to be effective. Square-groove joints and joints with heavy root faces are preferred to prevent melt-through. Weld joint designs and the associated procedures are listed in **Figure 9-22**. The values given for a butt joint can also be used to produce a fillet weld.

Since the arc does not provide any cleaning action, the joint must be very clean. If the joint is not clean, any gases formed in the weld pool will remain trapped in the weld deposit, creating porosity.

DCEN produces a narrow, deep weld. Because the best results on aluminum are obtained with very short arc lengths, DCEN is best suited for automatic

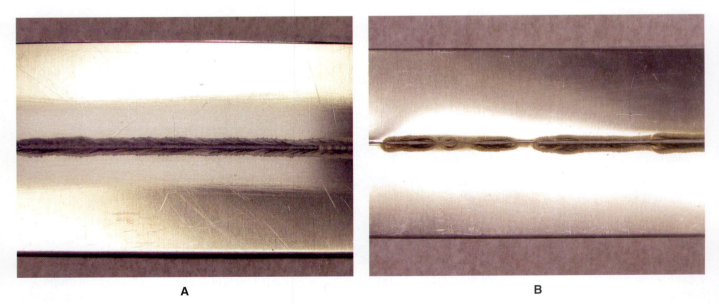

A   B

Figure 9-19. Weld penetration. A—Properly made welds have good, even penetration. B—The penetration of this weld did not fuse together due to the lack of cleaning of the mating surfaces. (Mark Prosser)

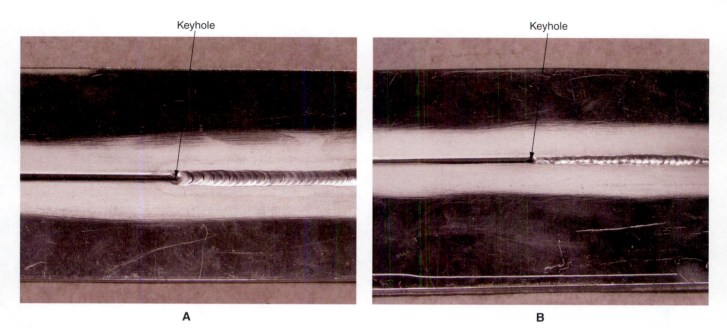

Figure 9-20. The keyhole method provides full penetration on thicker materials. A—A keyhole-shaped gap at the leading edge of the weld indicates full penetration has been achieved. B—Full penetration resulting from the keyhole method is evident on the back side of the weld. (Mark Prosser)

Common Problems Associated with GTAW on Aluminum		
Problems	**Causes**	**Corrective Measures**
Impurities and porosity	Oil, dirt, rust, mill scale, or other surface impurities	Use proper cleaning techniques Use proper cleaning compounds Use proper abrasives Use proper torch angles
Insufficient penetration (especially on heavy sections)	Too low temperature Inadequate joint preparation Improper fitup	Use recommended amperage Use recommended joint preparation Use correct fitup
Temperature too cold	Wrong polarity Arc length too long Too much torch angle Travel speed too fast Current too low	Use correct polarity Use correct arc length Decrease torch angle Lower travel speed Increase current
Temperature too hot	Travel speed too slow Current too high Current incorrect for base metal	Increase travel speed Lower current Use correct current setting for base metal
Gas entrapment	Weld pool cooling too quickly Too much heat Travel speed too fast	Use supplemental heat Use correct current setting Reduce travel speed
Alloying problems	Incompatible base and filler metals	Use compatible filler and base metal Follow all manufacturer's recommendations
Cracking	Excessive heat input Weld solidifying too fast Incompatible filler metal Crack-prone alloys Discontinuous welds, welds that intersect, or repair welds Cold-working materials	Lower current Increase travel speed Reduce current at end of weld Slow cooling processes Use compatible filler metal Use less crack-prone alloy Preheat materials

Figure 9-21. Common problems encountered during GTAW of aluminum and their possible solutions.

Material Thickness (inch)	Joint Design	Current DC (amperes) [1]	Volts	Diameter of Electrode [2] (inch)	Helium Gas Flow [3] (cfh)	Travel (ipm)	Filler Rod or Wire Diameter (inch)	Number of Passes
0.010	Standing edge	10–15		0.020	20–50		.01	
0.020	Square butt	15–30		0.020	20–50		.020	1
0.030	Square butt	20–50		0.020 or 0.047				
0.032	Square butt	65–70	10	3/32	20–50	52	None	1
0.040	Square butt	25–65		3/64	20–50		.047 (3/64)	1
0.050	Square butt	35–95		3/64	20–50		3/64	1
0.050	Square butt	70–80	10	3/32	20–50	36	None	1
0.060	Square butt	45–120		3/64 or 1/16	20–50		3/64 or 1/16	1
0.070	Square butt	55–145		1/16	20–50		1/16	1
0.080	Square butt	80–175		1/16	25–50		1/16	1
0.090	Square butt	90–185		1/16	20–50		1/16	1
1/8	Square butt	120–220		1/8	20–50		1/8	1
1/8	Square butt	180–200	12.5	1/8	20–50	24	None	1
1/4	Square butt	230–340		1/8	25–60		1/8 or 3/16	1
1/4	Square butt	220–240	12.5	1/8	25–60	22	None	1
1/2	60°V-bevel, 1/4″ root face	300–450		3/16	25–60		1/8 or 1/4	1
1/2	Square butt	260–300	13	5/32	25–60	20	None	2
3/4	60°V- or Double-V-bevel, 3/16″ root face	300–450		3/16	25–60		1/8 or 1/4	3 for Single-V; 2 for Double-V
3/4	Square butt	450–470	9.5	3/16	40–60	6	None	2
1	60° V- or Double-V-bevel, 3/16″ root face	300–450		3/16	25–60		1/8 or 1/4	4 for Single-V; 2 for Double-V
1	Square butt	550–570	9.5	1/4	40–60	5	None	2

[1] Automatic welding is required for the higher amperages. Manual welding can be done at the lower amperages.

[2] In lighter gauges of material, it is common to use larger diameter electrodes than recommended and to taper the tip.

[3] It is possible to substitute helium-argon mixtures. In automatic welding, the arc can be started in argon and the helium added to the shielding gas when welding begins. The best ratio of He-A is usually determined by experimentation. The gas flow is dependent in part on the welding speed.

Figure 9-22. Recommended practices for welding aluminum in the flat position using DCEN polarity.

welding. Arc lengths are usually about 1/16″ (1.6 mm) for manual welding and as little as 1/64″ (0.4 mm) for machine welding. Therefore, control of the arc length is critical. It is somewhat difficult to control arc length in manual DCEN gas tungsten arc welding.

Gases and Electrodes for Welding Aluminum Using DCEN

Helium-rich mixtures (minimum 75%) or pure helium should be used with the DCEN process. Argon (maximum 25%) can be used to help stabilize the arc. The helium-rich mixture requires an increased gas flow, since helium is lighter than argon.

Different types of electrodes can be used in DCEN processes. In all cases, the electrode must have sufficient capacity to carry the welding current used. The electrode should be tapered.

Techniques for Welding Aluminum Using DCEN

High-frequency voltage is not needed to maintain a stable DCEN arc, but it can be used for starting the arc. The arc can also be started by scratching the electrode to the workpiece.

The high weld current will melt the base material immediately after the arc is struck. Therefore, the filler metal must be fed into the molten pool immediately. Do not withdraw the welding rod from the shielding gas area. Move the welding rod forward into the weld pool to supply the required filler metal for the weld.

The helium gas used for shielding will leave a black sooty deposit on the top of the weld. Remove this deposit by wire brushing. Always remove this deposit after each pass when making multiple pass welds. Fill in the crater at the end of the weld.

To produce satisfactory welds on aluminum when using DCEN, follow these steps:

- Select the proper procedure from those shown in the chart in **Figure 9-22**. To avoid cross contamination of the aluminum, always use a dedicated stainless steel brush for cleaning the metal and removing the oxides.
- Because aluminum is hot short, provide support for the base material near the weld area. Otherwise, the metal will collapse.
- When manual welding, maintain the torch and welding rod angles as shown in **Figure 9-15**.
- Remember that the weld pool does not change color as it changes temperatures.
- If the electrode contacts the filler metal or the weld pool, stop and replace or regrind the electrode. (If this is not done, the aluminum on the electrode will burn away and contaminate the weld.)
- Fill the crater at the end of the weld.

Corrective measures for GTAW on aluminum are listed in the chart shown in **Figure 9-21**. Study this information.

Welding Aluminum Using DCEP (Reverse) Polarity

DCEP polarity is used to produce shallow penetration welds. Therefore, the joint thickness is usually below 1/16″ (1.6 mm). DCEP polarity is generally suitable for welding only edge-type joints or joints with very small gaps, such as those shown in **Figure 9-23**.

The cleaning action of reverse polarity is very good, since the electron flow is from the workpiece to the electrode. However, the base metal should still be thoroughly cleaned before the joint is welded.

Using DCEP polarity on thinner gauges of material requires good tooling if the joint is to stay in alignment during welding. The backing bar must have a relief for the penetration to drop into.

Gases and Electrodes for Welding Aluminum Using DCEP

Pure argon should always be used for welding aluminum using DCEP polarity. The use of thoriated tungsten electrodes is recommended because they can carry more heat for a given size than any other type of electrode. Electrode diameters for various welding currents are listed in **Figure 9-23**.

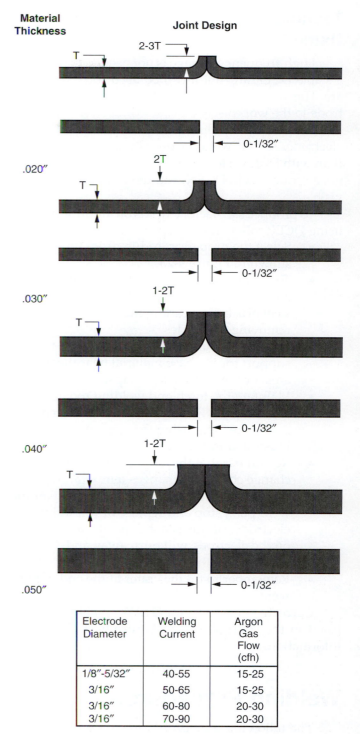

Electrode Diameter	Welding Current	Argon Gas Flow (cfh)
1/8″–5/32″	40-55	15-25
3/16″	50-65	15-25
3/16″	60-80	20-30
3/16″	70-90	20-30

Figure 9-23. Joint designs and machine settings recommended for welding aluminum using DCEP polarity.

The electrode should have a slight radius on the tip. Prior to welding, the current is applied until a "ball" is formed. If the "ball" elongates at the required current level, a larger electrode should be used.

Techniques for Welding Aluminum Using DCEP

High-frequency voltage is not needed to maintain a stable DCEP arc, but it can be used for starting the arc. The arc can also be started by scratching the electrode to the workpiece.

Most of the welding heat is concentrated on the electrode, so the welding operation is much slower than with DCEN. However, control of the arc length is not as critical as in DCEN, and the weld pool is highly visible.

To produce satisfactory welds on aluminum when using DCEP, follow these steps:

- Select the proper procedure from those shown in **Figure 9-23**.
- To avoid cross-contamination of the aluminum, always use a dedicated stainless steel brush for cleaning the metal and removing the oxides.
- Because aluminum is hot short, provide support for the base material near the weld area, or the metal will collapse.
- Maintain the torch and welding rod angles, as shown in **Figure 9-15**. Move the torch across the joint, using the step technique illustrated in **Figure 9-16**.
- Remember that the weld pool does not change color as it changes temperatures.
- If the electrode contacts the filler metal or the weld pool, stop and replace or regrind the electrode. (If this is not done, the aluminum on the electrode will burn away and contaminate the weld.)
- Remember to fill the crater at the end of the weld.

Corrective measures for GTAW on aluminum are listed in the chart shown in **Figure 9-21**. Study this information.

Welding Procedures

The following four procedures have been developed for GTAW practice and production. Any scrap aluminum material can be used for practice. The weld joint should fit properly for good results. All of the various materials should be cleaned before tack welding and welding.

Gas nozzle sizes are not specified because of the many variables involved. Always use the largest possible size available that does not obstruct your view of the weld pool.

Welding Procedure Number 9-1

Weld joint type:	Square-groove butt
Position:	Flat, horizontal, vertical
Material type:	Pure aluminum
Thickness:	3/32″ (2.4 mm)
Filler metal:	Pure aluminum
Diameter:	3/32″ (2.4 mm)
Machine setup:	ACHF
Shielding gas:	Argon
CFH:	15–25
Tungsten type:	Lanthanated, ceriated, zirconiated, or pure
Diameter:	3/32″ (2.4 mm) (tapered and balled)

Procedure:

Remove the oxide film on the surface of aluminum immediately prior to welding. Use a dedicated stainless steel brush or a chemical etch.

1. Tack weld the two pieces of material at each end.
   A. Add welding rod to each tack weld.
   B. Keep the joint tight (do not gap).
2. Position the torch and welding rod as shown in **Figure 9-15** and begin welding. The technique for welding aluminum is different from the technique for steel, because the weld is made in steps. This allows good penetration and heat control over the weld pool. The operation is shown in **Figure 9-16** and includes the following steps:
   A. Form a weld pool with penetration through the joint.
   B. Immediately add welding rod to stop the weld pool from falling. Continue to add filler metal to form a crown.
   C. Withdraw the welding rod.
   D. Move the torch and welding rod forward approximately 1/4″ (6.5 mm) and stop.
   E. Repeat steps A–D until the weld is complete.
   F. Increase the travel speed as the material heats up.

Problem Areas and Corrections:

1. Incomplete penetration.
   Possible cause: Amperage too low.
   Solution: Increase amperage.
   Possible cause: Insufficient heat in the weld pool.
   Solution: Wait longer before adding welding rod.
2. Penetration too heavy.
   Possible cause: Amperage too high.
   Solution: Decrease amperage.
   Possible cause: Too much heat in the weld pool.
   Solution: Add the welding rod sooner to cool the weld pool faster.

3. Uneven crown.
   Possible cause: Filler metal deposited inconsistently.
   Solution: Increase or decrease the amount of filler metal deposited into the weld pool as needed to maintain a consistent crown size.
   Possible cause: Inconsistent travel speed.
   Solution: Maintain a consistent travel speed.

Welding Procedure Number 9-2

Weld joint type:	60°–75° V-groove butt
Position:	All
Material type:	Pure aluminum
Thickness:	1/4″ (6.5 mm)
Filler metal:	Pure aluminum
Diameter:	3/32″ (2.4 mm)
Machine setup:	ACHF
Shielding gas:	Argon
CFH:	20–30
Tungsten type:	Lanthanated, ceriated, zirconiated, or pure
Diameter:	3/32″–1/8″ (2.4 mm–3.2 mm)

Procedure:

Prepare the bevel angles by machining. Clean the metal as described in Welding Procedure Number 9-1.
1. Tack weld the two pieces of metal.
   A. Do not gap the plates.
   B. Add welding rod to each tack weld.
2. Move the plates into the desired welding position.
3. Preheat the plates.
4. Weld the joint using the torch and welding rod positions shown in **Figure 9-15**.
   A. Maintain the preheat temperature range for all passes to obtain even penetration and weld pool control.
   B. Use the welding technique specified in Welding Procedure Number 9-1.
   C. Fill the weld joint with additional passes as required.

Problem Areas and Corrections:

The major problem in welding aluminum over 1/8″ (3.2 mm) thick is insufficient heat. Preheating is critical if the molten weld pool is to move correctly.
1. Incomplete penetration.
   Possible cause: Amperage too low.
   Solution: Increase amperage.
   Possible cause: Insufficient heat in the weld pool.
   Solution: Wait longer before adding welding rod.
   Possible cause: Insufficient preheating of thick base material.
   Solution: Preheat material to the proper temperature.
2. Cold shuts between passes.
   Possible cause: Insufficient preheating of thick base material.
   Solution: Insufficient preheating of thick base material.
3. Sluggish weld pool.
   Possible cause: Insufficient preheating of thick base material.
   Solution: Insufficient preheating of thick base material.
4. Penetration too heavy.
   Possible cause: Amperage too high.
   Solution: Decrease amperage.
   Possible cause: Too much heat in the weld pool.
   Solution: Add the welding rod sooner to cool the weld pool faster.
5. Uneven crown.
   Possible cause: Filler metal deposited inconsistently.
   Solution: Increase or decrease the amount of filler metal deposited into the weld pool as needed to maintain a consistent crown size.
   Possible cause: Inconsistent travel speed.
   Solution: Maintain a consistent travel speed.

Welding Procedure Number 9-3

Weld joint type:	T-joint
Position:	All
Material type:	Pure aluminum
Thickness:	1/16″–3/32″ (1.6 mm–2.4 mm)
Filler metal:	Pure aluminum
Diameter:	1/16″–3/32″ (1.6 mm–2.4 mm)
Machine setup:	ACHF
Shielding gas:	Argon
CFH:	15–25
Tungsten type:	Lanthanated, ceriated, zirconiated, or pure
Diameter:	3/32″ (2.4 mm) (tapered and balled)

Procedure:

Clean the metal as described in Welding Procedure Number 9-1.
1. Align the plates as required.
2. Tack weld the two plates at approximately right angles.
   A. Add welding rod to each tack weld.
   B. Keep the joint tight.

3. Weld the first pass using the torch and welding rod angles shown in **Figure 9-24**. Maintain the proper torch and welding rod angles throughout the welding process. To achieve penetration into the corner, the following technique must be used:
   A. Apply heat to start the weld pool on the lower plate.
   B. Apply a small amount of welding rod.
   C. Move the torch and welding rod forward approximately 1/8″ (3.2 mm) and stop.
   D. Add welding rod to achieve the desired crown height.
   E. Remove the welding rod from the weld pool.
   F. Continue along the weld joint using this technique. Develop a rhythm when adding the filler metal to the weld pool as the torch moves forward.

Problem Areas and Corrections:

1. Lack of penetration.
   Possible cause: Insufficient heat.
   Solution: Increase amperage.
   Solution: Use a smaller diameter welding rod.
   Solution: Avoid making too large a weld.
   Possible cause: Improper torch angle.
   Solution: Keep the electrode pointed to the corner.
   Solution: Adjust the torch angle to direct more heat into the bottom plate.

If the problem persists, check the chart of recommended procedures in **Figure 9-12**. Then, review the measures listed in the *Techniques for ACHF Welding of Aluminum* section of this chapter.

Welding Procedure Number 9-4

Weld joint type:	T-joint
Position:	All
Material type:	Pure aluminum
Thickness:	1/4″ (6.5 mm)
Filler metal:	Pure aluminum
Diameter:	3/32″ (2.4 mm)
Machine setup:	ACHF
Shielding gas:	Argon
CFH:	20–25
Tungsten type:	Lanthanated, ceriated, zirconiated, or pure
Diameter:	3/32″–1/8″ (2.4 mm–3.2 mm) (tapered and balled)

Procedure:

Clean the metal as described in Welding Procedure Number 9-1.

1. Tack weld the two plates at approximately right angles.
   A. Add welding rod to each tack weld.
   B. Keep the joint tight.
2. Move the plates into the desired welding position.
3. Preheat the assembly.
4. Weld the first pass using the required torch and welding rod angles as shown in **Figure 9-24**. Maintain the proper torch and welding rod angles through the entire pass. To achieve penetration into the corner, the following technique must be used:
   A. Apply heat to start the weld pool.
   B. Apply small amount of welding rod to the weld pool.
   C. Move the torch and welding rod forward approximately 1/4″ (6.5 mm) and stop.

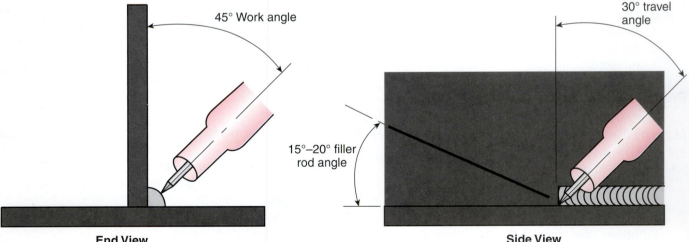

45° Work angle

15°–20° filler rod angle

30° travel angle

End View   **Side View**

Figure 9-24. Torch and welding rod positions for welding multipass T-joints.

   D. Add filler metal to achieve the desired weld crown height.
   E. Remove the welding rod from the weld pool.
   F. Continue along the weld joint using this technique. Develop a rhythm when adding the filler metal to the weld pool as the torch moves forward.
5. Complete additional passes as required to build the weld to the proper size.

Problem Areas and Corrections:

1. Lack of penetration.
   Possible cause: Insufficient heat.
   Solution: Increase amperage.
   Solution: Use a smaller diameter welding rod.
   Solution: Avoid making too large a weld.
   Solution: Wait longer before adding welding rod.
   Possible cause: Improper torch angle.
   Solution: Keep the electrode pointed to the corner.
   Solution: Adjust the torch angle to direct more heat into the bottom plate.

Summary

Aluminum is very light compared to steels, has a good strength-to-weight ratio, and is highly resistant to corrosion. Aluminum requires some slightly different welding techniques than those used on steel. The oxide film on aluminum will not melt before the base aluminum, so it must be removed or it will contaminate the weld.

Aluminum and aluminum alloys are identified by their composition, thermal treatment, and work-hardening characteristics. Wrought materials, which are those made by processes other than casting, are identified by a numbering system established by the Aluminum Association. Aluminum castings are grouped by major alloying elements. The Aluminum Association Temper Designation System applies to all forms of wrought and cast aluminum and aluminum alloys except ingot. Aluminum is classified as nonheat-treatable or heat treatable.

It is important to choose the correct filler metal for welding different aluminum alloys. The base metal and filler metal must match each other as closely as possible. The filler metal must be kept clean to prevent contamination that causes weld porosity.

Prior to welding, joint edges with a heavy oxide film must be thoroughly cleaned to prevent dross inclusion and porosity in the weld. Dross produced during cutting processes needs to be removed from the base metal by grinding. Joint edges should be sharp without tears or ridges, which can trap dirt and oil.

Joint preparation for welding aluminum includes using weld backing, preheating, and tack welding. Two types of weld backing can be used for welding aluminum groove welds: integral backing (or a backing ring) and a removable backing bar. It is often necessary to preheat thick pieces of aluminum to achieve proper penetration and bead contour. Tack welds should be kept as small as possible, and the filler metal should be the same as that used for the main welding procedure.

For ACHF welding of aluminum, argon gas is used as a shielding gas for thicknesses up to 1/8" (3.2 mm). The 75Ar-25He combination provides better penetration on all types of aluminum joints. Although pure tungsten electrodes have been used for a long time, zirconiated, ceriated, and lanthanated tungsten electrodes are frequently recommended for ac welding of aluminum. For welding of aluminum using DCEN polarity, helium-rich mixtures (minimum 75%) or pure helium must be used. The electrode must have sufficient capacity to carry the welding current used and should be tapered. For DCEP welding of aluminum, pure argon should be used. The use of thoriated tungsten electrodes is recommended.

Aluminum has low strength at high temperatures. There must be support near the weld area or the metal will collapse.

Review Questions

Write your answers on a separate sheet of paper. Do not write in this book.

1. At what temperature does pure aluminum melt?
2. Why must oxide film be removed before welding?
3. Wrought materials are those made by processes other than _____.
4. Aluminum castings are grouped by _____.
5. What does a *T6* temper designation indicate?
6. Aluminum can be classified into what two categories?
7. The strength of nonheat-treatable materials can be increased by _____ methods, such as stretching, drawing, and swaging.
8. Filler metal contaminants that most often cause weld porosity are _____ and _____.
9. What are the two types of backing used for welding aluminum groove welds?
10. What are the functions of preheating aluminum?
11. Aluminum is generally preheated to _____°F.
12. A tack weld on aluminum should be made as _____ as possible.
13. If oxide cleaning action is insufficient during ACHF welding, the _____ should be checked.
14. When thicker sections of aluminum are welded manually, a gas mixture of _____ and _____ can be used to obtain deeper penetration.
15. Why are thoriated tungsten electrodes *not* used for ac welding of aluminum?
16. Is the electrode used with alternating current pointed or blunted on the end?
17. What type of brush should be used to remove oxides from the surface of the material or the weld?
18. What does *hot short* mean?
19. What type of shielding gas is used for DCEN welding of aluminum?
20. What type of electrode is recommended for welding aluminum using DCEP polarity?

Chapter 10

Manual Welding of Stainless Steel

Objectives

After completing this chapter, you will be able to:

❐ Recall how stainless steels are identified.

❐ Identify characteristics and properties of stainless steel base materials.

❐ Recall the problems associated with welding stainless steel, including carbide precipitation, cracking, and distortion.

❐ Summarize the filler metal choices for various stainless steels.

❐ Recall methods of preparing stainless steel, including preweld cleaning, weld backing, chill bars, preheating, and tack welds.

❐ Select the correct power source, shielding gases, and electrodes for welding stainless steels.

❐ Apply correct procedures for welding stainless steel using DCEN.

❐ Recall the steps involved in postweld heat treatment.

Key Terms

annealing
austenitic stainless
carbide precipitation
chill bars
ferritic stainless
martensitic stainless
precipitation-hardening
ultra-high-purity ferritic steels
weld backing

Introduction

Stainless steels are highly corrosive-resistant, have good strength-to-weight ratios, and can be alloyed to produce specialty materials. These alloys are used in a wide variety of applications, ranging from the food service industry to aerospace. Stainless steels are welded in much the same way as regular steels, with a few exceptions. Stainless is a cleaner material than steel because of the addition of chromium and nickel to the alloy. In general, stainless steel is a very weldable material.

171

Base Materials

The stainless steel families include the following groupings of materials:

- **Chromium-nickel manganese stainless steel**—types 201 and 202. These grades are austenitic, nonmagnetic, and nonhardenable.
- **Chromium-nickel stainless steel**—types 301, 302, 303, 304, 305, 308, 309, 310, 312, 314, 316, 317, 321, 347, 349. These grades are austenitic, nonmagnetic, and nonhardenable.
- **Chromium stainless steel**—types 403, 410, 414, 416, 420, 431, 440, 501, 502. These grades are martensitic, magnetic, and hardenable.
- **Chromium stainless steel**—types 405, 430, 446. These grades are ferritic, magnetic, and nonhardenable.
- **Precipitation-hardenable stainless steel**—types 15-5 PH, PH 15-7 MO, 17-4 PH, 17-7 PH, AM 350, AM 355, A 286. These stainless steels are not identified by the American Iron and Steel Institute code system.

Wrought Material Identification

Commercially wrought stainless steels have been classified with type numbers by the American Iron and Steel Institute (AISI). **Figure 10-1** lists the type numbers and chemical analysis of commonly used stainless steels.

Cast Material Identification

Commercial stainless steels suitable for castings have been classified with type numbers by the Alloy Casting Institute (ACI). **Figure 10-2** lists the type numbers and chemical analysis of commonly used stainless steels.

Special Ferritic Stainless Steels

These alloys have been developed by various companies and are specified by trade names and/or chemical content or composition. **Figure 10-3** lists the identification numbers and the chemical analysis of common alloys in this group.

Effects of Alloying Elements

Alloying can make a significant difference in the base material. Many properties of the material, such as strength, hardness, weldability, corrosion resistance, and rigidity can be manipulated by alloying steel with different materials. The use of various alloying elements and their effects on the base material properties are shown in **Figure 10-4**.

Commercially Wrought Stainless Steel Identification (AISI)								
Composition, Percent[a]								
Type	C	Mn	Si	Cr	Ni	P	S	Others
302	0.15	2.00	1.00	17.0–19.0	8.0–10.0	0.045	0.03	
302B	0.15	2.00	2.0–3.0	17.0–19.0	8.0–10.0	0.045	0.03	
303	0.15	2.00	1.00	17.0–19.0	8.0–10.0	0.20	0.15 min	0–0.6 Mo
303Se	0.15	2.00	1.00	17.0–19.0	8.0–10.0	0.20	0.06	0.15 Se min
304	0.08	2.00	1.00	18.0–20.0	8.0–10.5	0.045	0.03	
304L	0.03	2.00	1.00	18.0–20.0	8.0–12.0	0.045	0.03	
305	0.12	2.00	1.00	17.0–19.0	10.5–13.0	0.045	0.03	
308	0.08	2.00	1.00	19.0–21.0	10.0–12.0	0.045	0.03	
309	0.20	2.00	1.00	22.0–24.0	12.0–15.0	0.045	0.03	
309S	0.08	2.00	1.00	22.0–24.0	12.0–15.0	0.045	0.03	
310	0.25	2.00	1.50	24.09–26.0	19.0–22.0	0.045	0.03	
310S	0.08	2.00	1.50	24.0–26.0	19.0–22.0	0.045	0.03	
314	0.25	2.00	1.5–3.0	23.0–26.0	19.0–22.0	0.045	0.03	
316	0.08	2.00	1.00	16.0–18.0	10.0–14.0	0.045	0.03	2.0–3.0 Mo
316L	0.03	2.00	1.00	16.0–18.0	10.0–14.0	0.045	0.03	2.0–3.0 Mo
317	0.08	2.00	1.00	18.0–20.0	11.0–15.0	0.045	0.03	3.0–4.0 Mo
317L	0.03	2.00	1.00	18.0–20.0	11.0–15.0	0.045	0.03	3.0–4.0 Mo
321	0.08	2.00	1.00	17.0–19.0	9.0–12.0	0.045	0.03	5 x %C Ti min
329	0.10	2.00	1.00	25.0–30.0	3.0–6.0	0.045	0.03	1.0–2.0 Mo
330	0.08	2.00	0.75–1.5	17.0–20.0	34.0–37.0	0.04	0.03	
347	0.08	2.00	1.00	17.0–19.0	9.0–13.0	0.045	0.03	C
348	0.08	2.00	1.00	17.0–19.0	9.0–13.0	0.045	0.03	0.2 Cu[b,c]
384	0.08	2.00	1.00	15.0–17.0	17.0–19.0	0.045	0.03	

a. Single values are maximum unless indicated otherwise. b. (Cb + Ta) min — 10 x %C. c. Ta — 0.10% max.

Figure 10-1. Common stainless steel compositions.

		Composition, Percent[a]					
Alloy Designation	Similar Wrought Type[b]	C	Si	Cr	Ni	Mo[c]	Other
CE-30	312	0.30	2.00	26–30	8–11	—	—
CF-3	304L	0.03	2.00	17–21	8–12	—	—
CF-3M	316L	0.003	1.50	17–21	9–13	2.0–3.0	—
CF-8	304	0.08	2.00	18–21	8–11	—	—
CF-8C	347	0.08	2.00	18–21	9–12	—	d
CF-8M	316	0.08	1.50	18–21	9–12	2.0–3.0	—
CF-12M	316	0.12	1.50	18–21	9–12	2.0–3.0	—
CF-16F	303	0.16	2.00	18–21	9–12	1.5	0.20–0.35 Se
CF-20	302	0.20	2.00	18–21	8–11	—	—
CG-8M	317	0.08	1.50	18–21	9–13	3.0–4.0	—
CH-20	309	0.20	2.00	22–26	12–15	—	—
CK-20	310	0.20	2.00	23–27	19–22	—	—
CN-7M	—	0.07	1.50	18–22	27.5–30.5	2.0–3.0	3–4 Cu
HE	—	0.2–0.5	2.0	26–30	8–11	—	—
HF	304	0.2–0.4	2.0	19–23	9–12	—	—
HH	309	0.2–0.5	2.0	24–28	11–14	—	0.2 N
HI	—	0.2–0.5	2.0	26–30	14–18	—	—
HK	310	0.2–0.6	2.0	24–28	18–22	—	—
HL	—	0.2–0.6	2.0	28–32	18–22	—	—
HN	—	0.2–0.5	2.0	19–23	23–27	—	—
HP	—	0.35–0.75	2.0	24–28	33–37	—	—
HT	330	0.35–0.75	2.5	15–19	33–37	—	—
HU	—	0.35–0.75	2.5	17–21	37–41	—	—

a. Single values are maximum.
   Manganese—1.50% maximum in CX-XX types; 2.0% maximum in HX types.
   Phosphorous—0.04% maximum except for Cf-16F–0.17% maximum.
   Sulfur—0.04% maximum.
b. Compositions are not exactly the same.
c. Molybdenum in HX types is 0.5% maximum.
d. Cb – 8 × %C (1.0% maximum), or Cb + Ta – 9 × %C (1.1% maximum).

Figure 10-2. Casting stainless steel compositions.

Stainless Steel	Typical Compositions					Strength, lb/in^2 (MPa)		Elongation, %
	Element (percent)					Tensile	Yield	
	C	N	Cr	Mo	Ni			
E-Brite 26-1	0.002	0.010	26	1	—	70,000 (480)	50,000 (345)	30
29-4	0.005	0.013	29	4	—	90,000 (620)	75,000 (520)	25
29-4-2	0.005	0.013	29	4	2	95,000 (650)	85,000 (590)	22
Carpenter 20 Cb-3	0.04	—	20	2.5	36			

Figure 10-3. Special stainless steel compositions.

Austenitic Stainless Steel Properties

The *austenitic stainless* steel family (AISI 200 and 300 series) consists of 18% chromium steel to which at least 8% nickel is added. Austenitic stainless steels are the most commonly used of all the stainless families.

The basic grain structure of austenitic stainless steel is austenitic at *all* temperatures. Therefore, the materials are not hardenable by heat treatment. Some hardening can be obtained by cold working. Cold working does not change the basic austenitic grain structure.

Element	Types of Steels	Effects
Carbon	All types	Strongly promotes the formation of austenite. Can form a carbide with chromium that can lead to intergranular corrosion.
Chromium	All types	Promotes formation of ferrite. Increases resistance to oxidation and corrosion.
Nickel	All types	Promotes formation of austenite. Increases high temperature strength, corrosion resistance, and ductility.
Nitrogen	XXXN	Very strong austenite former. Like carbon, nitrogen is thirty times as effective as nickel in forming austenite. Increases strength.
Columbium	347	Primarily added to combine with carbon to reduce susceptibility to intergranular corrosion. Acts as a grain refiner. Promotes the formation of ferrite. Improves creep strength.
Manganese	2XX	Promotes the stability of austenite at or near room temperature but forms ferrite at high temperatures. Inhibits hot shortness by forming MnS.
Molybdenum	316, 317	Improves strength at high temperatures. Improves corrosion resistance to reducing media. Promotes the formation of ferrite.
Phosphorous, selenium, or sulfur	303, 303Se	Increases machinability but promotes hot cracking during welding. Lowers corrosion resistance slightly.
Silicon	302B	Increases resistance to scaling and promotes the formation of ferrite. Small amounts are added to all grades for deoxidizing purposes.
Titanium	321	Primarily added to combine with carbon to reduce susceptibility to intergranular corrosion. Acts as a grain refiner. Promotes the formation of ferrite.
Copper	CN-7M	Generally added to stainless steels to increase corrosion resistance to certain environments. Decreases susceptibility to stress-corrosion cracking and provides age-hardening effects.

Figure 10-4. Effects of the various elements in stainless steels.

The high chromium percentage in austenitic stainless steel imparts good resistance to corrosion, heat, and acids. Nickel improves the material's ductility from room temperatures into cryogenic temperatures, which are temperatures below –250°F (–157°C).

The carbon content of austenitic stainless steel is held to a low percentage to reduce carbide precipitation and intergranular corrosion (these problems will be discussed later in this chapter). Grades 304L and 316L, as denoted by the letter L, have a lower percentage of carbon than the standard grades. Grades 321 and 347 are stabilized with columbium and tantalum or titanium to prevent carbide precipitation.

Martensitic Stainless Steel Properties

Martensitic stainless steels consist of low-carbon steel to which 11.5–18% chromium is added. The lower grades of the family, with approximately 12% chromium, provide moderate corrosion resistance up to approximately 1100°F (593°C). The carbon content of the grades varies, and the grade used depends on the hardness requirement for the finished part. Lower carbon grades, such as the 410 series, are weldable with certain precautions. Higher carbon grades with high chromium are not considered weldable materials.

With the combination of carbon and chromium, martensitic stainless steel responds to heating and air cooling for hardening. The specific physical characteristics of the treated materials depend on the ratio of alloying elements. A slight variation in carbon content can have a significant effect on the characteristics of the material after it is treated.

Precipitation-Hardening Stainless Steel Properties

Precipitation-hardening stainless steels are a group of steels that are alloyed with chromium, nickel, and smaller quantities of other elements such as copper, titanium, columbium, and aluminum. The combination of chromium and nickel gives the metal good corrosion and oxidation resistance. The other elements are added to promote hardening by precipitation throughout the structure during the heat-treating process.

Depending on the weldment design, fabrication procedures, and final use of the part, various types of heat treatments may be required to impart the desired properties to the metal. All of the precipitation-hardening stainless steels, except an alloy known as A-286, can be welded in all conditions. A-286 stainless steel is very difficult to weld due to hot cracking in the heat-affected zone.

Ferritic Stainless Steel Properties

Ferritic stainless steels contain iron, chromium, carbon, and additional elements such as aluminum, columbium, molybdenum, and titanium. These added elements prevent transformation of the grain structure during heating. Therefore, ferritic stainless steels remain ferritic and nonhardenable.

The higher chromium and carbon grades are susceptible to carbide precipitation and intergranular corrosion. Since these materials cannot be obtained with low carbon, they must be annealed after welding to redissolve the carbides and restore corrosion resistance.

Special Ferritic Stainless Steel Properties

These alloys are considered *ultra-high-purity ferritic steels* because of their very low carbon and nitrogen content. They have good mechanical properties and corrosion resistance (particularly to stress corrosion, cracking, and crevice corrosion due to chlorides). The low carbon and nitrogen content reduces the possibility of carbide precipitation and intergranular corrosion without thermal treatment of the material after welding.

Material Problems Associated with Stainless Steel

Welding stainless steel presents several potential problems, most of which result from contamination by atmospheric gases. Special steps, such as back purging of butt welds, must be taken when welding certain joint types. Care must be taken to preserve the desirable characteristics of stainless steel when it is welded. Problems include *carbide precipitation*, cracking, and distortion.

Carbide Precipitation

During the welding of unstabilized austenitic stainless steels, the weld and immediate area rise into and above the 800°–1600°F (427°–871°C) range. Within this range, chromium and carbon combine and precipitate into the grain boundaries of chromium-rich carbides. When this occurs, the areas next to the grain boundaries have insufficient chromium to produce a protective film. As a result, these areas become sensitive to corrosion by certain materials. The sequence of events is as follows:

1. **Sensitization.** The weld area is subjected to temperatures of 800°–1600°F (427°–871°C) or above.
2. **Carbide precipitation.** Chromium moves out of the grains and into the grain boundary to combine with carbon that has precipitated from the solid solution, forming chromium-rich carbides.

3. **Intergranular corrosion.** The corrosion takes place when the media corrodes the grains that have insufficient chromium.

Cracking

Cracking of austenitic stainless steel welds usually occurs in restrained joints. Cracking is caused by the weld metal's lack of ductility during the freezing of the molten metal. This problem may be caused by insufficient ferrite (crystalline form of iron), an improper welding procedure, or both.

Distortion

Austenitic stainless steels have very low thermal conductivity. Therefore, they retain heat from welding very well. The heat remains in the weld area instead of being dispersed throughout the material. As a result, austenitic stainless steels tend to distort more than carbon steels. For this reason, weld joints and tooling should be designed to remove welding heat from the completed weld as soon as possible. Stringer beads are favored over weave beads to reduce the amount of heat required to make the weld.

Filler Metals

Stainless steel filler metals are selected for use depending on the chemical analysis of the base material. In most cases, the mechanical properties of the filler metal matches or is superior to those of the base alloy. The American Welding Society specification AWS A5.9 *Specification for Bare Stainless Steel Welding Electrodes and Rods* lists the composition of various filler metals and the filler metals required to weld various austenitic, martensitic, and ferritic stainless steels. The chemical compositions of various filler metals are shown in **Figure 10-5**. The table in **Figure 10-6** lists the recommended filler metals for welding precipitation-hardening materials to each other and to other precipitation-hardening materials. The table in **Figure 10-7** lists the various filler metals that can be used to weld common austenitic and martensitic stainless steels.

Several types of filler metals are made with low carbon content. These filler metals are identified with an *L* at the end of the AWS classification number. Extra-low carbon filler metals are identified with *ELC* at the end of the AWS classification number. The *ELC* grade can be used as a substitute in all areas without difficulty.

A clean, high-quality filler metal must be used in order to attain maximum weld quality. Welding rods

AWS Classification[a]	C	Cr	Ni	Mo	Cb plus Ta	Mn	Si	P	S	N	Cu
ER209[c]	0.05	20.5–24.0	9.5–12.0	1.5–3.0	—	4.0–7.0	0.90	0.03	0.03	0.10–0.30	0.75
ER218	0.10	16.0–18.0	8.0–9.0	0.75	—	7.0–9.0	3.5–4.5	0.03	0.03	0.08–0.18	0.75
ER219	0.05	19.0–21.5	5.5–7.0	0.75	—	8.0–10.0	1.00	0.03	0.03	0.10–0.30	0.75
ER240	0.05	17.0–19.0	4.0–6.0	0.75	—	10.5–13.5	1.00	0.03	0.03	0.10–0.20	0.75
ER307	0.04–0.14	19.5–22.0	8.0–10.7	0.5–1.5	—	3.3–4.75	0.30–0.65	0.03	0.03	—	0.75
ER308[d]	0.08	19.5–22.0	9.0–11.0	0.75	—	1.0–2.5	0.30–0.65	0.03	0.03	—	0.75
ER308H	0.04–0.08	19.5–22.0	9.0–11.0	0.75	—	1.0–2.5	0.30–0.65	0.03	0.03	—	0.75
ER308L[d]	0.03	19.5–22.0	9.0–11.0	0.75	—	1.0–2.5	0.30–0.65	0.03	0.03	—	0.75
ER308Mo	0.08	18.0–21.0	9.0–12.0	2.0–3.0	—	1.0–2.5	0.30–0.65	0.03	0.03	—	0.75
ER308MoL	0.04	18.0–21.0	9.0–12.0	2.0–3.0	—	1.0–2.5	0.30–0.65	0.03	0.03	—	0.75
ER309[c]	0.12	23.0–25.0	12.0–14.0	0.75	—	1.0–2.5	0.30–0.65	0.03	0.03	—	0.75
ER309L	0.03	23.0–25.0	12.0–14.0	0.75	—	1.0–2.5	0.30–0.65	0.03	0.03	—	0.75
ER310	0.08–0.15	25.0–28.0	20.0–22.5	0.75	—	1.0–2.5	0.30–0.65	0.03	0.03	—	0.75
ER312	0.15	28.0–32.0	8.0–10.5	0.75	—	1.0–2.5	0.30–0.65	0.03	0.03	—	0.75
ER316[d]	0.08	18.0–20.0	11.0–14.0	2.0–3.0	—	1.0–2.5	0.30–0.65	0.03	0.03	—	0.75
ER316H	0.04–0.08	18.0–20.0	11.0–14.0	2.0–3.0	—	1.0–2.5	0.30–0.65	0.03	0.03	—	0.75
ER316L[d]	0.03	18.0–20.0	11.0–14.0	2.0–3.0	—	1.0–2.5	0.30–0.65	0.03	0.03	—	0.75
ER317	0.08	18.5–20.5	13.0–15.0	3.0–4.0	—	1.0–2.5	0.30–0.65	0.03	0.03	—	0.75
ER317L	0.03	18.5–20.5	13.0–15.0	3.0–4.0	—	1.0–2.5	0.30–0.65	0.03	0.03	—	0.75
ER318	0.08	18.0–20.0	11.0–14.0	2.0–3.0	8 × C min. to 1.0 max.	1.0–2.5	0.30–0.65	0.03	0.03	—	0.75
ER320	0.07	19.0–21.0	32.0–36.0	2.0–3.0	8 × C min. to 1.0 max.	2.5	0.60	0.03	0.03	—	3.0–4.0
ER320LR[c]	0.025	19.0–21.0	32.0–36.0	2.0–3.0	8 × C min. to 0.40 max.	1.5–2.0	0.15	0.015	0.020	—	3.0–4.0
ER321[f]	0.08	18.5–20.5	9.0–10.5	0.75	—	1.0–2.5	0.30–0.65	0.03	0.03	—	0.75
ER330	0.18–0.25	15.0–17.0	34.0–37.0	0.75	—	1.0–2.5	0.30–0.65	0.03	0.03	—	0.75
ER347[d]	0.08	19.0–21.5	9.0–11.0	0.75	10 × C min. to 1.0 max.	1.0–2.5	0.30–0.65	0.03	0.03	—	0.75
ER349[g]	0.07–0.13	19.0–21.5	8.0–9.5	0.35–0.65	1.0–1.4	1.0–2.5	0.30–0.65	0.03	0.03	—	0.75
ER410	0.12	11.5–13.5	0.6	0.75	—	0.6	0.50	0.03	0.03	—	0.75
ER410NiMo	0.06	11.0–12.5	4.0–5.0	0.4–0.7	—	0.6	0.50	0.03	0.03	—	0.75
ER420	0.25–0.40	12.0–14.0	0.6	0.75	—	0.6	0.50	0.03	0.03	—	0.75
ER430	0.10	15.5–17.0	0.6	0.75	—	0.6	0.50	0.03	0.03	—	0.75
ER630	0.05	16.0–16.75	4.5–5.0	0.75	0.15–0.30	0.25–0.75	0.75	0.04	0.03	—	3.25–4.00
ER26-1	0.01	25.0–27.5	h	0.75–1.50	—	0.40	0.40	0.02	0.02	0.015	0.20[h]
ER16-8-2	0.10	14.5–16.5	7.5–9.5	1.0–2.0	—	1.0–2.5	0.30–0.65	0.03	0.03	—	0.75

Table heading: **Composition, Percent[b]**

a. Refer to AWS A5.9—*Specification for Bare Stainless Steel Welding Electrodes and Rods.*
b. Single values shown are maximum percentages except where otherwise specified.
c. Vanadium – 0.10–0.30%
d. These grades are available in high silicon classifications that have the same chemical composition requirements as tabulated here with the exception that the silicon content is 0.65 to 1.00%. These high silicon classifications are designated by the addition *Si* to the standard classification designations in the table. The fabricator should consider carefully the use of high silicon filler metals in highly restrained fully austenitic welds. A discussion of the problem is presented in the Appendix to the specification.
e. Carbon shall be reported to the nearest 0.01% except for the classification E320LR for which carbon shall be reported to the nearest 0.005%.
f. Titanium – 9 × C min. to 1.0 max.
g. Titanium – 0.10 to 0.30%; tungsten–1.25 to 1.75%.
h. Nickel, max. – 0.5 minus the copper content, percent.

Figure 10-5. Austenitic, martensitic, and ferritic stainless steel filler metal compositions follow specifications for corrosion resistance.

Material Type	First Choice	Second Choice
17-4 PH or 15-5 PH	AMS 5826 or 17-4 PH or ER 308	ER 309 or ER 309 CB
Stainless W	AMS 5805C or A-286 or ER NiMo-3	ER NiMo-3 or ER 309
17-7 PH	AMS 5824A or 17-7 PH	ER 310 or ERNiCr-3
PH 15-7 Mo	AMS 5812C (15-7 Mo)	ER 309 or ER 310
AM 350	AMS 5774B (AM 350)	ER 308 or ER 310
AM 355	AMS 5780A (AM 355)	ER 308 or ER 309
A-286	ER NiMo-3	ER 309 or ER 310

Figure 10-6. Precipitation-hardening stainless steels can be welded together using the material listed in the First Choice column. When dissimilar precipitation-hardening materials are being welded together, the filler metal listed in the Second Choice column should be used.

	Weld Filler Metal									
	308	308L	309	310	316	316L	317	317L	330	347
Base Material	301 302		309	310						321
	304	304L 310S	309S		316	316L	317	317L	330	347
	304N				316N					348
	308		384							

Figure 10-7. This chart lists filler metals for common austenitic and martensitic stainless steels. Filler metals listed for each column can be used to weld the materials in the same column together or in combination.

should be stored in plastic containers in a clean, dry, dust-free area. Clean gloves should be worn to handle welding rods. Welding rods must be clearly and individually identified so that they can be distinguished from each other if they get mixed together. Flag tags work well for identifying welding rods.

Preparation of Stainless Steel

The base metal must be clean if the weld is to have satisfactory properties. Oil and grease must be removed from all types of stainless steels prior to welding. Acetone, alcohol, or any commercial degreaser can be used. Any oxidation should be removed mechanically, as described in Chapter 7, *Weld Preparation and Equipment Setup*. Also, any tooling that will be used in the welding operation should be properly cleaned before use.

Weld Backing

Various types of *weld backing* are used when welding stainless steel. Types of weld backing include integral, removable backing bar, commercial flux, and nitrogen gas.

Integral backing. An integral backing is the overlapping portion of a joint like the one shown in **Figure 10-8**. Another method of weld backing is to place a backing strip on the back side of the weld, as shown in

Figure 10-9. The main problem with this design is the fit of the backing bar. A loose fit allows expulsion of metal at the weld root and contamination of the penetration. These problems are shown in **Figure 10-10**.

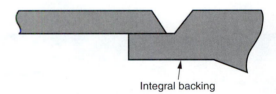

Integral backing

Figure 10-8. Integral backing bar designs are often used for pipe to fit welds where excess stock for machining is available.

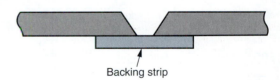

Backing strip

Figure 10-9. Backing strips must fit tightly to prevent gaps between the back of the joint and the strip.

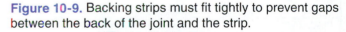

Gap

Figure 10-10. If the backing plate does not fit tightly to the joint, air can enter through the gap to contaminate the weld. Also, filler metal can be expelled out through the gap.

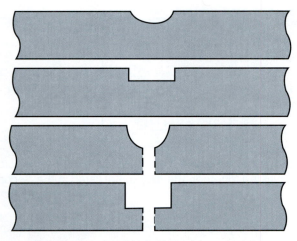

Figure 10-11. Types of weld backing grooves for full-penetration groove welds. The bottom two designs have passages for delivering backing gas to the back side of the weld.

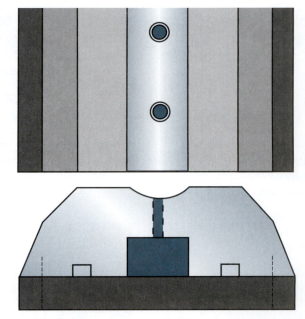

Figure 10-12. Backing bar design for admitting gas to the penetration side of the weld joint.

Removable backing bar/ purge blocks. These types of backing bars are found in stakes and seamers, or they may be a part of the assembly tooling. The groove in the bar may be designed for casting, or forming, the penetration, as shown in **Figure 10-11**. Purge blocks, like the one shown in **Figure 10-12**, flood the back side of the weld with argon, preventing oxidization. Purge blocks are used when high quality must be maintained on the back side of the weld. Dimensions for the grooves of backing bars and purge blocks are shown in **Figure 10-13**.

Commercial flux. This type of flux, **Figure 10-14**, is supplied as a powder and is mixed with acetone or alcohol to form a paste that is applied to the part by brushing. When the alcohol or acetone in the paste evaporates, the powder remains on the joint, as shown in **Figure 10-15**. During welding, the powder melts and forms a barrier that prevents air from reaching the molten metal in the weld pool. If necessary, wire brushes and hot water can be used to remove the residue after welding.

Backing Bar Groove Dimensions			
Metal Thickness	**Weld Type**	**Groove Dimensions for Casting Penetration**	**Groove Dimensions for Gas Backup Penetration**
.005″–.012″	Autogenous	.040″ wide .010 deep	.040″ wide .125″ deep
	Filler added	-----	
.013″–.020″	Autogenous	.063″ wide .010″ deep	
	Filler added	-----	
.021″–.032″	Autogenous	.093″ wide .010″ deep	.187″ wide .100″ deep
	Filler added	.125″ wide .020″ deep	
.033″–.040″	Autogenous	.125″ wide .020″ deep	
	Filler added	.187″ wide .025″ deep	
.041″–.050″	Autogenous	.125″ wide .020″ deep	
	Filler added	.187″ wide .025″ deep	
.051″–.062″	Autogenous	.187″ wide .020″ deep	
	Filler added	.250″ wide .040″ deep	
.063″–.072″	Autogenous	.250″ wide .020″ deep	.250″ wide .100″ deep
	Filler added	.250″ wide .040″ deep	
.073″–.125″	Autogenous	.250″ wide .020″ deep	
	Filler added	.312″ wide .040″ deep	
.126″–.250″	Autogenous	.312″ wide .020″ deep	.312″ wide .100″ deep
	Filler added	.375″ wide .050″ deep	
.251″–.375″	Autogenous	-----	
	Filler added	-----	

Figure 10-13. Backing bar groove dimensions for full-penetration welds.

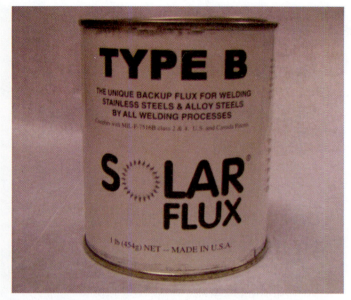

Figure 10-14. Commercial flux is often used when backing bars or tooling cannot be used. (Mark Prosser)

Figure 10-15. The flux powder remains on the part after the fluid evaporates. (Mark Prosser)

Nitrogen gas. Nitrogen gas is not an inert gas. However, it can be used in areas where it is not in contact with the molten metal, such as the areas shown in **Figure 10-16A** and **Figure 10-16B**. Nitrogen cannot be used as a backing gas for full-penetration welds, **Figure 10-16C**.

Chill Bars

Chill bars can be used for welding all types of stainless steels. For welding austenitic stainless steels, chill bars assist in removing heat from the joint area and reduce distortion. For welding martensitic steels, the bars are used to retain the preheat and slow down the cooling rate to reduce cracking.

Preheating

Preheating is used to slow down the cooling rate and prevent the formation of hard and brittle welds and heat-affected zones. The following list describes

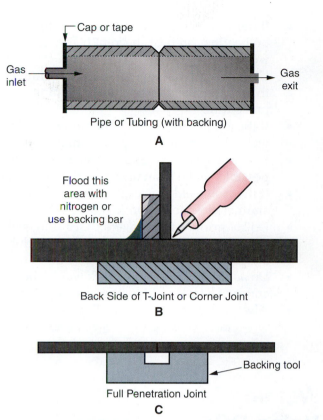

Figure 10-16. Nitrogen can be used as backing gas as long as it does not contact molten metal. A—Nitrogen can be used as a backing gas for pipe and tubing welds made with integral backing or a backing ring. B—Nitrogen can also be used in place of a backing bar on the back side of a T-joint or corner joint. C—Nitrogen must not be used as a backing gas for full-penetration welds, like a butt weld in plate.

the preheating requirements for various chrome and chrome-nickel stainless steels:

- Austenitic, chromium-nickel manganese stainless steel—no preheat.
- Austenitic, chromium-nickel stainless steel—no preheat.
- Martensitic, chromium stainless steel (hardenable)—preheat the parts to 300°–500°F (149°–260°C).
- Ferritic, chromium stainless steel (nonhardenable)—preheat the parts to approximately 300°F (149°C) on welds with high restraint in grades 430, 434, 442, 446.
- Precipitation-hardenable stainless steel—no preheat.

Tack Welds

When tack welds must be made to maintain joint alignment for welding, observe the following guidelines:

- If the joint welding procedure requires preheating, preheat the joint before tack welding.
- Use the same filler metal for tack welding that will be used for the main welding procedure.
- If the joint requires full penetration, the tack weld also requires full penetration.
- Keep the tack weld as small as possible. If possible, make it smaller than the main weld.
- If the tack weld cracks, leave it and make another close to the broken one.
- Do *not* grind out tack welds. Grinding grit in a joint makes it almost impossible to achieve a high-quality weld.
- Do *not* leave a crater in a tack weld. The crater of a tack weld is its weakest point.
- Always use an inert gas or flux backing when tack welding full-penetration type joints.

Welding Stainless Steel Using DCEN (Straight) Polarity

The power source selected for welding stainless steel should be a motor generator or a rectifier with a high-frequency arc start. An inverter-type machine is desirable, but not mandatory.

Gases for Welding Stainless Steel

Argon gas can be used for welding all thicknesses of stainless steel. For thicknesses over 1/8" (3.2 mm), adding helium to the argon creates a hotter arc and increases the penetration. Pure helium gas can be used for automatic welding of stainless steels.

The following are standard argon-helium mixes:
- 75% argon—25% helium
- 50% argon—50% helium
- 25% argon—75% helium

Argon-hydrogen gas mixtures are often used for making cleaner welds. However, an excessive amount of hydrogen in the mixture can result in weld porosity. Refer to the Chapter 4 section *Argon-Hydrogen Mixtures*.

Stainless steels sometimes require more shielding gas than regular steels in order to provide adequate weld protection and good coloring of the weld. The weld should have a straw color, which indicates the weld is in the correct temperature range. Adequate

protection is achieved by using a slightly larger cup size, which enables more gas flow to the weld area.

Electrodes for Welding Stainless Steel

Thoriated (1% or 2%), ceriated, or lanthanated electrodes should be used. The electrodes should be tapered to a sharp point. Always use an electrode large enough to carry the required amperage.

Procedure for Welding Stainless Steel

1. Clean any oxide film on the weld joint and surrounding area down to bright metal.
2. Prior to welding, grind all groove weld joints prepared by plasma arc cutting or carbon arc gouging to a bright metal finish. See **Figure 10-17**.
3. Remove grease, oil, dirt, and other contaminants with alcohol or acetone.
4. Select a filler metal. Refer to **Figure 10-6** or **Figure 10-7**.
5. Complete the weld, keeping the following guidelines in mind:
   - If martensitic and ferritic alloys are being welded, properly preheat the weldment, maintain the proper interpass temperature, and postheat the weldment as specified.
   - Use the largest gas nozzle possible.
   - Use the largest diameter welding rod possible.
   - Always trim contaminated metal from the used end of a welding rod, as shown in **Figure 10-18**.

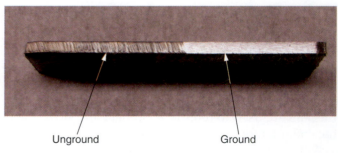

Unground   Ground

Figure 10-17. Plasma arc cut material has a severe oxide film on the surface that must be removed down to bright metal prior to welding. (Mark Prosser)

Figure 10-18. Filler metal that is removed from the torch shielding gas while it is still hot will be contaminated. Always cut off the colored end of the welding rod. (Mark Prosser)

- Use minimum welding heat. For most joints, deep penetration into the base material is not required. Full penetration is required only on a butt weld.
- Make stringer beads; do *not* make weave beads. Stringer beads minimize carbide precipitation.
- If austenitic stainless steels are being welded, use chill bars to cool the weld area as fast as possible to prevent distortion.
- When stopping a weld, always maintain gas coverage from the torch nozzle until the weld cools to a color. Proper gas coverage from the torch will result in a light blue, straw, or gold-colored weld.
- When wire brushing base metal or a completed weld, always use an uncontaminated austenitic stainless steel wire brush.
- Tack welds, root passes, and weld ends are susceptible to cracking because the weld metal cross section is thinner. To prevent cracks, make welds slightly convex and do not leave craters.

6. Evaluate the welding results.

Figures 10-19 to **10-23** illustrate stainless steel butt welds and fillet welds made under various conditions. The butt weld in **Figure 10-19** was made using chill bar tooling to remove welding heat. Note the good gold and light blue coloration. Welds made without the use of chill bar tooling are shown in **Figures 10-20** and **10-21**. Note the purple and magenta hues surrounding the butt weld, and the enlarged heat-affected zone in both joints. The butt weld in **Figure 10-22** was made without any type of shielding on the penetration,

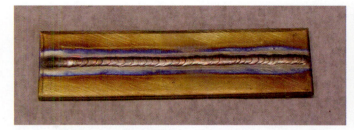

Figure 10-19. Chill bars remove the welding heat rapidly, resulting in a weld with a good color and contour. (Mark Prosser)

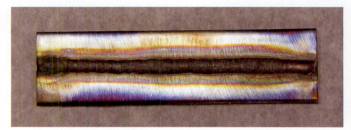

Figure 10-20. Without chill bars, the welding heat moves out into the plate material. The crown has no color and is flatter. (Mark Prosser)

Figure 10-21. A fillet weld made without chill bars has a wide heat-affected zone. However, the weld has good colors due to proper gas coverage within the T-joint. (Mark Prosser)

Figure 10-22. The unprotected penetration weld side of this joint is badly oxidized from the atmosphere. The weld has a poor bead contour and will not do the job intended. The oxidized material must be completely removed and the joint must be rewelded. (Mark Prosser)

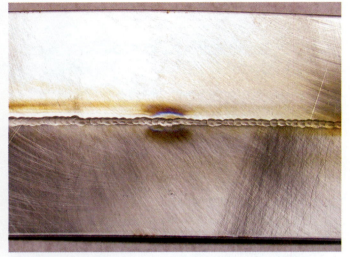

Figure 10-23. The penetration side of this weld was protected by an inert gas (argon). As a result, the weld has good color, even penetration, and will do the job intended. (Mark Prosser)

resulting in oxidation. **Figure 10-23** shows a butt weld that was made with gas shielding on the penetration. This weld has good color and even penetration.

Postweld Heat Treatment

During the welding of hardenable martensitic stainless steels, some martensite will form, regardless of what preheat is used. If the weldment will not be fully hardened after welding, *annealing* is recommended. The annealing procedure is as follows:

1. After the welding is completed, heat the weldment to 1550°–1650°F (843°–899°C) and hold that temperature for two hours.
2. Furnace-cool the weldment to 1100°F (593°C) at a rate not to exceed 100°F (38°C) per hour.
3. Allow the weldment to cool to room temperature.

Specialty stainless steels and precipitation-hardening alloys should be thermally treated to the manufacturer's recommendations to achieve full mechanical values.

Summary

Wrought materials and cast materials are identified by type numbers. Special ferritic stainless steels are specified by trade names and/or chemical content or composition. The use of various alloying elements affects properties such as strength, hardness, weldability, corrosion resistance, and rigidity.

Austenitic stainless steels consist of 18% chromium steel to which at least 8% nickel is added. The high chromium content imparts good resistance to corrosion, heat, and acids. Austenitic stainless steels are not hardenable by heat treatment, but some hardening can be obtained by cold working.

Martensitic stainless steels consist of low-carbon steel to which 11.5–18% chromium is added. Higher carbon grades with high chromium are not considered weldable. Heating and air cooling can be used for hardening of martensitic stainless steels.

Precipitation-hardening steels are alloyed with chromium, nickel, and smaller quantities of other elements. The combination of chromium and nickel provides good corrosion and oxidation resistance. The other elements are added to promote hardening.

Ferritic stainless steels contain iron, chromium, carbon, as well as additional elements that prevent transformation of the grain structure during heating. The higher chromium and carbon grades are susceptible to carbide precipitation and intergranular corrosion. Special ferritic stainless steels (ultra-high-purity ferritic steels) have very low carbon and nitrogen content, which helps prevent carbide precipitation and intergranular corrosion without thermal treatment of the material after welding.

Material problems associated with stainless steel include carbide precipitation, cracking, and distortion. Weld joints and tooling should be designed to remove welding heat from the completed weld as soon as possible.

Specification AWS A5.9 lists the composition of various filler metals and the filler metals required to weld various types of stainless steels. Several types of filler metals are made with low carbon content. Care should be taken to keep welding rods clean and properly identified.

Preparation of stainless steel for welding includes proper cleaning of the material and tooling, use of weld backing and chill bars, preheating, and tack welding. Types of weld backing include integral, removable backing bar, commercial flux, and nitrogen gas. Preheating requirements vary for the various chrome and chrome-nickel stainless steels.

Argon gas can be used when welding all thicknesses of stainless steel. Stainless steels sometimes require more shielding gas than regular steels. Thoriated (1% or 2%), ceriated, or lanthanated electrodes should be used. The electrodes should be tapered to a sharp point.

Review Questions

Write your answers on a separate sheet of paper. Do not write in this book.

1. List the five groupings in the stainless steel families.
2. Commercially wrought stainless steels have been classified by the _____ Institute.
3. How are the special ferritic stainless steels identified?
4. What metal is used in the austenitic stainless steels to provide corrosion resistance?
5. Nickel improves the _____ of stainless steel.
6. Austenitic stainless steels have a low carbon content in order to reduce the problems of _____ and intergranular corrosion.
7. What two materials are used in series 410 stainless steel? Will this steel harden?
8. List four materials that can be added to the precipitation-hardening stainless steels to promote hardening.
9. The special ferritic stainless steels are also called _____ ferritic steels.
10. The lower carbon and _____ content of special ferritic stainless steels reduces the possibility of intergranular corrosion.
11. In what temperature range does sensitization occur in the weld area of unstabilized austenitic steels?
12. Why do austenitic stainless steels distort more than carbon steel during welding?
13. The letters *ELC* at the end of an AWS classification number identify a(n) _____ filler metal.
14. Commercial flux is a powder that is mixed with _____ or _____ to form a paste that is brushed onto the part.
15. _____ gas can be used as a backing gas in areas where it is not in contact with the molten metal.
16. Which hardenable stainless steel family requires preheat prior to welding?
17. Why is helium added to argon to weld thicker sections of aluminum?
18. Pure _____ gas can be used when automatic welding stainless steel.
19. Why are stringer beads preferred over weave beads when welding the austenitic stainless steels?
20. In the annealing procedure, the weldment should be cooled at a rate not to exceed _____ per hour.

Stainless steels are very weldable materials and are used in a wide variety of applications. (senlektomyum/Shutterstock)

Chapter

Manual Welding of Magnesium

Objectives

After completing this chapter, you will be able to:

- ❏ Identify the characteristics of magnesium.
- ❏ Identify wrought material and tempers by their designations.
- ❏ Summarize the filler metal choices for magnesium and its alloys.
- ❏ Recognize common joint designs for welding magnesium with ac and continuous high-frequency voltage.
- ❏ Recall joint preparation techniques, including preweld cleaning, weld backing, tooling, preheating, and tack welds.
- ❏ Select the correct power source, shielding gases, and electrodes for welding magnesium using DCEN and DCEP.
- ❏ Apply correct procedures for welding magnesium using ACHF.
- ❏ Summarize the postweld heat treatment needed for magnesium.
- ❏ Apply correct procedures for welding magnesium using DCEN.
- ❏ Apply correct procedures for welding magnesium using DCEP.

Key Terms

intermetallic compounds
preheating
spitting
tempers

Introduction

Magnesium is a very lightweight material that is similar to aluminum, but with a better strength-to-weight ratio. Magnesium's rigidity and corrosion resistance characteristics make it a desirable choice for many applications. Magnesium also has good weldability if the proper procedures are followed. Caution is needed when grinding, cutting, and welding magnesium due to its flammability. The grinding dust from magnesium can cause a flash fire.

Base Materials

Magnesium is a nonferrous material that is readily welded with GTAW. The major characteristics of magnesium include the following:

- extreme lightness
- high strength-to-weight ratio
- good thermal conductivity
- good electrical conductivity
- high resistance to corrosion
- nonsparking
- nontoxicity
- does not change color as it is heated

Pure magnesium melts at approximately 1200°F (649°C). Magnesium alloys melt at a slightly lower temperature. Welding of magnesium requires approximately two-thirds of the amount of heat required to weld the same thickness of aluminum. The oxide film on the surface must be removed prior to welding.

Wrought Material Identification

Magnesium alloys are designated by the ASTM International system. The designations are based on chemical composition. The first two letters represent the two alloying elements present in the greatest amount. These are arranged in decreasing percentages, or alphabetically if they have equal percentages. The letters are followed by the respective percentages rounded off to whole numbers. These are followed by a serial letter that indicates some variation in composition.

The letters that designate various alloying elements include:

A—Aluminum
E—Rare earths
H—Thorium
K—Zirconium
L—Lithium
M—Manganese
Q—Silver
S—Silicon
T—Tin
Z—Zinc

The chemical compositions of wrought magnesium alloys are listed in **Figure 11-1**.

Tempers

Magnesium materials can be further identified by the condition in which they are supplied. Designations used for *tempers* of magnesium mill products are:

—F	As extruded
—T, —T51	Artificially aged
—T8, —T81	Solution heat-treated, strain-hardened, then artificially aged
—0	Fully annealed
—H24, —H26	Strain-hardened, then partially annealed

Filler Metals

Choosing the correct filler metal for welding a specific magnesium alloy is very important. The metal produced in the weld pool is a combination of the filler metal and the base material. Weld metal must have the strength, ductility, and cracking and corrosion resistance required by the application. Using

Magnesium Alloy Chemical Composition (Percent)												
ASTM Designation Alloy	Aluminum	Manganese (minimum)	Zinc	Zirconium	Rare Earths	Thorium	Calcium	Silicon (maximum)	Copper (maximum)	Nickel (maximum)	Iron (maximum)	Other Important (maximum)	Magnesium
AZ31B	2.5–3.5	0.20	0.7–1.3	–	–	–	0.04 maximum	0.30	0.05	0.005	0.005	0.30	Balance
AZ31C	2.4–3.6	0.15	0.5–1.5	–	–	–	0.04 maximum	0.30	0.10	0.03	–	0.30	Balance
AZ61A	5.8–7.2	0.15	0.4–1.5	–	–	–	–	0.30	0.05	0.005	0.005	0.30	Balance
AZ80A	7.8–9.2	0.15	0.2–0.8	–	–	–	–	0.30	0.05	0.005	0.005	0.30	Balance
HK31A	–	0.15 maximum	–	0.45–1.0	–	2.5–4.0	–	–	–	–	–	0.30	Balance
HM21A	–	0.45–1.1	–	–	–	1.5–2.5	–	–	–	–	–	0.30	Balance
HM31A	–	1.20	–	–	–	2.5–3.5	–	–	–	–	–	0.30	Balance
M1A	–	1.20	–	–	–	–	0.08–0.14	0.05	0.05	0.01	–	0.30	Balance
ZK60A	–	–	4.8–6.2	0.45 minimum	–	–	–	–	–	–	–	0.30	Balance

Figure 11-1. Wrought magnesium alloy chemical compositions.

the correct filler alloy eliminates or significantly reduces the formation of intermetallic compounds and low ductility in magnesium welds. *Intermetallic compounds* are combinations of two or more metals that differ from the base metal in composition and structure.

Figure 11-2 lists the various filler metals used for joining magnesium alloys to each other and to other magnesium alloys. The AWS specifications for electrodes and welding rods are published in AWS A5.19 *Specification for Magnesium Alloy Welding Electrodes and Rods.*

Maximum weld quality can only be achieved if the filler metal is clean. Filler metal contaminants that most often cause weld porosity are hydrocarbons, such as oil, and hydrated oxides, which are a combination of an oxide and water. The heat of welding releases hydrogen from these contaminants, which causes porosity in the weld.

Filler metal should be handled with the utmost care. Clean gloves should be put on before touching the filler metal. This will help keep the filler metal clean and ready to use.

Weld Joint Design

Weld joint designs vary slightly, depending on the polarity that will be used to weld the joint. Common joint designs for magnesium that will be welded with alternating current and continuous high-frequency voltage are shown in **Figure 11-3**. Joint designs used with DCEN and DCEP welding are specified in the individual welding procedures.

Magnesium Joint Preparation

Joint edges prepared by plasma arc cutting or carbon arc gouging have a heavy oxide film on the surface. This surface should be thoroughly cleaned prior to welding to prevent dross and porosity in the final weld. Joint edges prepared by shearing should be sharp, without tearing or ridges. Otherwise, dirt and oil can become entrapped, resulting in faulty welds.

Preweld Cleaning

Magnesium is generally received from the mill with a lightly oiled surface. The material should be degreased and cleaned using the same procedures used to clean aluminum.

Weld Backing

Weld backing bars designed for welding of aluminum are recommended for use with magnesium. Using an inert gas on the root side of the weld helps ensure the good penetration and defect-free root reinforcement required for high-quality welds.

Tooling

Magnesium tends to distort more than aluminum. Using hold-down bars and chill bars in conjunction with the backing bars will greatly reduce distortion.

Preheating

Thin structures of magnesium do not require *preheating*. Material over 1/4″ (6.4 mm) can be preheated to a maximum of 300°F (149°C) to increase the depth of penetration. Magnesium castings requiring welding should be preheated to about 300°F (149°C) to prevent shrinkage cracks.

Chart of Filler Metals									
Alloy	**AZ10A**	**AZ31B**	**AZ61A**	**AZ80A**	**HK31A**	**HM21A**	**HM31A**	**ZE10A**	**ZK60A**
AZ10A	AZ261A*	–	–	–	–	–	–	–	–
AZ31B	AZ261*	AZ261*	–	–	–	–	–	–	–
AZ61A	AZ261*	AZ261*	AZ261*	–	–	–	–	–	–
AZ80A	AZ261*	AZ261*	AZ261*	AZ261*	–	–	–	–	–
HK31A	AZ261*	AZ261*	AZ261*	AZ261*	EZ33A	–	–	–	–
HM21A	AZ261*	AZ261*	AZ261*	AZ261*	EZ33A	EZ33A	–	–	–
HM31A	AZ261*	AZ261*	AZ261*	AZ261*	EZ33A	EZ33A	EZ33A	–	–
ZK60A	NR	NR	NR	NR	NR	NR	NR	NR	NR

*Use AZ61A for best weldability of wrought alloys and greatest economy. AZ92A, AZ101A or AZ61A welding rod.

NR = Not recommended

Figure 11-2. Filler metals used to weld magnesium and alloys. (Dow Chemical Co.)

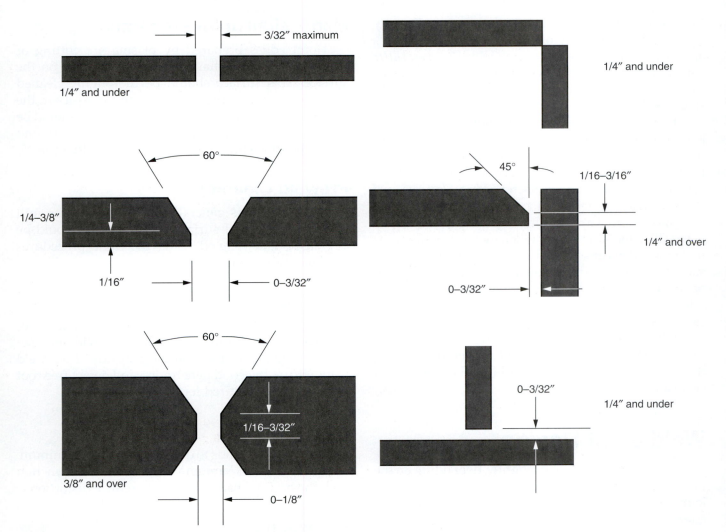

Figure 11-3. Weld joint designs used for welding magnesium with ACHF current.

Temperature-indicating paints, crayons, and pellets can be used to verify the actual preheat temperature. The temperature-indicating material should not be allowed to contact the molten metal in the weld pool.

Tack Welds

When tack welds must be made to maintain joint alignment for welding, the following considerations apply:

- If the joint welding procedure requires preheating, the joint must be preheated before being tack welded.
- Use the same filler metal as will be used for the main welding procedure.
- If the joint requires full penetration, the tack weld must have full penetration.
- Keep the tack weld as small as possible. If possible, make it smaller than the main weld.
- If the tack weld cracks, leave it and make another close to the broken one.

- Do *not* grind out tack welds. Grinding grit in a joint makes it almost impossible to achieve a high-quality weld.
- Do *not* leave a crater in the tack weld. The crater of a tack weld is its weakest point.

If a tack-welded assembly needs straightening, a soft mallet can be used to even the joint prior to welding. Do not get the mallet material into the open joint. The joint must be kept clean.

Welding Magnesium Using AC High-Frequency (ACHF) Polarity

1. When using a transformer or a combination ac/dc power source *not* designed for GTAW, do not exceed the derated capacity of the unit.
2. Ensure that the welding machine is properly grounded to prevent high-frequency radiation.

3. If the power source has a wave balancer control, set the control to normal (or 50%) to start welding. This adjusts the amount of current produced during the negative (straight) polarity and positive (reverse) polarity phases of alternating current cycle. Once a stable arc is established, adjust the control to provide the desired penetration or cleaning action.

4. If the oxide cleaning action is lost or insufficient during welding, check the high-frequency point gap. The high-frequency spark is needed to maintain the start of each straight and reverse polarity part of the cycle. Without the reverse polarity, the cleaning action is lost.

Gases for Welding Magnesium Using ACHF

Argon gas is used to weld material thicknesses up to 1/8" (3.2 mm) on all types of joints. For materials over 1/8" (3.2 mm) thick, a 75He-25Ar mixture can be used. This mixture provides better penetration on all types of joints. Using a gas mixture with larger amounts of helium creates problems with arc starting. The gas used should be high quality, and there should be no leaks in the supply system.

Electrodes for Welding Magnesium Using ACHF

Pure, zirconiated, or lanthanated tungsten electrodes are recommended for welding with alternating current because they have the good "balling" capability required for making high-quality magnesium welds. Thoriated tungstens do not have this capability, and as a result will "spit" after brief use. *Spitting* refers to a breakdown of the electrode, in which small particles of tungsten are separated and transferred into the weld pool by the arc stream.

Techniques for Welding Magnesium Using ACHF

To produce satisfactory welds on aluminum when using ACHF, follow these steps:

1. Select the electrode diameter from those listed in **Figure 11-4**.
2. Taper the electrode as shown in **Figure 11-5**.
3. Blunt the electrode on the end. Start an arc on scrap material and increase amperage until a radius is formed, as shown in **Figure 11-5**.
4. Extend the electrode beyond the end of the cup a distance not greater than the inside diameter of the cup.
5. Maintain the torch and welding rod angles shown in **Figure 11-6**.

Magnesium Welding Parameters							
Material Thickness (inches)	Number of Passes	Current (A*a*)	Electrode Diameter (inches)	Weld Rod Diameter (inches)	Argon Flow (ft³/h)	Joint Design*b*	Filler Metal Consumption (lb/ft of weld)
0.040	1	35	1/16	3/32	12	A	0.004
0.063	1	50	3/32	3/32	12	A	0.005
0.080	1	75	3/32	3/32	12	A	0.006
0.100	1	100	3/32	3/32	12	A	0.008
0.125	1	125	3/32	1/8	12	A	0.009
0.190	1	160	1/8	1/8	15	A	0.011
0.250	2	175	5/32	1/8	20	B	0.026
0.375	3	175	5/32	5/32	20	B	0.057
0.375	2	200	3/16	1/8	20	C	0.024
0.500	3	175	5/32	5/32	20	B	0.106
0.500	2	250	3/16	1/8	20	C	0.047

a Magnesium alloys containing thorium will require 20% higher current. The use of helium as the shield gas will reduce the weld current required by 20–30 amperes.

b A = Square-groove butt joint, zero root opening.

B = Single-V 60° bevel butt joint, 1/16 inch root face, zero root opening.

C = Double-V 60° bevel butt joint, 3/32 inch root face, zero root opening.

Figure 11-4. The welding parameters in this table are used with ACHF welding current. (International Magnesium Assoc.)

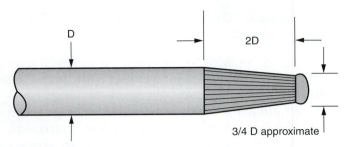

Figure 11-5. Electrodes used for alternating current should be ground as shown. The ball or rounded tip will form as the welding current is increased.

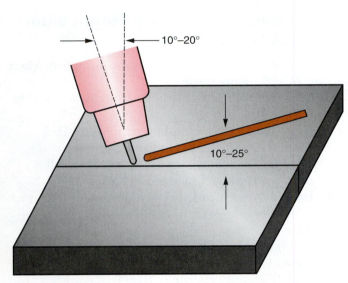

Figure 11-6. Torch and welding rod positions for welding groove welds in the flat (1G) position.

6. Use Welding Procedure 9-1 described in Chapter 9. Move the torch across the joint as shown in **Figure 9-16**. The arc should be a yellowish color. A magnesium weld pool does not change color as it changes temperature.
7. If the electrode contacts the weld pool, stop and replace the electrode. Otherwise, the magnesium on the electrode will burn away and contaminate the weld.

8. Fill the crater at the end of the weld.
9. If an orange-colored deposit forms on the tip of the electrode during welding, remove it from the electrode by rubbing it with light sandpaper between weld passes.
10. Always use a dedicated stainless steel brush to remove oxides from the surface of the material or the weld.

Postweld Treatment

Magnesium alloys containing more than 1.5% aluminum are susceptible to stress cracking. They must be stress-relieved after welding. The stress-relief times and temperatures required for various alloys are shown in **Figure 11-7**.

Welding Magnesium Using DCEN (Straight) Polarity

DCEN welding is used to produce deeply penetrating welds. **Figure 11-8** shows the various types of joints used with this technique. The variables and techniques listed in the Chapter 9 Welding Aluminum Using DCEN (Straight) Polarity section generally apply to welding magnesium as well. Postweld treatment as specified in **Figure 11-7** also applies.

Welding Magnesium Using DCEP (Reverse) Polarity

DCEP welding is used to produce shallow penetration welds. **Figure 11-9** shows the various types of joints that are used with this technique. The variables and techniques listed in Chapter 9 in the Welding Aluminum Using DCEP (Reverse) Polarity section generally apply to welding magnesium as well. Postweld treatment as specified in **Figure 11-7** also applies.

Stress-Relief					
Sheet			Castings		
Alloy	Temperature (°F)	Time (minutes)	Alloy	Temperature (°F)	Time (minutes)
AZ31B-0	500 (260°C)	15	AM100A	500 (260°C)	60
AZ31B-H24	300 (149°C)	60	AZ63A	500 (260°C)	60
HK31A-H24	550 (288°C)	30	AZ81A	500 (260°C)	60
HM21A-T8	700 (371°C)	30	AZ91C	500 (260°C)	60
HM21A-T81	750 (399°C)	30	AZ92A	500 (260°C)	60

Figure 11-7. Stress-relief treatment schedule.

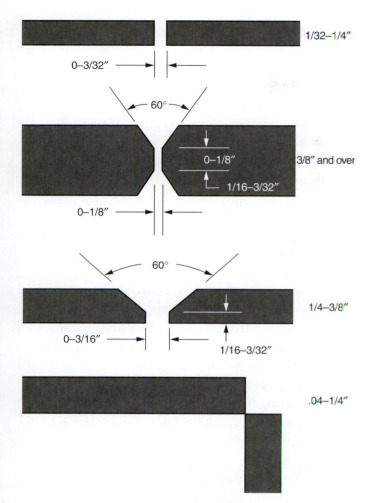

Figure 11-8. Weld joint designs used when welding magnesium with DCEN current.

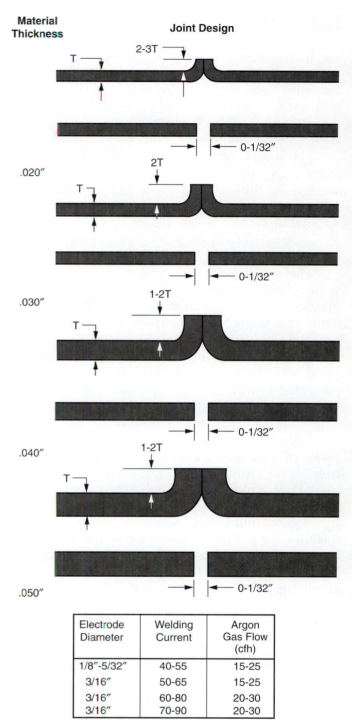

Figure 11-9. Joint configurations and settings recommended for DCEP welding of magnesium.

Electrode Diameter	Welding Current	Argon Gas Flow (cfh)
1/8″-5/32″	40-55	15-25
3/16″	50-65	15-25
3/16″	60-80	20-30
3/16″	70-90	20-30

Summary

Magnesium is found in many applications. It has many desirable characteristics such as light weight, excellent strength-to-weight ratio, good corrosion resistance, and easy weldability. Magnesium alloys are designated based on chemical composition. Magnesium materials are further identified by tempers.

The use of the correct filler alloy eliminates or reduces the formation of intermetallic compounds and low ductility in magnesium welds. AWS specifications for welding rods are published in AWS A5.19 *Specification for Magnesium Alloy Welding Electrodes and Rods*. Gloves should be worn when handling filler metal to keep it clean.

The use of weld backing bars designed for the welding of aluminum is recommended for welding magnesium. Magnesium tends to distort more than aluminum. Using hold-down bars and chill bars along with the backing bars will greatly reduce distortion.

Thin structures of magnesium do not require preheating. Material over 1/4″ (6.5 mm) can be preheated to increase the depth of penetration. The same considerations that apply to tack welding for aluminum also apply to tack welding for magnesium.

Argon gas is used to weld material thicknesses up to 1/8″ (3.2 mm) on all types of joints. For materials over 1/8″ (3.2 mm) thick, a 75He-25Ar mixture can be used. Pure, zirconiated, or lanthanated tungsten electrodes are recommended for welding with ac.

Because aluminum and magnesium have similar qualities, the techniques used to weld magnesium are very similar to those used to weld aluminum. Magnesium alloys containing more than 1.5% aluminum must be stress-relieved after welding.

Review Questions

Write your answers on a separate sheet of paper. Do not write in this book.

1. List five characteristics of magnesium.
2. At what temperature does pure magnesium melt?
3. Which requires more heat to melt, magnesium or aluminum?
4. Magnesium alloy designations conform to the _____ system.
5. Using the correct filler alloy eliminates or reduces the formation of _____ in magnesium welds.
6. What two contaminants most often cause porosity in magnesium welds?
7. Magnesium from the mill has a(n) _____ surface.
8. What combination of tooling reduces the distortion of magnesium?
9. What thickness of magnesium requires preheating before it can be welded?
10. How can a tack-welded joint in magnesium be straightened prior to being welded?
11. What is the purpose of a wave balancer on an ac welding machine?
12. If the cleaning action is lost during ACHF welding, the _____ should be checked.
13. Thoriated tungsten electrodes are not used for welding magnesium. Why?
14. During welding, how far should the electrode extend from the gas nozzle?
15. Magnesium alloys containing more than 1.5% _____ are susceptible to stress cracking.

Chapter

Manual Welding of Copper and Copper Alloys

Objectives

After completing this chapter, you will be able to:

- ❏ Identify characteristics of copper and copper alloys.
- ❏ Identify copper and copper alloys by their identification numbers.
- ❏ Summarize the filler metal choices for copper and copper alloys.
- ❏ Recall joint preparation techniques, including preweld cleaning, weld backing, preheating, and tack welds.
- ❏ Select the correct power source, shielding gases, and electrodes for welding copper and copper alloys using DCEN.
- ❏ Apply correct procedures for welding copper and copper alloys using DCEN.
- ❏ Identify the types of copper materials that require postweld treatment.

Key Terms

brass
bronze
phosphor bronze

Introduction

The use of copper dates back at least ten thousand years. Its early use was mostly for decoration. One of the world's largest copper reserves was discovered in Copper Harbor, Michigan in the late 1800s. This area of Michigan's Upper Peninsula is known as having the cleanest and most pure copper in the world.

Copper is a very ductile metal with excellent thermal and electrical conductivity. Copper oxidizes and eventually turns a greenish color when exposed to the elements.

Base Materials

Copper and copper alloys are nonferrous materials. Their major characteristics include the following:
- high thermal conductivity
- good electrical conductivity
- excellent corrosion resistance
- good formability and ductility
- high hardness and good wearing properties (some alloys)

Copper melts at approximately 1980°F (1082°C). Copper and most of its alloys require preheating and high rates of heat input during welding. The preheating is necessary because copper conducts the heat away from the weld joint very rapidly. However, certain copper alloys may not require preheating.

The following are three of the most common alloys of copper:
- *Brass*. Brass is an alloy of copper and zinc. It is available in a number of different copper-to-zinc ratios for different applications. Brass is very desirable for decoration because it looks like gold and can be highly polished. Brass can also be made to have a very smooth surface. It is commonly used in applications where little friction is desired, such as bearing surfaces and bushings.
- *Bronze*. Bronze is an alloy of copper and tin. It is available with different ratios of copper to tin, which result in bronzes with different characteristics. Even though there is a distinct difference between brass and bronze, some brasses are commonly called bronze and some bronzes are referred to as brass.
- *Phosphor bronze*. This bronze is an alloy of copper, tin, and phosphorus. The added phosphorus works as a deoxidizing agent to help clean the material and create a tough, strong material with low-friction characteristics.

Copper and Copper Alloy Identification

The Copper Development Association has assigned identification numbers to copper and copper alloys. These numbers are listed in the table in **Figure 12-1**. Also included in the table are gas tungsten arc weldability ratings. The ratings are as follows:

G– Good
F– Fair
NR– Not recommended
E– Excellent

Note that numbers 314 through 385 contain lead, which boils at welding temperatures, causing porosity and cracking.

Filler Metals

Filler metals used to join copper and copper alloys are listed in **Figure 12-2**. Filler metal specifications are found in AWS A5.7 *Specification for Copper and Copper-Alloy Bare Welding Rods and Electrodes.*

Weld Joint Design and Fitup

Copper weld joint designs are generally more open than steel designs. This is due to copper's higher thermal conductivity and the resulting loss of preheat from the weld joint. Some common joint designs are shown in **Figure 12-3**.

As a copper joint heats up, it expands more than other types of metal. Also, heat transfers and disperses through the weldment faster than with other types of metals. The areas of the metal that are in contact with the tooling will be cooler than areas surrounded by air because the tooling acts like a heat sink. This is true in all metals, but because of high thermal conductivity, it is more critical for copper and its alloys. The joint design must be carefully considered when tooling is designed for the weld. Similarly, the type of tooling used will have a significant impact on the technique used to weld the joint. Whereas the thickness of a backing block would have little effect on a steel weldment, it could significantly change the heat input requirements for a weld in copper alloy.

Because copper expands more than most materials, a copper weldment must be preheated to the proper temperature before the root openings can be set. If groove openings are not established at the preheat temperature, the joint will close during the preheat cycle due to expansion.

Joint Preparation for GTAW of Copper

Joint edges that are prepared by plasma arc cutting or carbon arc gouging have a heavy oxide film on the surface. This surface should be thoroughly cleaned prior to welding to prevent dross and porosity in the final weld. Joint edges prepared by shearing should be sharp without tearing or ridges. Where these conditions exist, dirt, oil, and grease can become trapped and result in faulty welds.

Copper and Copper Alloys							
Copper Alloy Number	Trade Name	Nominal Composition (%)	GTAW	Copper Alloy Number	Trade Name	Nominal Composition (%)	GTAW
102	Oxygen-free copper (OF)	Cu, 100.	G	443–445	Inhibited admirality	Cu 71. Sn 1. Zn 28.	G
110	Electrolytic tough pitch (ETP)	Cu + 0, 100.	F	464–467	Naval brass	Cu 60. Sn .75 Zn 39.25	F
113, 114 116	Silver-bearing tough pitch (STP)	Cu + Ag +0, 100.	F				
122	Deoxidized, high residual phosphorous (DHP)	Cu + P, 100.	G	485	Leaded naval brass	Cu 60. Sn .7 Pb 1.8 Zn 37.5	NR
210	Gilding, 95%	Cu 95. Zn 5.	G	505	Phosphor bronze, 1.25% E	Cu 98.7 Sn 1.3.	G
220	Commercial bronze, 90%	Cu 90. Zn 10.	G	510	Phosphor bronze, 5% A	Cu 95. Sn 5.	G
226	Jewelry bronze, 87.5%	Cu 87.5 Zn 12.5	G	521	Phosphor bronze, 8% C	P .2 Cu 92. Sn 8.	G
230	Red brass, 85%	Cu 85. Zn 15.	G	524	Phosphor bronze, 10% D	Cu 90. Sn 10.	G
240	Low brass, 80%	Cu 80. Zn 20.	G				
260	Cartridge brass, 70%	Cu 70. Zn 30.	F	544	Free-cutting phosphor bronze	Cu 88. Pb 4. Sn 4. Zn 4.	NR
268, 270	Yellow brass	Cu 65. Zn 35.	F	613, 614	Aluminum bronze D	Cu 91. Al 7. Fe 2.	E
280	Muntz metal	Cu 60. Zn 40.	F				
314	Leaded commercial bronze	Cu 89. Pb 1.9 Zn 9.1.	NR	651	Low-silicon bronze B	Cu 98.5 Si 1.5	E
330	Low-leaded brass tube	Cu 66. Pb .5 Zn 33.5.	NR	655	High-silicon bronze A	Cu 97. Si 3.	E
332	High-leaded brass tube	Cu 66. Pb 1.6 Zn 32.4	NR	675	Manganese bronze A	Cu 58.5 Fe 1.4 Mn .1 Sn 1. Zn 39.0	F
335	Low-leaded brass	Cu 65. Pb .5 Zn 34.5	NR				
340	Medium-leaded brass	Cu 65. Pb 1. Zn 34.	NR	687	Aluminum brass	Cu 77.5 Al 2. Zn 20.5	F
342, 353	High-leaded brass	Cu 64.5 Pb 2. Zn 33.5	NR	706	Copper nickel, 10%	Cu 88.6 Fe 1.4 Ni 10.	E
356	Extra-high-leaded brass	Cu 62. Pb 2.5 Zn 35.5	NR	715	Copper nickel, 30%	Cu 69.5 Ni 30. Fe .5	F
360	Free-cutting brass	Cu 61.5 Pb 3. Zn 35.5	NR	745	Nickel silver, 65–10	Cu 65. Ni 10. Zn 25.	F
365–368	Leaded Muntz metal	Cu 60. Pb .6 Zn 39.4	NR	752	Nickel silver, 65–18	Cu 65. Ni 18. Zn 17.	F
370	Free-cutting Muntz metal	Cu 60. Pb 1. Zn 39.	NR	754	Nickel silver, 65–15	Cn 65. Ni 15. Zn 20.	F
377	Forging brass	Cu 60. Pb 2. Zn 38.	NR	757	Nickel silver, 65–12	Cu 65. Ni 12. Zn 23.	F
385	Architectural bronze	Cu 57. Pb 3. Zn 40.	NR	770	Nickel silver, 55–18	Cu 55. Ni 18. Zn 27.	F

Figure 12-1. Copper and copper alloy identification.

Filler Metals for Copper		
Filler Metal *a*	**Common Name**	**Base Metal Applications**
RCu	Copper	Coppers
RCuSi-A	Silicon bronze	Silicon bronzes, brasses
RCuSn-A	Phosphor bronze	Phosphor bronzes, brasses
RCuSn-C	Phosphor bronze	Phosphor bronzes, brasses
RCuNi	Copper-nickel	Copper-nickel alloys
RCuAl-A2	Aluminum bronze	Aluminum bronzes, brasses, silicon bronzes, manganese bronzes
RCuAl–A3	Aluminum bronze	Aluminum bronzes
RCuNiAl	–	Nickel-aluminum bronzes
RCuMnNiAl	–	Manganese-nickel-aluminum bronzes
RBCuZn-A	Naval brass	Brasses, copper
RCuZn-B	Low fuming brass	Brasses, manganese bronzes
RCuZn-C	Low fuming brass	Brasses, manganese bronzes

a. See the specification AWS A5.7, *Copper and Copper Alloy Bare Welding Rods and Electrodes.*

Figure 12-2. Filler metal applications for copper and some copper alloys.

Preweld Cleaning

Before a copper alloy can be properly welded, oil, grease, and other contaminants must be removed with acetone or alcohol. Any corrosion or scale must also be removed before the material is welded. Commercial copper cleaners or acid baths can be used for this purpose.

Weld Backing

Groove welds that are welded from one side of the joint where 100% penetration is required should have a grooved copper backing. The backing groove retains argon from the torch and protects the penetration area from oxidation. A typical backing bar configuration is shown in **Figure 12-4**.

Preheating Copper

If preheating is required, heat the weldment with an oxyacetylene torch with a neutral flame. Temperature-indicating crayons or liquids can be used to determine

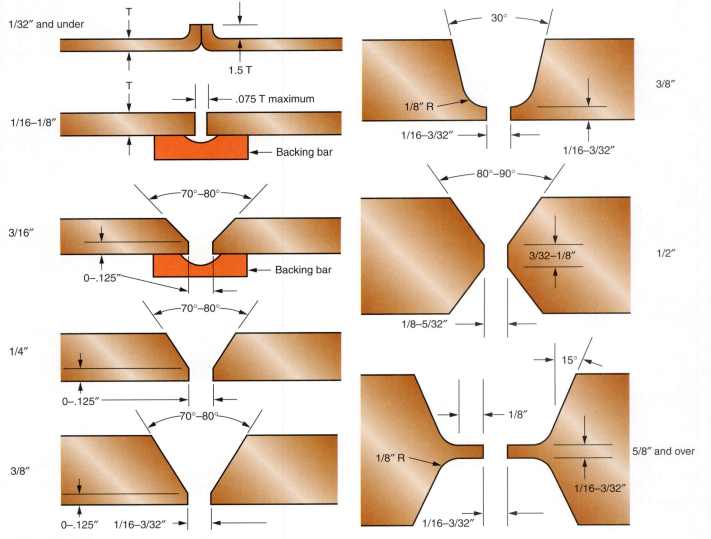

Figure 12-3. Copper and copper alloy joint designs.

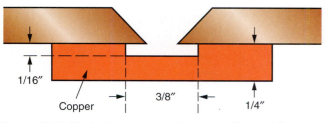

Figure 12-4. Typical copper and copper alloy weld backing bar design.

the correct temperature. Do not allow the temperature-indicating material to flow into the weld joint. Heat the joint and the backing bar (if used) thoroughly.

Remember that the heat rapidly flows away from the joint. This means the preheat operation must be constantly and carefully monitored. Failure to preheat the weldment properly can cause poor fusion in the joint and incomplete penetration. Some common preheat temperatures are listed in **Figure 12-5**.

Tack Welds

When tack welds must be made to maintain joint alignment for welding, the following considerations apply:

- If the welding procedure calls for preheating, the weldment must be at the proper preheat temperature before it can be tack welded.
- Use the same filler metal as will be used for the main welding procedure.
- If the joint requires full penetration, the tack weld also requires full penetration.
- Keep the tack weld as small as possible. If possible, make the tack weld smaller than the main weld.
- If the tack weld cracks, leave it and make another tack weld close to the broken one.

Thickness	Preheat
1/32″	No preheat
1/32–1/16″	No preheat
1/16–1/8″	100°F (38°C)
3/16″	100°F (38°C)
1/4″	200°F (93°C)
3/8″	450°F (232°C)
1/2″	650°F (343°C)
5/8″	750°F (399°C)

Figure 12-5. Recommended preheats for copper.

- Do *not* grind out tack welds. If one must be removed, use a rotary file. Grinding grit in a joint makes a good quality weld almost impossible.
- Do *not* leave a crater in a tack weld. The crater of a tack weld is its weakest point.

Welding Copper Using DCEN (Straight) Polarity

Power source types used in welding copper or copper alloys include motor generators, rectifiers, and inverter-type machines. High-frequency arc start is required to prevent copper contamination of the electrode. If the electrode were scratched across the soft copper base metal, some of the copper would be transferred to the electrode, causing contamination.

Gases for Welding Copper

Argon gas can be used for welding copper up to 1/16″ (1.6 mm) thick. If thicker copper is being welded, helium must be added to the argon for improved weld pool control and penetration. The addition of helium increases the intensity and the heat of the arc. Standard gas mixes that can be used are:

- 75% argon–25% helium
- 50% argon–50% helium
- 25% argon–75% helium

Pure helium gas creates severe arc starting problems. Helium gas is usually used only in automatic welding operations.

Electrodes for Welding Copper

A 1% or a 2% lanthanated, thoriated, or ceriated electrode should be used to weld copper. The electrode should be tapered to a sharp point. The electrode diameter must be large enough to carry the required amperage.

Copper Welding Techniques

1. Select a joint design from **Figure 12-3**.
2. Use RCu filler metal. This filler metal is deoxidized copper made specifically for welding. Do *not* use other types of copper for filler metals.
3. Select the proper shielding gas for the thickness of metal being welded.
4. Preheat the base metal as specified.
5. Weld with stringer beads only. Do *not* oscillate the weld pool. Copper is a hot short material and is prone to cracking. On multipass welds, the first pass must be large enough to prevent cracking.

Copper-Zinc Alloy (Brass) Welding Techniques

1. Select joint design from **Figure 12-3**.
2. Use RCuSi filler metal. This filler metal has a high zinc content and will fume during the melting of the material. This can cause porosity in the weld.
3. Select the proper shielding gas for the thickness of metal being welded.
4. Preheat is required only to reduce current input. Do *not* overheat the copper-zinc alloys. Whitish formations on the weld pool indicate it is too hot.
5. Position the filler metal in the arc. The intense heat of the arc will burn the zinc alloy base material as explained in step 2.

   Weld with stringer beads only. Do *not* oscillate the weld pool. Zinc alloys are hot short and prone to cracking. On multipass welds, the first pass must be large enough to prevent cracks.

Copper-Tin Alloy (Phosphor Bronze) Welding Techniques

Copper-tin alloys have a tendency to crack during welding and cooling. For this reason, GTAW is usually limited to the welding of light-gauge materials and to minor repairing or surfacing.

1. Use RCuSn filler metal.
2. Use argon shielding gas.
3. Preheat the base metal to approximately 400°F (204°C).
4. Maintain the interpass temperature during welding.
5. Keep the weld pool as small as possible.
6. Slowly cool the metal to room temperature after welding.

Copper-Aluminum Alloy (Aluminum Bronze) Welding Techniques

GTAW can be used to weld the copper-aluminum alloys listed in **Figure 12-6**.

1. Joint designs up to 1/8″ (3.2 mm) thick can be square-groove with an approximately 1/16″ (1.6 mm) root opening. On joints over 1/8″ (3.2 mm) thick, use a 60°–70° V-groove and weld from one side.
2. Use argon shielding gas.
3. Alloys containing up to 10% aluminum can be preheated to remove residual moisture. Interpass temperature should not exceed 300°F (149°C). After the metal is welded, allow it to cool in still air.

Copper Alloy (Aluminum Bronze) Number	
612	623
613	624
614	625
618	626
619	628
620	630
622	

Figure 12-6. Copper alloy (aluminum bronze) numbers that can be gas tungsten arc welded.

4. A commercial fluoride-type flux can be applied to straight butt edges of copper-aluminum alloy sheet material. This will improve the flow and wetting of the weld pool.
5. Use RCuAl filler metal.

Copper-Silicon Alloy (Silicon Bronze) Welding Techniques

The silicon bronzes are the easiest to weld of all the copper alloys. They have low thermal conductivity and preheating is not required.

1. Joint designs up to 1/8″ (3.2 mm) thick can be square-groove. Joints over 1/8″ (3.2 mm) thick should be prepared with a 60°–70° V-groove and welded from one side.
2. Use either DCEN or ACHF polarity.
3. Use argon shielding gas.
4. Use RCuSi filler metal.
5. To prevent centerline cracking of the weld, allow the heat to fully penetrate material at the start of weld. Add sufficient material at the start of weld to provide strength at the higher temperature. Slow the welding speed.
6. A film of silica forms on the top of the weld during welding. Remove the film between passes by wire brushing with a stainless steel wire brush.
7. The interpass temperature should not exceed 200°F (93°C).

Copper-Nickel Alloy Welding Techniques

The copper-nickel alloys are very weldable with GTAW. Since their thermal conductivities are about the same as steel, no preheat is required.

1. Joint design up to 1/8″ (3.2 mm) thick can be square-groove. Joints over 1/8″ (3.2 mm) thick should be prepared with a 60°–70° V-groove and welded from one side.
2. Use argon shielding gas.
3. Use RCuNi filler metal.
4. Surface contaminants can cause cracking in the heat-affected zone. To prevent this condition, clean the base material thoroughly before welding.
5. Maintain a short arc and weld only with stringer beads.

Welding Castings of Copper

Many rebuilding, repairing, surfacing, and joining jobs involve the welding of copper alloy castings. The various alloys and filler metals used are listed in **Figure 12-2**.

The area to be welded should be cleaned by grinding or machining until bright metal is visible. This removes the hard scale on the part of the casting that came into contact with the casting mold. All grease, oil, dirt, and scale adjacent to the weld area must be removed in order to prevent contamination of the weld.

If the area to be welded extends into the casting, the groove must be tapered as shown in **Figure 12-7** to avoid cold shuts or lack of fusion.

Brasses

1. Preheat the metal to approximately 400°F (204°C).
2. Use 75He-25Ar shielding gas.
3. Use direct current electrode negative (DCEN) polarity.

Silicon Bronzes

1. Do not preheat the metal.
2. Use argon shielding gas.
3. Interpass temperature should not exceed 150°F (66°C).
4. Keep the weld pool as small as possible.
5. Mechanically clean between passes.

Phosphor Bronzes

1. Preheat the metal to 300°–400°F (149°–204°C).
2. Use 100% argon or 75Ar-25He shielding gas.
3. Maintain an interpass temperature of 400°F (204°C).
4. Keep the weld pool as small as possible.
5. Allow the metal to cool slowly after welding.

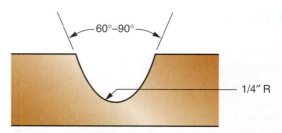

Figure 12-7. The crack or defect has been ground out of the solid piece. Prepare castings for welding with a generous radius at the bottom of the groove. Taper side walls of the groove to avoid cold shuts at the edge of the weld.

Aluminum Bronzes

1. Preheat the metal to approximately 300°F (149°C).
2. Use argon, helium, or an argon-helium mixture for shielding gas.
3. Interpass temperature should not exceed 300°F (149°C).
4. Allow the metal to cool in still air after welding.

Copper-Nickel Alloys

1. Do not preheat the metal.
2. Clean the surface very well before welding.
3. Use argon shielding gas.
4. Maintain a short arc.
5. Weld with stringer beads.

Postweld Treatment of Castings

High-copper alloys, copper-aluminum alloys, and some copper-nickel alloy castings are susceptible to stress cracking in some service applications. To prevent stress cracking, the part should be heat-treated. Heat treating reduces the stresses within the part. Furnace times and temperatures are listed in **Figure 12-8**.

Material	Temperature
Red Brasses	750°F (399°C)
Copper Alloy 443–445	750°F (399°C)
Copper Alloy 464–467	850°F (454°C)
Copper Alloy 614	1100°F (593°C)
Copper Alloy 628	1200°F (649°C)
Copper Alloy 655	850°F (454°C)
Copper Nickel Alloys	1000°F (538°C)

Heat slowly to temperature.
Hold for one hour minimum.

Figure 12-8. Stress-relief times and temperatures for materials susceptible to stress cracking.

Summary

Copper and copper alloys have many desirable characteristics, including high thermal conductivity, good electrical conductivity, excellent corrosion resistance, and good formability and ductility. Copper and many copper alloys require preheating because they conduct heat away from the weld joint very rapidly. However, some copper alloys do not require preheating. Three common copper alloys are brass, bronze, and phosphor bronze.

The Copper Development Association has assigned identification numbers to copper and copper alloys. Filler metals used to join copper and copper alloys are specified in AWS A5.7 *Specification for Copper and Copper-Alloy Bare Welding Rods and Electrodes*.

Copper weld joint designs are generally more open than steel designs. Groove openings must be established at the preheat temperature, or the joint will close during the preheat cycle due to expansion.

Preweld cleaning involves removing oil, grease, and other contaminants with acetone or alcohol. Any corrosion or scale should be removed with commercial copper cleaners or acid baths.

Copper weldments are preheated using an oxyacetylene torch with a neutral flame. The preheat operation must be carefully monitored. Failure to preheat the weldment properly can cause poor fusion in the joint and incomplete penetration.

High-frequency arc start is required to prevent copper contamination of the electrode. Argon gas can be used for welding copper up to 1/16" (1.6 mm) thick. For welding thicker copper, helium is added to the argon for improved weld pool control and penetration. A 1% or a 2% lanthanated, thoriated, or ceriated electrode should be used. The electrode should be tapered to a sharp point.

Different welding procedures and filler metals are required for welding copper and its various alloys. The area of a casting to be welded should be cleaned by grinding or machining until bright metal is seen in order to remove the hard scale on the part of the casting that came into contact with the casting mold.

Review Questions

Write your answers on a separate sheet of paper. Do not write in this book.

1. List five major characteristics of copper and the copper alloys.
2. At what temperature does copper melt?
3. Why is copper difficult to get hot enough to weld?
4. What are the benefits of the phosphorus in phosphor bronze?
5. Copper and copper alloys are assigned identification numbers by the _____.
6. Why are copper alloy numbers 314 through 385 almost impossible to weld?
7. Copper weld joint designs are generally more _____ than steel designs.
8. Why are groove openings specified at preheat temperatures rather than at room temperatures?
9. Any corrosion or scale must be removed before welding using _____ or _____.
10. Groove welds that are welded from one side of the joint where 100% penetration is required should have a(n) _____ backing.
11. When using an oxyacetylene torch for preheating, what type of flame is used?
12. Failure to properly preheat the weldment can cause poor _____ in the joint and incomplete _____.
13. Why is high-frequency arc start required when welding copper?
14. Argon gas can be used for welding copper up to _____ thick.
15. What type of electrodes can be used when welding with DCEN polarity?
16. What does RCu filler metal consist of?
17. Whitish formations that appear on the surface of copper-zinc alloys when welding indicate that the weld pool is _____.
18. Which are the easiest copper alloys to weld?
19. When welding silicon bronze, a film of _____ forms on the top of the weld.
20. Why do castings require grinding or machining before welding?

Manual Welding of Nickel, Nickel Alloys, and Cobalt Alloys

Objectives

After completing this chapter, you will be able to:

- Identify the characteristics of nickel, nickel alloys, and cobalt alloys.
- Explain how nickel alloys and cobalt alloys are specified.
- Identify material groups and uses for nickel, nickel alloys, and cobalt alloys.
- Recognize and resolve major problems that can occur when welding nickel and cobalt alloys.
- Summarize the filler metal choices for nickel, nickel alloys, and cobalt alloys.
- Recall joint preparation techniques, including preweld cleaning, weld backing, and tack welds.
- Select the correct power source, shielding gases, and electrodes for welding nickel, nickel alloys, and cobalt alloys.
- Apply correct procedures for welding nickel, nickel alloys, and cobalt alloys.

Key Terms

cobalt
nickel

Introduction

Nickel is an element that has been used by humans for thousands of years, although until the nineteenth century it was mistaken for other metals. Nickel has a silvery-white appearance with a hint of gold coloring and can be highly polished. It is used in a variety of applications, including making coins and in the plating industry. Nickel is used as an alloying agent in a variety of materials because of its desirable characteristics. Nickel increases the strength, toughness, and corrosion resistance of steels and is very easily welded. Nickel and chromium are important alloying elements in stainless steel.

Cobalt is a metallic element acquired as a by-product of mining copper and nickel. It has excellent wear and heat resistance properties. Cobalt also produces high-strength alloys when added to other materials.

Base Materials

There are approximately 100 different materials within the nickel, nickel alloy, and cobalt alloy groups. The materials are made by different companies and, for the most part, are identified by registered trade names. Some of the alloys are listed by SAE International's Aerospace Material Specifications (AMS), ASTM International, ASME (American Society of Mechanical Engineers), and the American Welding Society (AWS). The chemical composition of various nickel alloys is shown in **Figure 13-1**.

Material Groups and Uses

- **Pure nickel** is used in the fabrication of food processing equipment, vessel liners, and piping for the food industry.
- **Nickel-copper alloys** are used in the fabrication of processing equipment, refining equipment, and heat exchangers.
- **Nickel-chromium alloys** are used in many applications involving corrosive media, with a wide range of service temperatures.
- **Nickel-iron-chromium alloys** have good high-temperature strength and resistance to oxidation and corrosion.
- **Nickel-molybdenum alloys** have good corrosive resistance at low temperatures.
- **Nickel-chromium-molybdenum alloys** have good corrosion resistance at room temperature. They also resist oxidation at elevated temperatures.
- **Pure cobalt** has few uses of interest to welders.
- **Cobalt alloys** provide good corrosion resistance and high-temperature service.

Material Problems Associated with Nickel and Cobalt Alloys

The major problems that occur when nickel and cobalt alloys are welded are porosity and cracking within the welds. Both of these problems can be avoided by proper weld joint design, appropriate preweld cleaning techniques, and the use of correct welding procedures.

The main point to remember is nickel and cobalt alloys can withstand exposure to extreme temperatures and corrosive media. For this reason, they are commonly found on parts designed for use in harsh environments. The service environment must be considered when placing each weld. Welds that are questionable at room temperature in a normal atmosphere will usually fail rapidly in service.

To prevent porosity and cracking, follow these guidelines:

- Use the proper base material for the intended service.
- Use the correct filler metal.
- Design the weld joint with larger groove angles, since the molten metal does not "wet" readily. See **Figure 13-2**.
- Use full penetration joints wherever possible. Fillet and edge joints promote cracking from the unwelded area.
- Design the weld joint for minimum restraint during welding. Allow the joint to move.
- Design tooling to provide maximum chilling of the weld area.
- Use the proper cleaning techniques.
- Use the proper welding techniques.

Filler Metals

Filler metals for welding nickel, nickel alloys, and cobalt alloys are selected based on the chemical composition of the base materials. When two pieces of the same alloy are being joined, the filler metal should generally match the base material. When two pieces made from dissimilar alloys are being joined, the filler metal should match the higher-alloy base material.

Filler metals developed by individual material companies are often listed by trade names and numbers. In some cases, they do not meet any current specification. The manufacturer of the base material can be contacted regarding the proper filler metal to be used. For joining nickel alloys to dissimilar metals, the materials specified in **Figure 13-3** can be used.

Filler metal specifications for nickel and nickel alloys include the following:

- AWS A5.14
- MIL-E-21562
- SAE-AMS 5838, 5786, 5679, 5794, 5675, 5832, 5837

Filler metal specifications for cobalt alloys include the following:

- AMS 5789, 5796, 5801

Maximum weld quality can only be obtained if clean, high-quality filler metal is used. Store wire in plastic containers in a clean area. Always handle filler rods with clean gloves. Any wire that has been handled with bare hands must be cleaned before use. Use alcohol or acetone and a clean rag to remove any contamination. Loose rods should be individually identified so they can be distinguished from each other if they get mixed together.

Common designation[a]	Ni[b]	C	Cr	Mo	Fe	Co	Cu	Al	Ti	Cb[c]	Mn	Si	W	B	Other
Composition, Percent															
Commercially Pure Nickels															
Nickel 200	99.5	0.08			0.2		0.1				0.2	0.2			
Nickel 201	99.5	0.01			0.2		0.1				0.2	0.2			
Nickel 205	99.5	0.08			0.1		0.08		0.03		0.2	0.08			0.05Mg
Solid Solution Types															
Monel 400	66.5	0.2			1.2		31.5				1.	0.2			
Monel 404	54.5	0.08			0.2		44.	0.03			0.05	0.05			
Monel R-405	66.5	0.2			1.2		31.5				0.1	0.02			
Hastelloy F	47.	0.05	22.	6.5	17.	2.5				2.	1.5	1.	1.		
Hastelloy X	47.	0.10	22.	9.	18.	1.5					1.	1.	0.6		
Nichrome V	76.	0.1	20.		1.						2.	1.			
Nichrome	57.	0.1	16.		25.						1.	1.			
Hastelloy G	44.	0.1	22.	6.5	20.	2.5	2.			2.	1.5	1.	1.		
IN 102	68.	0.06	15.	3.	7.			0.4	0.6	3.			3.	0.005	0.03Zr, 0.02Mg
RA 333	45.	0.05	25.	3.	18.	3.				1.	1.5	1.2	3.		
Inconel 600	76.	0.08	15.5		8.		0.2				0.5	0.2			
Inconel 601	60.5	0.05	23.		14.			1.4			0.5	0.2			
Inconel 625	61.	0.05	21.5	9.	2.5			0.2	0.2	3.6	0.2	0.2			
Carpenter 20Cb3	36.	0.04	20.	2.5	36.		3.5			0.5	1.	0.5			
Incoloy 800	32.5	0.05	21.		46.			0.4	0.4		0.8	0.5			0.02Mg
Incoloy 825	42.	0.03	21.5	3.	30.		2.25	0.1	0.9		0.5	0.25			
Hastelloy B	61.	0.05	1.	28.	5.	2.5					1.	1.			
Hastelloy C	54.	0.08	15.5	16.	5.	2.5					1.	1.	4.		
Hastelloy D	82.	0.10			1.	1.5	3.				1.	9.			
Hastelloy N	70.	0.06	7.	16.5	5.						0.8	0.5			
Hastelloy W	60.	0.12	5.	24.5	5.5	2.5					1.	1.			
Precipitation Hardenable Types															
Duranickel 301	96.5	0.15			0.3		0.13	4.4	0.6		0.25	0.5			
Monel K-500	66.5	0.10			1.		29.5	2.7	0.6		0.8	0.2			
Waspaloy	58.	0.08	19.5	4.		13.5		1.3	3.					0.006	0.06Zr
Rene 41	55.	0.10	19.	10.	1.	10.		1.5	3.		0.05	0.1		0.005	
Nimonic 80A	76.	0.06	19.5					1.6	2.4		0.3	0.3		0.006	0.06Zr
Nimonic 90	59.	0.07	19.5			16.5		1.5	2.5		0.3	0.3		0.003	0.06Zr
M 252	55.	0.15	20.	10.		10.		1.	2.6		0.5	0.5		0.005	
Udimet 500	54.	0.08	18.	4.		18.5		2.9	2.9		0.5	0.5		0.006	0.05Zr
Alloy 713C[d]	74.	0.12	12.5	4.				6.	0.8	2.				0.012	0.10Zr
Inconel 718	52.5	0.04	19.	3.	18.5			0.5	0.9	5.1	0.2	0.2			
Inconel X750	73.	0.04	15.5		7.			0.7	2.5	1.	0.5	0.2			
Inconel 706	41.5	0.03	16.		40.			0.2	1.8	2.9	0.2	0.2			
Alloy 901	42.5	0.05	12.5		36.	6.		0.2	2.8		0.1	0.1		0.015	
Ni-Span-C 901	42.2	0.03	5.3		48.5			0.6	2.6		0.4	0.5		0.014	
IN 100[d]	60.	0.18	10.	3.		15.		5.5	4.7						0.06Zr, 1.0V
Dispersion Strengthened Types															
TD Nickel	98.														2ThO$_2$
TD Ni Cr	78.		20.												2ThO$_2$

a. Several of these are registered trade names. Alloys of similar compositions may be known by other common designations or trade names.
b. Includes small amount of cobalt if cobalt content is not specified.
c. Includes tantalum also.
d. Casting alloys.

Figure 13-1. Nickel, nickel alloy, and cobalt alloy chemical compositions.

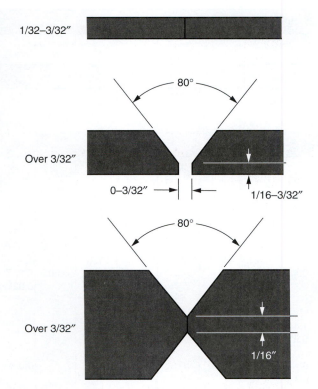

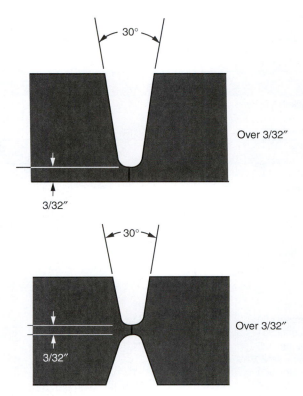

Figure 13-2. Common weld joint designs for nickel and cobalt alloys.

Nickel Alloy	Stainless Steel	Low Carbon and Stainless Steel	5-9 Nickel Steels	Copper	Copper Nickel
Nickel 200-201	Inconel 82	Inconel 82	Inconel 82	Monel 60	Monel 60
Monel Alloys 400, K-500, 502	Inconel 82	Nickel 61	Nickel 61	Monel 60	Monel 60
Inconel Alloys 600-601	Inconel 82	Inconel 82	Inconel 82	Nickel 61	Nickel 61
Inconel Alloys 625, 706, 718, X-750	Inconel 82	Inconel 82	Inconel 625	Nickel 61	Nickel 61
Incoloy Alloy 800	Inconel 82	Inconel 82	Inconel 82	Nickel 61	Nickel 61
Incoloy Alloy 825	Inconel 625	Inconel 625	Inconel 625	Nickel 61	Nickel 61

Figure 13-3. Filler metals used to join nickel alloys to dissimilar materials.

Weld Preparation

Cleaning procedures must be established before any welding is done on the actual weldment. These procedures must be followed exactly or the weld can be ruined.

Cleanliness is absolutely critical for nickel and cobalt alloys. Surface oxides, grease, machining coolant, lead, paint, copper, and shop dirt are contaminants that can lead to weld failure. Any foreign material that contacts the joint or filler metal must be considered a contaminant unless tested and proven otherwise.

Follow these precautions for preweld cleaning and setup:

- Remove all oxide film and cleaning solvent residue for a distance of at least 2″ (51 mm) from the joint edge by sanding, grinding, machining, grit blasting, or abrading with emery cloth.

- Thermally cut metal must be machined to remove *all* material within the heat-affected zones.
- Identification marking made with pens, inks, crayons, or paints may contain lead or sulfur. These areas must be cleaned. Use a solvent to remove the marks, and then grind or rub the area with emery cloth to remove any residue.
- *Never* use sanding discs, wheels, or belts that have been used on other materials.
- *Never* use brass or carbon steel wire brushes to clean the metal. Use only stainless steel wire brushes that have not been used on any other type of material.
- Do *not* use copper, brass, or bronze in any part of the welding fixture that could come into contact with the weld zone.
- Use only clean solvents. Use mechanical cleaning methods to remove any residue left by the cleaning solvents.
- Wear clean gloves and use clean rags.
- Do *not* handle parts with bare hands.
- Do *not* use shop air for blowing washed parts dry. This air may contain oil.
- Draw file all sheared edges with clean files.
- Any tooling that contacts the weld area must be washed with alcohol or acetone before being used.

Weld Backing

The root side of all full-penetration joints must be protected from the atmosphere. Tooling that directs an inert gas to the root side of the joint must be used. The tooling should have radius-type backing grooves, if possible, to prevent notches on the penetration face. Casting-type backing grooves with square edges often cause notches on the penetration face. See **Figure 13-4.**

If pipe is being welded, consumable ring inserts of the types shown in the chapter on pipe welding are preferred over backing rings. Consumable ring inserts and backing rings will be discussed in Chapter 16.

The interior penetration side of a joint being welded on pipe must also be shielded by an inert gas. The pipe ends can be closed off and the space filled with inert gas, purging the atmosphere from the pipe. This technique will be described in greater detail in Chapter 16.

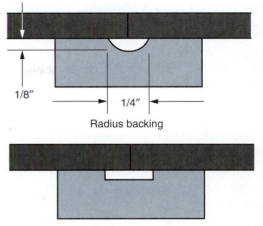

Figure 13-4. Correct and incorrect designs for backing tooling grooves.

Tack Welds

When tack welds must be made to maintain joint alignment for welding, the following considerations apply:

- Use the same filler metal for tack welding as will be used for the main welding procedure.
- If the joint requires full penetration, the tack weld requires penetration.
- Keep the tack weld as small as possible. If possible, make it smaller than the main weld.
- If the tack weld cracks, leave it and make another close to the broken one.
- Do *not* grind out tack welds. Grinding grit in a joint makes it almost impossible to achieve a high-quality weld.
- Do *not* leave a crater in a tack weld. The crater of a tack weld is its weakest point.
- Always use an inert gas backing when tack welding full-penetration joints.

Welding Nickel, Nickel Alloys, and Cobalt Alloys Using DCEN (Straight) Polarity

A motor generator or rectifier power source should be used to weld nickel, nickel alloys, or cobalt alloys. A high-frequency arc start is required.

Gases for DCEN Welding of Nickel or Cobalt

Argon gas can be used for all thicknesses. For material over 1/8" (3.2 mm) thick, adding helium to the argon will increase the penetration. The following are the standard mixes that may be used:

- 75% argon—25% helium
- 50% argon—50% helium
- 25% argon—75% helium
- 95% argon—5% hydrogen (This mix will leave a cleaner, brighter weld crown.)

Pure helium gas can be used for automatic welding of nickel or cobalt.

Electrodes for DCEN Welding of Nickel or Cobalt

Use 1% or 2% lanthanated or thoriated electrodes. Taper the electrodes to a sharp point. Always use an electrode large enough to carry the required amperage.

Procedure for Welding Nickel and Cobalt

1. Clean the weld areas as previously described.
   - When wire brushing material or the weld, always use a dedicated austenitic stainless steel wire brush.
2. Select a filler metal with the desired chemical composition.
   - Use the largest possible diameter filler metal.
   - Always trim contaminated wire from the used end of welding wire before using it.
3. Preheat the base material only if the temperature is below 60°F (16°C).
4. Weld the material. Follow these guidelines:
   - Maintain the torch in as vertical a position to the weld as possible. Use only enough amperage to flow the material. (Remember that full-penetration joints should be used with nickel and cobalt. Fillet welds are *not* recommended.) Deep penetration into the side walls of a joint is not required.
   - Use the shortest possible arc length.
   - Create stringer beads. Do *not* use a weave pattern.
   - Use chill bars to cool the weld area as fast as possible to prevent distortion.
   - Use the largest possible torch gas nozzle.

- When stopping a weld, always maintain gas coverage from the torch nozzle until the weld cools without oxide film. Set the torch gas postflow for 5–10 seconds longer than normal.
- Tack welds, root passes, and weld ends are susceptible to cracking because the weld metal cross section is thinner in those spots. To prevent cracks, make welds convex and do not leave craters.
- Consult the base material manufacturer regarding postweld heat treatment.

Summary

Nickel is used as an alloying agent in many materials because it is easily welded and increases the strength, toughness, and corrosion resistance of the metal to which it is added. Cobalt produces high-strength alloys when added to other metals. For the most part, the alloys are identified by registered trade names.

Porosity and cracking within the welds are problems that occur when nickel and cobalt alloys are welded. Proper weld joint design, appropriate preweld cleaning techniques, and correct welding procedures help prevent these problems. The service environment, which may involve extreme temperatures and corrosive media, must also be considered.

Filler metals are selected based on the chemical composition of the base materials. The filler metal should match the base material. If two dissimilar alloys are being joined, the filler metal should match the higher-alloy base material. Filler rods should be handled with clean gloves.

Cleanliness is more critical for nickel and cobalt alloys than for other materials. Cleaning procedures must be followed exactly or the weld can be ruined. Surface oxides, grease, machining coolant, lead, paint, copper, shop dirt, and any other foreign material can lead to weld failure if not removed. The root side of all full-penetration joints and the interior penetration side of a pipe joint must be shielded by an inert gas.

A motor generator or a rectifier with a high-frequency arc start should be used to weld nickel, nickel alloys, or cobalt alloys. Argon shielding gas can be used for all thicknesses. However, for material over 1/8" (3.2 mm) thick, adding helium to the argon will increase the penetration. A lanthanated or thoriated electrode should be used. The electrode should be tapered to a sharp point.

The base material should be preheated only if the temperature is below 60°F (16°C). When welding, use only stringer beads. Chill bars should be used to cool the weld area as fast as possible to prevent distortion. When stopping a weld, maintain gas coverage from the torch nozzle until the weld cools without oxide film. Consult the base material manufacturer regarding postweld heat treatment.

Review Questions

Write your answers on a separate sheet of paper. Do not write in this book.

1. Nickel increases the strength, toughness, and _____ of steels.
2. Nickel alloys and cobalt alloys are made by different companies and are often identified by _____.
3. List two major problems that occur when nickel and cobalt alloys are welded.
4. The weld joint should be designed for _____ restraint during welding.
5. Filler metals for welding the nickel and cobalt alloys are selected based on the _____ of the base materials.
6. How should filler metal be stored?
7. How should the weld joint in nickel or cobalt alloys be designed, since the filler metal does not "wet" to the base readily?
8. Why are fillet and edge joints not recommended when nickel and cobalt alloys are welded?
9. List three methods that can be used to remove oxide film and clean solvent residue.
10. How are thermally cut joints prepared for welding?
11. How can identification markings made with crayons, inks, and pencils be removed?
12. What type of brush should be used to clean nickel or cobalt weld joints and welds?
13. What type of shielding gas is added to argon to achieve deeper penetration when using manual DCEN?
14. The base material should be preheated only if the temperature is below _____.
15. Welds in nickel and cobalt alloys should be laid down as _____ beads.

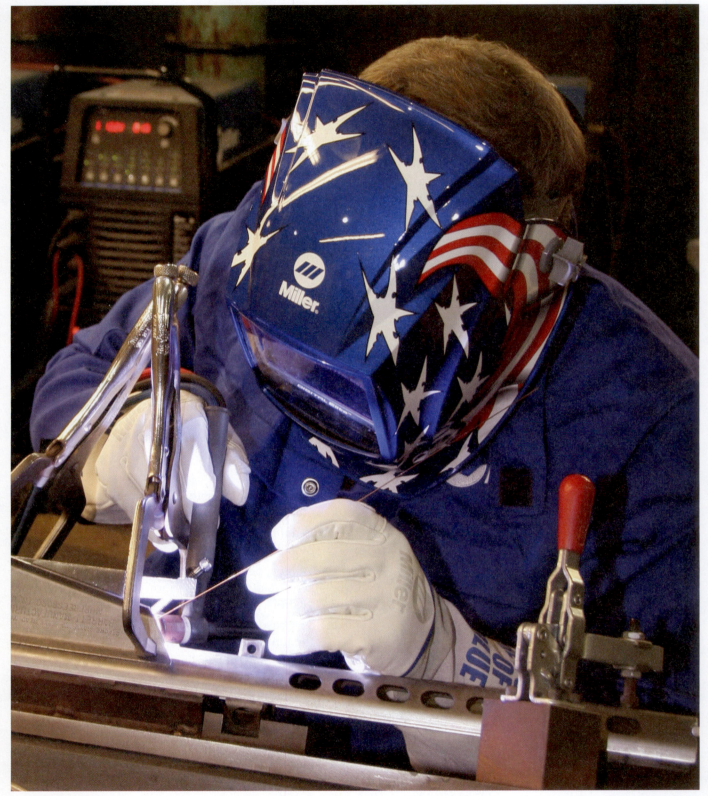

The gas tungsten arc welding process can be used to create visually pleasing welds on nickel-based alloys. (Miller Electric Mfg. Co.)

Chapter

Manual Welding of Titanium

Objectives

After completing this chapter, you will be able to:

- ❏ Identify the characteristics of titanium.
- ❏ Recall the material identification system and commercial specifications for titanium and titanium alloys.
- ❏ Identify the four groups of titanium.
- ❏ Summarize the filler metal choices for titanium and titanium alloys.
- ❏ Recall methods of preweld preparation, including cleaning, weld shielding, and tack welds.
- ❏ Select the correct power source, shielding gases, and electrodes for welding titanium and titanium alloys.
- ❏ Recall the proper techniques for welding titanium.
- ❏ Identify sources of contamination of titanium.

Key Terms

beta alloys
interstitials
notch toughness
titanium
trailing shield

Introduction

Titanium is a metallic element with good corrosion resistance, light weight, and high strength. It is very expensive and is often referred to as an exotic metal. The physical characteristics of titanium make it desirable in many manufacturing sectors, including jewelry making, the aviation industry, and the medical industry.

Proven procedures must be used to weld titanium. There is no shortcut method that yields acceptable welds. Procedures that work on many materials are not acceptable on titanium. The most important aspect of welding titanium is making sure it is properly prepared and cleaned. When the correct procedures are followed and the base metal is properly cleaned, titanium has excellent weldability.

Base Material

Titanium is a dull-silver metal with excellent corrosion resistance and a high strength-to-weight ratio. Pure titanium melts at approximately 3000°F (1649°C). The oxide film that forms on titanium's surface makes it resistant to corrosion. For this reason, titanium is very useful in chemical processing industries.

In the pure titanium grades, oxygen is used as a strengthening element. However, oxygen also decreases ductility. Many different alloying elements are added to titanium to improve its mechanical properties into the cryogenic temperature ranges, which are below –250°F (–157°C).

Material Identification

Base materials are identified by the letters *Ti* preceding the alloy content. The alloy content is listed with the major component first, followed by the lesser alloying components. For example, Ti-6Al-4V alloy is an alloy of titanium with 6% aluminum and 4% vanadium.

Specifications

Commercial titanium producers market titanium to commercial specifications and to the Aerospace Material Standards (AMS). All types of titanium listed in the AMS specification are located in the 4900 number series.

ASTM International specifications are used for identification of some types of titanium. These types are identified as grades. The ASTM grades are generally used for material used in the fabrication of weldments to ASME (American Society of Mechanical Engineers) specifications.

Types of Titanium

The titanium family consists of four groups of materials:

- commercial pure titanium
- alpha alloys
- alpha-beta alloys
- beta alloys

The names alpha, alpha-beta, and beta alloys refer to the type of grain structure in each metal. These names are often used in industry to describe the metals. The alpha alloys are not heat treatable and generally have good welding characteristics. Alpha-beta alloys are heat treatable and most are weldable. The **beta alloys** are heat treatable and generally weldable.

Filler Metals

Choosing the correct filler metal for gas tungsten arc welding of titanium and titanium alloys is very important. The metal produced in the weld pool is a combination of the filler wire and the base metal. The weld metal must have the strength, ductility, crack resistance, and corrosion resistance to withstand the service conditions for which the weldment is designed.

The filler alloy used will determine, to a large degree, the type of grain structure in the completed weld. This grain structure determines the physical properties of the completed weld.

In many cases, a nonalloyed filler metal is used to join alloy base material. This is often done where full 100% joint strength is not required, or the weldment is not heat-treated for strength purposes after welding. In these cases, the welds will have lower tensile strength and higher ductility.

Filler metal specifications are found in AWS A5.16 *Specification for Titanium and Titanium-Alloy Welding Electrodes and Rods.* **Figure 14-1** lists the filler metals available and their chemical compositions.

Filler metals used for the welding of titanium and titanium alloys must be free from contamination and defects, both on the inside and the outer surface of the wire. Some elements and gases are trapped within the wire when it is manufactured. Although they are not detrimental in other materials, these contaminants can cause failure in titanium welds. Carbon, oxygen, hydrogen, nitrogen, and iron are held to low limits in wire specifications. These elements are the main cause of a lack of weld toughness in titanium welds.

If high weld toughness is required, a filler metal with an extra-low level of impurities, or **interstitials**, should be selected. Welding filler metals with very low interstitials are called *ELI* materials (extra-low interstitial content). The specification listed in **Figure 14-1** lists these materials with a -1 following the material type:

ERTi-6Al-4V—Standard grade filler metal

ERTi-6Al-4V-1—ELI grade filler metal

Filler metal manufacturers clean titanium immediately prior to packaging. The packages are sealed to prevent dirt and contaminants from entering. Store the filler metal tightly sealed in these packages until it is needed. To prevent contamination, remove only the amount needed for immediate use and reseal the package immediately. Always handle the material with clean gloves; do not use bare hands. Wipe any material that needs to be put back with a clean, lint-free cloth that has been soaked in alcohol.

AWS Classification	Carbon Percent	Oxygen Percent	Hydrogen Percent	Nitrogen Percent	Aluminum Percent	Vanadium Percent	Tin Percent	Chromium Percent	Iron Percent	Molybdenum Percent	Columbium Percent	Tantalum Percent	Palladium Percent	Titanium Percent
RTi-1[a]	0.03	0.10	0.005	0.012	—	—	—	—	0.10	—	—	—	—	Remainder
RTi-2	0.05	0.10	0.008	0.020	—	—	—	—	0.20	—	—	—	—	Remainder
RTi-3	0.05	0.10–0.15	0.008	0.020	—	—	—	—	0.20	—	—	—	—	Remainder
RTi-4	0.05	0.15–0.25	0.008	0.020	—	—	—	—	0.30	—	—	—	—	Remainder
RTi-0.2Pd	0.05	0.15	0.008	0.020	—	—	—	—	0.25	—	—	—	0.15–0.25	Remainder
RTi-3Al-2.5V	0.05	0.12	0.008	0.020	2.5–3.5	2.0–3.0	—	—	0.25	—	—	—	—	Remainder
RTiA1-2.5V-1[a]	0.04	0.10	0.005	0.012	2.5–3.5	2.0–3.0	—	—	0.25	—	—	—	—	Remainder
RTi-5Al-2.5Sn	0.05	0.12	0.008	0.030	4.7–5.6	—	2.0–3.0	—	0.40	—	—	—	—	Remainder
RTi-5Al-2.5Sn-1[a]	0.04	0.10	0.005	0.012	4.7–5.6	—	2.0–3.0	—	0.25	—	—	—	—	Remainder
RTi-6Al-2Cb-1Ta-1Mo	0.04	0.10	0.005	0.012	5.5–6.5	—	—	—	0.15	0.5–1.5	1.5–2.5	0.5–1.5	—	Remainder
RTi-6Al-4V	0.05	0.15	0.008	0.020	5.5–6.75	3.5–4.5	—	—	0.25	—	—	—	—	Remainder
RTi-6Al-4V-1[a]	0.04	0.10	0.005	0.012	5.5–7.75	3.5–4.5	—	—	0.15	—	—	—	—	Remainder
RTi-8Al-1Mo-1V	0.05	0.12	0.008	0.03	7.35–8.35	.75–1.25	—	—	0.25	.75–1.25	—	—	—	Remainder
RTi-13V-11Cr-3Al	0.05	0.12	0.008	0.03	2.5–3.5	12.5–14.5	—	10.0–12.0	0.25	—	—	—	—	Remainder

a – This classification of filler metal restricts the allowable interstitial content to a low level in order that the high toughness required for cryogenic applications and other special uses can be obtained in the deposited weld metal.

Note 1—Analysis for interstitial content shall be made after the welding rod or electrode has been reduced to its final diameter.

Note 2—Single values are maximum percentages.

Figure 14-1. Titanium filler metal chemical compositions.

Preweld Preparation

Molten titanium has an affinity for (easily absorbs) carbon, oxygen, nitrogen, and hydrogen. If these elements are absorbed into the weld, the weld can become contaminated. Depending on the amount of contamination, porosity may form, or the weld may have a very low notch toughness value. *Notch toughness* is the ability of a metal to resist cracking failure at a notch during stress loading. Either porosity or low notch toughness can cause failure of the weld under a load at far below the rated value.

Contaminants are found in many places, such as oils, grease, tools, and body salts. Therefore, preweld cleaning must be done just prior to welding. The base materials and tooling must be thoroughly and completely cleaned. A sample cleaning operation includes the following steps:

1. Draw file all abutting edges. (Sheared edges must be filed until smooth.)
2. Wire brush the entire joint 1″ (25.4 mm) around the weld area. Use a stainless steel brush that has never been used to clean other types of materials. After use, thoroughly wash the wire brush with alcohol and store it in a sealed container. Clearly mark the container to indicate that the brush is for use on titanium only.
3. Wash the entire weld joint and adjacent area with a residue-free solvent. Alcohol and acetone are good degreasers. After washing the weld joint, allow it to dry thoroughly. Do not blow dry the metal with compressed air unless the air is completely filtered. An alternative to using alcohol or acetone is to wash or dip the parts in an alkaline solution. A dilute solution of sodium hydroxide or a commercial alkaline cleaner can be used. Rinse with clear water and dry.
4. If the parts are to be installed on a fixture or tool for welding, clean the fixture with alcohol or acetone before installing parts. If the parts cannot be installed in the fixture immediately, protect the parts by covering them with lint-free cloths or clean paper.
5. Always handle cleaned parts, fixtures, and tooling with lint-free gloves and cloths. Never use bare hands to handle cleaned titanium.

Weld Shielding

Titanium welds must be protected from the atmosphere during welding. Hot titanium absorbs carbon, oxygen, hydrogen, and nitrogen, which contaminate the weld. The degree of contamination can be determined by the color of the weld and the heat-affected zone. Welds and heat-affected zones that are silver in color, as shown in **Figure 14-2**, have sufficient protection and are considered good welds. Gold, straw, and light blue colors indicate progressive amounts of contamination. Depending on the final use of the weldment, the amount of contamination indicated by these colors may or may not be acceptable. The chart in **Figure 14-3** shows various weld colors that indicate acceptable welds and colors that indicate some contamination. Dark blue, light gray, or dark gray colors indicate excessive contamination, and the welds cannot be used for *any* purpose.

The size, configuration, and production quality of the weldment will usually dictate the type of shielding needed. In all cases, the shielding gas from the torch will not be sufficient to provide shielding for all of the hot material.

In order to produce a good weld, the welder must keep the weld and heat-affected zone shielded by gas until the metal is cooled below approximately 500°F (260°C). Small parts can be welded in a chamber like the one shown in **Figure 14-4**. The cleaned tools and parts are loaded into the chamber area, the top is installed, and argon is fed into the chamber. Welding can be started as soon as all of the air is removed from the chamber.

A test weld should be made on a scrap piece of material to test the atmosphere within the unit. A silver weld indicates a sufficient atmospheric protection. A straw or light blue color indicates contamination. An oxygen analyzer can also be used to determine proper gas coverage. For high production jobs, there may be an auxiliary chamber on the side of the main chamber. Parts are passed into and out of the main chamber without compromising the main chamber atmosphere.

Welding torches should always be fitted with a gas lens to reduce the turbulence of the gas leaving the nozzle. With a gas lens, the gas is delivered with a blanket effect, rather than a jet effect. *Trailing shields* protect the metal behind the torch as the torch is moved along the joint. **Figure 14-5** shows two types of trailing shield torches.

Backing fixtures used to locate parts should fit with virtually no gap between the part and the tool. If gaps are present, air can be drawn into the backing gas system, and contamination will result. Grooves

Figure 14-2. Good coloring is an indication that a weld was sufficiently protected from the atmosphere. (Miller Electric Mfg. Co.)

for the backing gas should always be as large as possible to reduce gas turbulence. **Figure 14-6** shows a hold-down clamp and a purge backing fixture for T-joint fillet welds. Tubing and cylindrical parts can be blocked by tape or dams for gas shielding the backside of the weld.

All of the connection fittings for argon gas must be tight, and the hoses must not leak. Any leak in the system will cause contamination of the material during welding. Since a considerable amount of gas is used to shield titanium, sufficient gas must always be available before starting the welding operation. For a long operation, several cylinders can be interconnected on a manifold system to provide a sufficient supply. These cylinders can be turned on all at once or used as needed.

Tack Welds on Titanium

The base metal should be tack welded to keep it in proper alignment for welding. The following considerations apply:

- Protect the joint from contamination.
- Use the same filler metal to make the entire weld.
- If the joint requires full penetration, the tack weld requires full penetration.
- Keep the tack weld as small as possible. If possible, make it smaller than the main weld.
- If the tack weld cracks, leave it and make another close to the broken one.
- Do *not* grind titanium. Use a rotary file only. Grinding grit in a joint makes it almost impossible to achieve a high-quality weld.
- Do *not* leave a crater in the tack weld. The crater of a tack weld is its weakest point.

Titanium Color Indicates Weld Quality

1: Silver

2: Straw

3: Brown

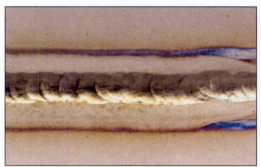

4: Brown-blue

5: Bright blue

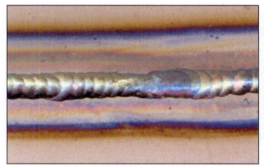

6: Green-blue

7: Dull salmon pink

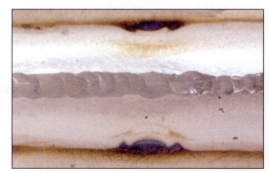

8: White oxide

Figure 14-3. Several different colors indicate acceptable welds. Example 1 is an acceptable weld. Examples 2 and 3 are acceptable welds, but the straw and brown colors may be deemed visually unacceptable. Examples 4 through 8 show discoloration that indicates some degree of contamination. These welds may be rejected, depending on the application. In all cases, discoloration must be removed between passes on a multipass weld. (Miller Electric Mfg. Co.)

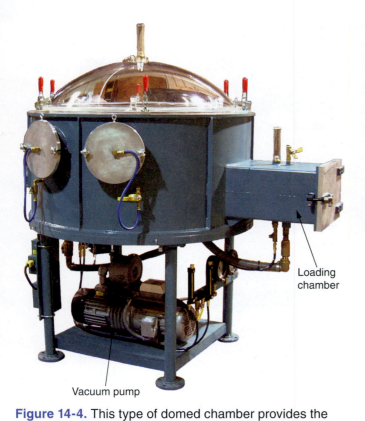

Figure 14-4. This type of domed chamber provides the best atmospheric protection for welding titanium. The level of gas purity can be checked using a gas analyzer after purging is complete. (Jetline Engineering)

Welding Titanium Using DCEN (Straight) Polarity

A motor generator or rectifier power source is recommended for GTAW on titanium. A high-frequency arc start is required.

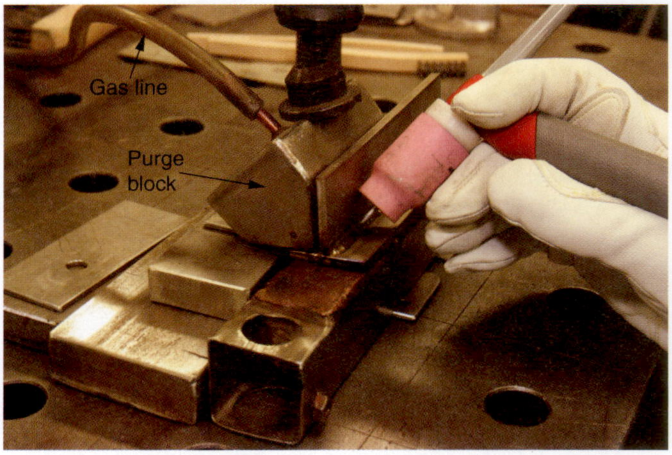

Figure 14-6. This purge block is being used to supply shielding gas to the back side of the weld. Note the gas line that supplies shielding gas to the block. (Miller Electric Mfg. Co.)

Gases for Welding Titanium

For all thicknesses of titanium, argon is generally recommended for the chamber, trailing shield, and backing gas. Helium can be used with argon as the torch shielding gas for automatic welding. However, trailing shield gases mix with the helium, which dilutes the effect of the helium gas.

Only gas that is known to be of the highest purity should be used. The gas purity, if questionable, should be tested before the gas is used. Because titanium requires absolute gas coverage, it is a good idea to use a larger cup and increase gas pressures slightly.

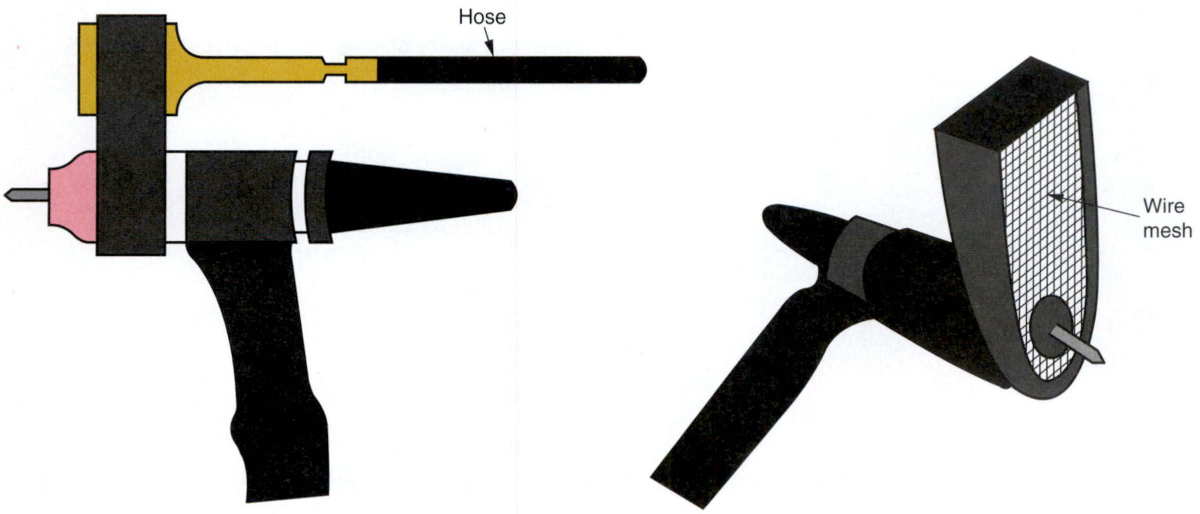

Figure 14-5. Two types of trailing torches. These trailing torches provide extra cover gas to provide a double protection of the weld.

Electrodes for Welding Titanium

A 1% or 2% lanthanated or thoriated electrode tapered to a sharp point is recommended for welding titanium and titanium alloys. An electrode large enough to carry the required amperage should be used.

Techniques for Welding Titanium and Titanium Alloys

1. Handle all tooling, parts, and welding wire with clean, lint-free gloves.
2. Check all purge gases and the torch gas prior to starting.
3. Always connect the work lead with a screw-type or locking-type clamp to make a tight connection to the part. Never use a spring clamp. Spring clamps have a tendency to arc out on the part, which can cause burn spots. High-frequency burns on titanium can cause internal contamination and possible failure of the weldment.
4. When using a chamber for welding, always make a test weld on a scrap block to determine if the purge is adequate.
5. If the weld has a silver color, the purge is good. A gold or blue color indicates that some contamination is present. Do not weld over welds with gold or blue coloring. If the weld is gray, the contaminated area must be completely removed by machining.
6. If the filler metal is removed from the weld pool and the wire end is contaminated, remove the contaminated portion. Do not put a contaminated wire into the weld pool.
7. Always fill craters.
8. Refer to the material manufacturer's specifications for proper postweld heat treatment procedures.

Sources of Contamination

Common causes of contamination of titanium include:

- Tools stored in an area where they absorb moisture.
- Chambers left open for several hours.
- Tools, wrenches, and other tools not cleaned before use in a chamber. (Backing gas tooling must always be cleaned to remove oil after manufacture. The oil will mix with the argon and cause severe contamination during welding.)
- Water leaks in a water-cooled welding torch.
- Argon hoses leaking argon.
- Argon connections loose.
- Improper fitup of assembly tools.
- Argon gas supplied at a too-high pressure.
- Impure argon gas.
- Holes in gloves in the welding chamber.
- Scratch starting an arc, which contaminates titanium with tungsten.
- Dirty gloves or rags.
- Dirty alcohol or acetone.
- Sticking the electrode into the work and welding over the broken piece.

Summary

Titanium is considered an exotic metal and is used in specialty applications in many different industries. It is one of the best materials for strength-to-weight ratio and corrosion resistance. The four groups of titanium materials are commercial pure titanium, alpha alloys, alpha-beta alloys, and beta alloys.

The type of filler alloy used is a significant factor in the grain structure of the completed weld. Where high weld toughness is needed, a filler metal with an extra-low level of interstitials should be used. Filler metals must be free from contamination and defects on both the inside and the outer surface of the wire.

Molten titanium easily absorbs carbon, oxygen, nitrogen, and hydrogen. These elements can contaminate the weld, resulting in porosity or very low notch toughness value. Working with titanium requires absolute cleanliness in all aspects. Titanium can be contaminated from many sources. Thorough cleaning of the base material and tooling must be done just prior to welding.

Titanium welds must be protected from the atmosphere during welding. Silver coloring in welds and heat-affected zones indicate sufficient shielding from the atmosphere. Gold, straw, and light blue colors indicate some contamination. Dark blue, light gray, or dark gray colors indicate excessive contamination, and welds with these colors are unusable.

Torches should be fitted with a gas lens. Backing fixtures used to locate parts should fit with virtually no gap between the part and the tool so that air is not drawn into the backing gas system. Small parts can be welded in a chamber into which argon gas is fed.

A motor generator or a rectifier power source with a high-frequency arc start should be used for welding titanium. Very high-purity argon is generally recommended for the chamber, trailing shield, and backing gas. Electrodes recommended for welding titanium are 1% or 2% lanthanated or thoriated electrodes tapered to a sharp point.

Review Questions

Write your answers on a separate sheet of paper. Do not write in this book.

1. List three physical characteristics of titanium.
2. What element is used to strengthen pure titanium?
3. What are ASTM grades of titanium generally used for?
4. What are the four groups of materials in the titanium family?
5. Which titanium alloy group is *not* heat treatable?
6. List five elements that are called *interstitials*.
7. What do the letters *ELI* mean? How are filler metals identified as *ELI* materials?
8. In order to produce a good weld, the welder must keep the weld and heat-affected zone shielded by gas until the metal is cooled below approximately _____°F (260°C).
9. What weld colors indicate a level of contamination that may or not make the weld unacceptable?
10. What weld colors indicate the weld is badly contaminated and cannot be used under any condition?
11. Welding torches can be fitted with a(n) _____ to reduce the gas flow turbulence from the gas nozzle.
12. What types of electrodes are recommended for welding titanium?
13. The ground lead should always be connected with a(n) _____ clamp.
14. What must be done after welding wire is removed from the weld pool and exposed to the air?
15. Why should backing tooling be cleaned after manufacture and before use in welding?

Chapter

Manual Welding of Dissimilar Metals

Objectives

After completing this chapter, you will be able to:
- ☐ Give examples of material combinations that can be joined using GTAW.
- ☐ Explain the buttering process.
- ☐ Explain the buildup process.
- ☐ Recall the uses of butt joints, clad-material joints, and built-up joints on dissimilar metals.
- ☐ Recall the process used to select the proper filler metal for welding dissimilar materials.
- ☐ Identify and select the proper method of joint preparation.
- ☐ Recall methods of joint cleaning.
- ☐ Recall the importance of joint fitup, heat input, and welding techniques for welding dissimilar metals.
- ☐ Recall welding techniques used to reduce dilution.

Key Terms

buildup
buttering materials
cladding
dilution
hardfacing

Introduction

It is often necessary to weld dissimilar materials when fabricating weldments. Welders also occasionally need to surface a base metal to prevent corrosion, oxidation from heat, and wear. Finally, welding of dissimilar materials may be required for maintenance or repair of worn parts. The ability to weld dissimilar metals is a highly useful skill.

Material Combinations/ Applications

Many material combinations can be joined using GTAW. Some examples are as follows:

- steel to steel
- steel to cast iron
- steel to stainless steel
- steel to nickel
- stainless steel to nickel
- stainless steel to Inconel
- copper to steel
- copper nickel to steel
- copper aluminum to steel
- silicon bronze to steel
- surfacing alloy to iron base metal

Weld joints are often made up of dissimilar metals. The filler metal used to join butted materials, as shown in **Figure 15-1**, can be different from one or both of the base metals. Depending on the desired alloy mix in the weld, the filler metal can be one of the base metals or a dissimilar metal, **Figure 15-2**. If a dissimilar metal is used as the filler metal, it must be compatible with both base metals.

Buttering materials, as shown in **Figure 15-3**, are used to join metals that are very different. Each metal must be "buttered" with a material that is compatible with the filler metal used to make the final joint.

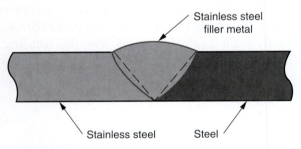

Figure 15-1. Stainless steel filler metal is used in many steel-to-stainless steel combinations where ductility is of prime importance.

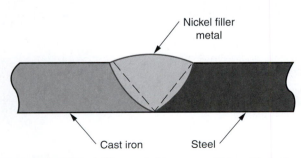

Figure 15-2. Nickel filler metal is compatible with both cast iron and steel.

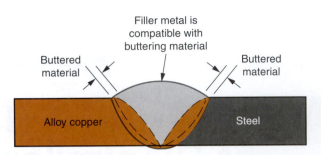

Figure 15-3. The buttering material is applied to each material before the joint filler metal is used.

Buttering is the process of applying a layer of compatible filler metal to the mating surface of each piece of base metal to be welded. Another type of filler metal is then used to weld the buttered surfaces together. The filler metal used must be compatible with the buttering material, but may be incompatible with the base metal itself. In other words, the filler metal used to butter the surface acts as a transitional layer between the base metal and the metal used to complete the weld joint.

Clad material, shown in **Figure 15-4**, is used extensively in the manufacture of processing equipment. The *cladding* is bonded to the base metal at the rolling mill. The thickness of the cladding will vary, depending on its final purpose. When clad materials are being joined, the filler metal must be compatible with both the cladding and the base metal.

Built-up materials, as shown in **Figure 15-5**, are similar to a clad or buttered joint; however, buildups are generally thicker. Unlike cladding, *buildup* is a process performed by the welder. The welder applies layers of weld beads to a base metal in order to give the surfaces the physical properties of the filler metal. In many cases, buildup is added to improve the wear resistance or chemical resistance of the base metal.

Usually, more than one layer of weld beads must be created. When a weld bead is created, the base metal mixes with the filler metal, diluting the filler metal and diminishing its desirable properties. *Dilution* is the mixing of filler alloy with base alloy. As additional layers of weld beads are added, and the buildup thickens, the amount of dilution decreases until the deposit contains only the filler metal chemistry. A GTAW buildup operation is shown in **Figure 15-6**.

Butt Joints of Dissimilar Metals

Butt joints with square edges, as shown in **Figure 15-7**, are used only where the material thickness can be welded in a single pass. This type of joint has considerable dilution between the base metals and the filler metals.

Butt joints with V-grooves welded from one side, as shown in **Figure 15-8**, are used to join thicker

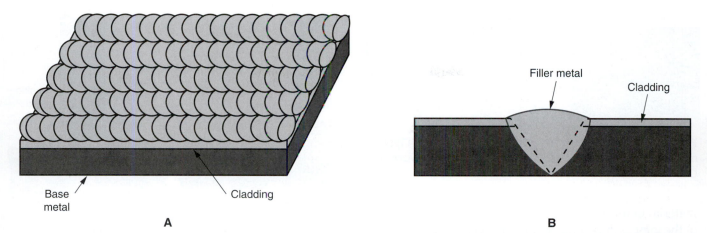

Base metal

Cladding

A

Filler metal

Cladding

B

Figure 15-4. Welding clad materials. A—A series of shallow surfacing welds have been added to the material. Since the welds do not penetrate into the base metal, the filler metal need only be compatible with the cladding. B—Welds into the heavier base metal require filler metals selected for strength, ductility, and other mechanical properties. The filler metal must be compatible with both the base metal and the cladding.

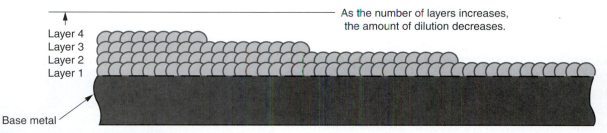

As the number of layers increases, the amount of dilution decreases.

Layer 4
Layer 3
Layer 2
Layer 1

Base metal

Figure 15-5. The number of weld layers required to achieve the desired chemistry is determined by testing a sample of the final joint design using chemical analysis.

Figure 15-6. The outside of this pipe is being built up. The special GTAW torch is oscillated during the operation to widen the weld bead. (CFM/VR-TESCO LLC)

Figure 15-7. Square-groove weld joint designs require a filler metal compatible with each of the base metals due to the considerable amount of dilution.

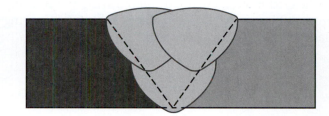

Figure 15-8. Stringer beads reduce penetration and dilution of V-groove welds. Weave beads should be avoided, if possible, to reduce heat input and the amount of dilution from the base metals.

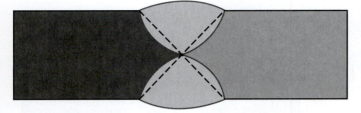

Figure 15-9. Double V-groove welds result in much less dilution of the base metals since less welding is involved. Use stringer beads to further reduce dilution.

material in multiple passes. Dilution is high at the edges of the joint and diminishes near the center of the joint.

Butt joints with double V-grooves are welded from both sides, **Figure 15-9**. Again, the weld should be made with stringer beads rather than weave beads. Very thick materials may require multiple passes on each side. Distortion and dilution of the base metal are minimized with this type of joint. Less filler metal is required to fill this type of joint than a single V-groove, which reduces heat input.

A butt joint with double-buttered edges is shown in **Figure 15-10**. The buttered material is applied in sufficient thickness so the edge of the joint will be composed of undiluted filler metal. Before the joint is welded, the buttered surface is cut or ground to form the desired type of groove for the joint. It is important that the material exposed by this operation be undiluted.

Clad Material Joints

Clad material often requires two joint designs—one for the base metal and one for the cladding. The base metal and the cladding are welded separately. The base metal joint is prepared and welded using standard practices. The cladding joint must be designed so the clad material is not diluted during the welding process. **Figure 15-11** shows joint designs for preparing clad materials and various weld applications.

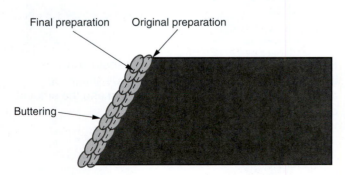

Figure 15-10. Buttering or buildup material must be applied thick enough to obtain the correct material chemistry.

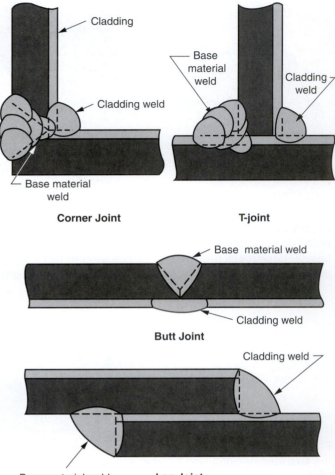

Figure 15-11. Common joint preparations used with clad base metals.

For groove welds in clad material, the base metal joint is welded first, using a filler metal that is compatible with the base metal. Next, a shallow weld bead known as a *weld overlay* is created at the root of the base metal weld. The filler metal used for the weld overlay pass must be compatible with both the base metal and the cladding material. Next, one or more beads are laid down on top of the weld overlay. The filler metal used for these passes must have the same chemical composition as the cladding.

Because cladding is not removed for a fillet weld, no overlay pass is required. The filler metal must have a chemical composition that is compatible with the cladding metal. Penetration beyond the cladding must be minimized. If the weld penetrates into the base metal, the filler metal can become diluted to unacceptable levels. In the case of a lap joint between clad materials, a small amount of dilution may be acceptable, but it should be kept to a minimum.

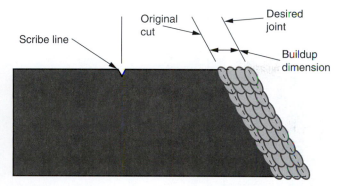

Figure 15-12. Reference lines or points are used to ensure proper buildup dimensions. It was calculated that this joint required four passes of buildup to eliminate dilution.

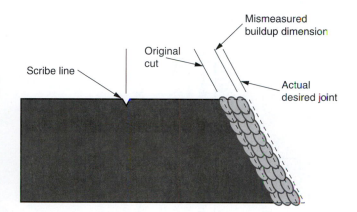

Figure 15-13. Insufficient buildup can result in incorrect weld metal chemistry. In this case, a miscalculated buildup dimension resulted in only three levels of buildup being applied. Four levels were required to eliminate dilution.

Built-Up Joints

Built-up joints require undiluted filler metal chemistry at the edge of the weld. To achieve this condition, the number of filler metal layers must be computed into the joint design, **Figure 15-12**. A groove prepared without enough filler metal, **Figure 15-13**, can result in improper chemistry in the final weld. If there is not enough weld metal at the edge of the joint, diluted filler metal will be exposed when the final edge is cut or ground. The results of a great many filler metal/base metal combinations have been analyzed by certifying institutions, and the recommended combinations have been published. The welder can refer to these resources when selecting a filler metal for a given base metal.

Filler Metals

The choice of filler metals for a weld joint requires analyzing the composition of the base metals, dilution percentages, and the final use of the joint. In many cases, a sufficient number of welds have been made to establish which filler metals can be used successfully.

Compatible stainless filler metals and base metals for the most common applications are listed in **Figure 15-14**. Proper buttering must be done in many instances to minimize dilution and avoid cracking problems with the weld. The table in **Figure 15-15** lists the types of surfacing materials to be used with specific base metal types, as well as the polarity to be used with the surfacing materials.

Figure 15-16 lists cladding materials and the proper filler metal selection for joining materials clad with those materials. **Figure 15-17** illustrates the weld overlay as a barrier between the steel weld metal and the alloy cladding metal. The weld overlay prevents dilution of the cladding weld. Cladding alloys, the proper filler alloys to use for the overlay pass on steel, and the proper filler metal alloys for welding the cladding pass(es) are listed in **Figure 15-18**. The table in **Figure 15-19** lists designations for common filler metals used for surfacing, as well as the Brinell hardness ranges, yield strengths, and tensile strengths.

Hardfacing refers to making a material stronger or more wear resistant by adding a layer of a different material to its surface. Cladding and buildup are forms of hardfacing. The preheat temperatures required for hardfacing various materials are shown in the chart in **Figure 15-20**. The base metals being hardfaced are listed in the left column, and the hardfacing alloys are listed in the header at the top of the chart. The numbers at the intersections of base metals and hardfacing materials are the proper preheating temperatures for those material combinations.

Joint Preparation

When preparing a joint, a welder should consider all of the methods available and choose the best one for the situation. Machining operations, like routing and milling, can create a very precise joint, and the part is ready for welding immediately after the machining operation. However, machining operations are expensive. Unless parts require the precision offered by machining operations, the joints are generally prepared by other means.

Base Alloy, Type	201, 202, 301, 302, 302B, 303[a], 304, 305, 308	304L	309, 309S	310, 310S, 314[a]	316	316L	317	317L	321, 347, 348	330[a]	402, 405, 410, 412, 414, 420[a]	430, 430F, 431, 440A, 440B, 440C[a]	446[d]	501, 502[c,d]	505[c,d]	Carbon Steels[c,d]	Cr-Mo Steels[c,d]
201, 202, 301, 302, 302B, 303[a], 304, 305, 308	E308	E308	E308	E308	E308	E308	E308	E308	E308	E309	E309	E309	E310	E309	E309	E309	E309
304L		E308L	E308	E308	E308	E308	E308	E308	E308	E309	E309	E309	E310	E309	E309	E309	E309
309, 309S			E309	E309	E309	E309	E309	E309	E309	E309	E309	E309	E310	E309	E309	E309	E309
310, 310S, 314[a]				E310	E316	E316	E317	E317	E308	E310	E309	E309	E310	E310	E310	E310	E309
316					E316[b]	E316	E316	E316	E308[b]	E309Mo	E309	E309	E310	E309	E309	E309	E309
316L						E316L	E316	E316L	E316L	E309Mo	E309	E309	E310	E309	E309	E309	E309
317							E317	E317	E316L	E309Mo	E309	E309	E310	E309	E309	E309	E309
317L								E317L	E308L	E309Mo	E309	E309	E310	E309	E309	E309	E309
321, 347, 348									E347	E309	E309	E309	E310	E309	E309	E309	E309
330[a]										E330	E309	E309	E310	E312	E312	E312	E312
403, 405, 410, 414, 416, 420											E410	E430[e]	E410[e]	E502[e]	E505[e]	E410[e]	E410[e]
430, 430F, 431, 440A, 440B, 440C												E430	E430	E502[e]	E505[e]	E430[e]	E430[e]
446													E446	E502[e]	E502[e]	E430[e]	E430[e]
501, 502														E502	E502[e]	E502[e]	E502[e]
505															E505	E505[e]	E505[e]

Notes: Grades shown are those most commonly selected for most applications; other combinations may be used. Wherever possible, recommendation is based upon the most available and lowest cost filler metal. Filler metal designations are those appearing in AWS Specification, A5.9 for bare filler wire.

a. These alloys are sensitive to weld cracks and fissures; for this reason, E312 filler metal is a frequently recommended alternative. It is preferred especially when thick sections or highly restrained joints are required. Buttering these metals with type 312 before joining is often desirable.

b. E16-8-2 is preferred for lower embrittlement danger in elevated temperature service.

c. When joining an austenitic steel, alternate choice is to butter carbon or chromium steel with E309 and join with E308 or with filler metal similar to austenitic base metal. E307 is also commonly used for welds between austenitic stainless steel and either carbon or low alloy steels.

d. ENiCrFe3 is preferred for elevated temperature service, except when sulfur compounds are present.

e. If austenitic weld metal is acceptable for service conditions, E309 or E310 is often employed.

Figure 15-14. Filler metals used for joining stainless steels and dissimilar metals.

Base Metal	Surfacing Material	Current Type Amps.	Rod Type	Deposit RC Hardness
Mild and stainless steels	Haynes Stellite alloys	ACHF ACHF ACHF ACHF ACHF	Stellite #1 Stellite #6 Stellite #12 Stellite #93 Hascrome	54 39 47 62 23–43
Copper	Stellite #6 alloy	DCEN 180–230 for 3/16″ material	Stellite #6	42
Steel, copper, and silicon bronze	Aluminum bronze	DCEN	Aluminum-bronze rods	
Mild steel and cast iron	Bronze and copper	ACHF or DCEN 150 for 1/2″ material	A1-bronze and copper rods	
Stainless steel	Silver	ACHF 160 for 1/2″ material		
Mild steel	Stainless steel	ACHF or DCEN		
Carbon and alloy tool steels	Tungsten carbide	DCEN 300–375	Tube of 8/15 mesh tungsten particles	

Figure 15-15. Base metal and surfacing material combinations.

Cladding Type	Filler Metal
405, 410, 410S, 429, 430	309, 310 Inconel A, B, 182, or equivalent Inconel 82 or equivalent 430
304	309, 310 309L
304L	309L 308L 309Cb 309CbL
321, 347	309Cb, 310Cb 309CbL
316	347 309Mo, 310Mo 316L
316L	309MoL 316L
317	318 309MoL, 310Mo, 309Mo
317L	317L 309MoL, 310Mo 317L
Incoloy 825	Incoloy 65 or equivalent Incoloy 135 or equivalent Inconel 625 or equivalent Inconel 112 or equivalent
Inconel 600	Inconel A, B, 182, or equivalent Inconel 82 or equivalent
Monel 400	Monel
70Cu-30Ni 90Cu-10Ni	Monel 70Cu-30Ni
Nickel	Nickel
Copper	Copper, Monel, nickel, Inconel

Figure 15-16. The cladding listed in the left hand column can be welded with the same alloy or with any of the materials listed on the right.

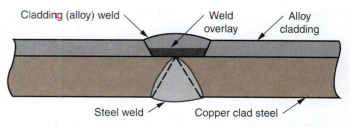

Figure 15-17. The proper sequence of welding clad steels is to weld the steel base metal, then add the weld overlay, and finally to make the cladding weld. The weld overlay acts as a buffer between the base metal and cladding, preventing dilution of the cladding.

Cladding Alloy	Alloy for Overlay on Steel	Alloy for Welding Cladding
Copper	RCu RCuAl-A2 RCuSi-A RNi-3	RCu
Copper-Zinc	RCuAl-A2	RCuAl-A2
Copper-Tin-Zinc	RCuSn-A	RCuSn-A
Copper-Aluminum	RCuAl-A2	RCuAl-A2
Copper-Silicon	RCuSi-A	RCuSi-A
Copper-Nickel	RCuNi	RCuNi

Figure 15-18. Filler metals used for welding clad steels.

Wire Designation	Brinell Hardness	Yield Strength (ksi)	Tensile Strength (ksi)
RCu	50–60	8	28
RCuSn	70–85		35
RCuSi-A	80–100	25	50–55
RCuNi	60–80	10–20	50
RCuAl-A1	100–150	25–30	55–65
RCuAl-A2	130–190	30–35	60–75
RCuAl-B	140–220	40–45	90–110
RCuAl-C	180–280	40–45	90–100
RCuAl-D	250–350	50–55	75–85
RCuAl-E	290–390	55–70	70–80
NiAl bronze	180–200	55–60	100

Figure 15-19. Common filler metals used for surfacing.

Base Metal ↓ / Hard Surfacing →		Cobalt Base Alloys						Nickel Base Alloys						Fe-Cr Alloys
		1	6	7	12	T-400	T-800	44	45	46	10XN	A	T-700	
Carbon steel	°F	600	300	200	500	900	900	200	400	900	400	70	900	70
.30C maximum	°C	325	150	95	275	480	480	95	205	480	205	20	480	20
Carbon steel	°F	600	400	300	500	900	900	200	400	900	400	100	900	200
.30–.50C	°C	325	205	150	275	480	480	95	205	480	205	40	480	95
Low-alloy steels up	°F	600	400	300	500	900	900	200	400	900	400	100	900	200
to 3% total alloys	°C	325	205	150	275	480	480	95	205	480	205	40	480	95
Medium-alloy steels	°F	600	400	400	500	900	900	200	400	900	400	200	900	300
3–10% total alloys	°C	325	205	205	275	480	480	95	205	480	205	95	480	150
High-alloy steels Martensitic e.g. Type 410	°F	600	400	400	500	900	900	200	400	900	400	200	900	300
	°C	325	205	205	275	480	480	95	205	480	205	95	480	150
High-alloy steels Ferritic e.g. Type 430	°F	600	300	200	500	900	900	200	400	900	400	100	900	100
	°C	325	150	95	275	480	480	95	205	480	205	40	480	40
High-alloy steels Austenitic e.g. Types 304, 316	°F	500	300	200	400	900	900	200	400	900	400	70	900	70
	°C	275	150	95	205	480	480	95	205	480	205	20	480	20
Nickel alloys e.g. Inconel* e.g. Monel*	°F	500	300	200	400	900	900	200	400	900	400	70	900	70
	°C	275	150	95	205	480	480	95	205	480	205	20	480	20

*Trade names of the International Nickel Co., Inc.

Figure 15-20. Preheating of the base metal reduces the amount of cracking in the surfacing alloy during the cooling period.

Shearing of sheet and plate is often the primary operation involved in joint preparation. In the thinner materials, the sheared edge may be used in the joint without any additional preparation. However, the sheared edge always contains a rough surface that could potentially trap dirt or other foreign material. Further processing, such as filing or grinding, may be required. See **Figure 15-21**.

Clad materials should always be sheared with the clad surface upward. This protects the cladding from scratches and nicks during the operation. The edge of the cladding will have a radius formed on the corner, as shown in **Figure 15-22**. The edge will need to be cut or ground to square it off and to ensure that the cladding is the proper thickness. See **Figure 15-23**.

The carbon and alloy content of steels should be considered before they are flame cut. Higher-alloy steels can harden during the operation due to the fast cooling of the metal. Improperly adjusted torches can introduce carbon onto the surface or oxidize the cut edge.

Steels that have stainless steel or nickel cladding can be flame cut if the clad is not more than 20% of the total thickness. The material is set up so the heavier steel side is placed on the top, as shown in **Figure 15-24**. The molten steel then penetrates the alloy, making the cut possible. Proper travel speeds must be maintained. If the cut is lost during the operation, it is almost impossible to restart. One solution would be to start the cut from the opposite direction and come back into the disrupted cut area. Since the cladding is on the bottom, the cladding must be protected to prevent scratches, nicks, and other marks. This can be done by using good clamping techniques and setting the cladding on a softer material.

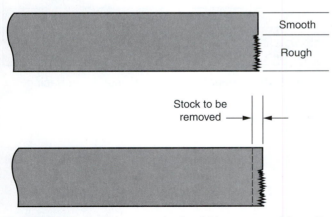

Smooth

Rough

Stock to be removed →

Figure 15-21. Sufficient stock should be removed from the edge of sheared material to eliminate all rough edges.

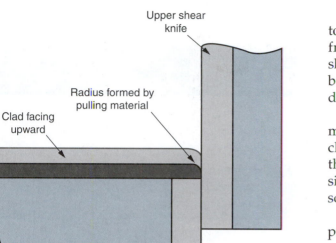

Figure 15-22. Due to the ductility of the clad material, the metal will stretch during the shear operation. This reduces the clad thickness at the sheared edge.

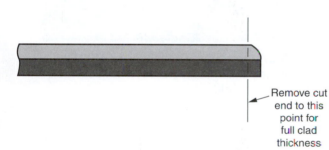

Figure 15-23. Remove sufficient stock from the sheared edge to ensure full clad thickness.

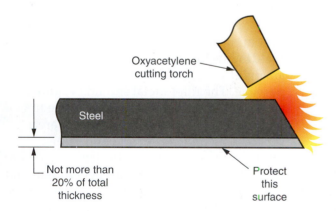

Figure 15-24. In order to establish cutting parameters (gas pressures, tip size, and travel speed), make a test cut on scrap clad material before making the full cut.

Steels with copper cladding are very difficult to flame cut because copper conducts the heat away from the joint very rapidly. Therefore, copper cladding should be removed from the steel where the cut is to be made, as shown in **Figure 15-25**. The copper cladding can be removed by grinding or chiseling.

Plasma arc cutting can be used to cut all types of metals, including clad metals. As **Figure 15-26** shows, clad metal can be plasma arc cut from either side of the plate. If the plate is positioned with the cladding side up, the cladding will be better protected from scratches, nicks, and other marks.

Air carbon arc gouging is frequently used to prepare U-groove weld joints. This process offers fast preparation time with minimum cost. Repairs, spot cladding, and irregularly shaped welds are prepared for welding using this process.

Joint Cleaning

Defect-free welds cannot be made if the weld joint is dirty or contaminated. To ensure the best possible welding conditions, carefully inspect the joint before starting to weld. Remove slag and scale from the part by grinding or machining the surfaces. Follow grinding with wire brushing to remove any particles left from the grinding wheel. Remove oil, grease, and pencil marks with solvent. Remove burrs and nicks to prevent foreign materials from getting trapped in the weld joint.

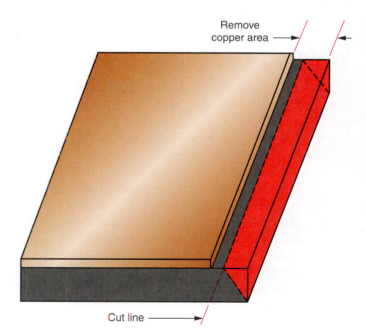

Figure 15-25. Cladding should be removed far enough from the edge or cut line so the copper is not melted into the cut.

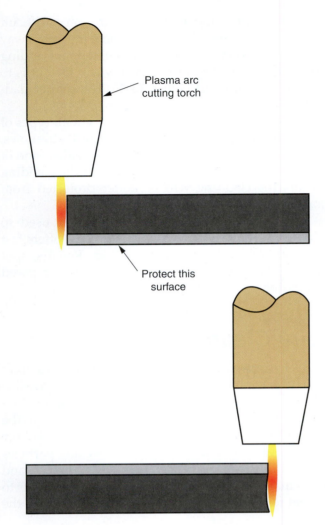

Figure 15-26. The plasma arc cutting process can be used to cut ferrous or nonferrous metals. Therefore, the cut can be made from either side.

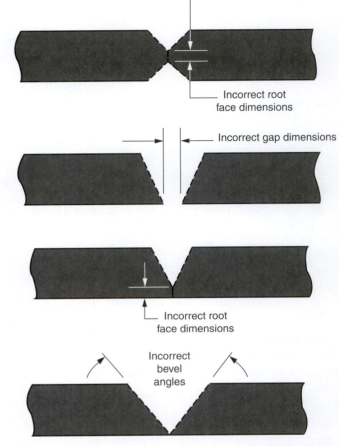

Figure 15-27. Incorrect joint dimensions affect the alloy mix within the joint. This may affect the mechanical properties of the weld.

Welding

Welding dissimilar metals requires careful consideration of the metals to be joined and the welding parameters:

- **Joint fitup.** Joints that are improperly fit up for welding often fail. Establish joint and assembly tolerances by completing test welds and determining the welding parameters. If changes in the welding parameters are required to correct improper fitup, the changes should be allowed only within a prescribed set of tolerances. **Figure 15-27** shows areas in which excessive tolerances can affect the welding operation and joint integrity.
- **Heat input.** Heat input is dependent on amperage, voltage, travel speed, and application of the filler metal. Insufficient heat input results in lack of fusion, voids, porosity, and cold shuts within the weld joint. Too much heat can cause excessive dilution within the joint, between the base metal and the filler metal. The use of pulsers to establish heating and cooling cycles will regulate amperage and reduce the overall heat input required to weld the joint.
- **Welding technique.** Hold the torch at an angle that melts as little base metal as possible. The filler metal should be added to the weld pool as rapidly as possible to provide a "cold pool." Large molten areas stir the pool and cause more dilution.

Out-of-position welding usually generates more heat in the base metal, causing more penetration and dilution. Welds made in these positions should be stringer beads, made with no oscillation.

Sizes of welds and sequencing of beads should be made to control dilution of one bead to another. Each weld should overlap the previous bead by one-half of the previous bead width, as shown in **Figure 15-28**.

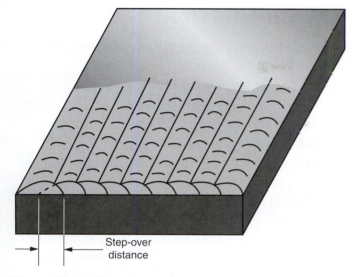

Figure 15-28. The amount of overlap of one weld to another is called the "step-over" distance. This dimension must be established during testing and maintained in production to control the alloy mix.

Each layer of weld should be flat, without valleys or high spots. High and low spots can lead to defects within the weld if not corrected prior to starting the next layer. To correct these conditions, grind all high spots even and fill or grind valleys even. See **Figure 15-29**.

Each weld should be cleaned and visually inspected for porosity, voids, slag, and silicon before the next bead is applied. The specified number of layers should always be deposited. Adding or reducing the number of layers can radically affect the dilution and composition of the final layer of the weld.

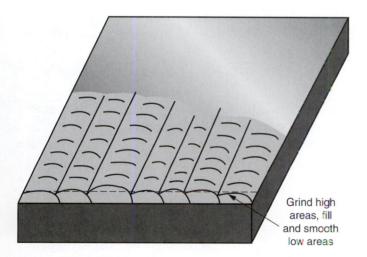

Figure 15-29. Welds with valleys or high spots do not have the proper alloy mix. These areas can cause defects such as lack of penetration, lack of fusion, and excessive dilution.

Summary

Filler metal used to join butted materials can be different from one or both of the base metals, but it must be compatible with both base metals. Buttering is the application of a filler metal to the surfaces of base metals. The filler metal acts as a transitional layer between the base metal and the metal used in the weld joint. Cladding is bonded to the base metal at the rolling mill. When clad materials are being joined, the filler metal must be compatible with both the cladding and the base metal.

Butt joints with square edges are used only where the material thickness can be welded in a single pass. Butt joints with V-grooves welded from one side are used to weld thicker material with multiple passes. Butt joints with double V-grooves are used to weld thicker material with one or more passes on each side. Distortion and dilution of the base metal are minimized in butt joints with double V-grooves.

Clad material often requires two joint designs. The base metal and the cladding are welded separately. The cladding joint must be designed so the clad material is not diluted during the welding process. For built-up joints, the number of layers of filler metal is computed into the joint design to ensure undiluted filler metal chemistry at the edge of the weld.

Hardfacing makes material stronger or more wear resistant by adding a layer of a different material to its surface. Cladding and buildup are forms of hardfacing.

Joint preparation involves shearing of sheet and plate. Clad materials should always be sheared with the clad surface upward to protect the cladding from damage.

The carbon and alloy content of steels should be considered before they are flame cut, because higher-alloy steels can harden during the operation. Steels with stainless steel or nickel cladding can be flame cut if the clad is not more than 20% of the total thickness. Steels with copper cladding are very difficult to flame cut because copper conducts the heat away from the joint very rapidly.

Welding dissimilar metals requires careful attention to joint fitup, heat input, and welding technique. Changes made to correct improper fitup should be allowed only within a prescribed set of tolerances. Insufficient or excessive heat input results in weld joint problems. Filler metal should be added to the weld pool as rapidly as possible to avoid large molten areas that cause increased dilution. Out-of-position welds should be made with stringer beads.

Sequencing of weld beads should be made to control dilution of one bead to another. Each weld should overlap the previous bead by one-half of the previous bead width. Each layer of weld should be flat, without valleys or high spots. Each weld should be cleaned and inspected before the next bead is applied.

Review Questions

Write your answers on a separate sheet of paper. Do not write in this book.

1. The filler metal used to weld buttered surfaces together must be compatible with the _____, but may be incompatible with the _____.
2. What does the term *hardfacing* mean?
3. Where is clad material made? Is all cladding the same thickness?
4. Which is usually thicker, a built-up or a clad surface?
5. Does a single V-groove butt weld result in more or less dilution than a double V-groove butt weld?
6. Why are clad materials sheared with the clad surface facing upward?
7. Steels with stainless or nickel cladding may be flame cut if the cladding is not more than _____% of the total thickness.
8. Which type of cladding metal must be thoroughly removed from steel before thermally cutting? Why?
9. Which thermal cutting process can be used to prepare dissimilar metal joints on all types of metals?
10. Which process is used to prepare U-groove weld joints?
11. Remove slag and scale from the part by _____ or _____ the surfaces.
12. What type of equipment can be used to reduce overall heat input into a weld?
13. When a surface is being built-up, why should filler metal be added to the weld pool as rapidly as possible?
14. Why should each layer of a surfacing weld be flat?
15. To control dilution of one bead to another, each weld should overlap the previous bead by _____ of the previous bead width.

Manual Welding of Pipe

Objectives

After completing this chapter, you will be able to:

- Summarize the factors to consider when selecting a joint design.
- Recall the various pipe joint designs.
- Recall the joint preparation processes for various joint designs.
- Explain reasons for purging pipe weld areas and describe criteria for selecting gases used for purging.
- Recognize the various types of purge dams.
- Explain the root purging operation.
- Summarize the methods and types of equipment used to determine the oxygen level of the purge area.
- Summarize fitup and tack welding procedures.
- Recall methods of torch manipulation.
- Distinguish between welding techniques for root passes on various types of pipe welds.

Key Terms

bladder dams
closed joint
gas purge tester
open joint
ovality
soluble dams
water-soluble films

Introduction

This chapter explains the general procedures for preparing and welding various types of pipe joints. Cleaning procedures, filler metals, preheating, interpass temperature, and postheating are covered in detail in the individual chapters that relate to welding of the applicable base metals.

Joint Designs

Factors to consider when selecting a joint design include the following:
- type of material and thickness
- quality of the weld
- position and accessibility of the weld joint
- internal flow restrictions
- welder skill
- production quantity

Socket-Type Joint

Socket-type joints use fittings machined or cast to accept a section of pipe. A shoulder in the socket determines the depth the pipe will be set into. The fitting may be an elbow, bushing, reducer, T, or a special design part. **Figure 16-1** shows a typical socket-type joint.

Lap Joint

Lap joints are used for joining ends of pipe together where a large gap exists between the ends. A coupling or sleeve is placed between the ends of the pipe to be joined. Depending on the application, the coupling or sleeve will either slide over or inside each end of the pipe. It is then welded to the pipes with fillet welds. Lap joint designs are shown in **Figure 16-2**.

Laterals, T-Joints, and Y-Joints

These types of joints are used for joining pipes with various diameters and intersections at various angles. There are many configurations, some of which are shown in **Figure 16-3**.

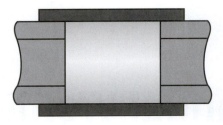
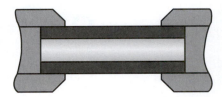

Figure 16-2. Lap joints are used when one pipe can be assembled inside of the other pipe, or if pipes are connected with a coupling or sleeve.

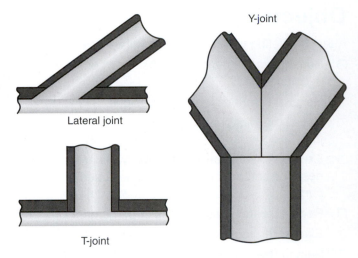

Figure 16-3. Various types of pipe joint connections.

Butt Joints

Butt joints are commonly used in industry. They are used to directly connect the ends of pipe. Several butt joint designs are shown in **Figure 16-4**. They include the following:
- standard butt joint with root face-closed joint
- standard butt joint with root face-open joint
- special machined joint configuration
- integral backing ring
- backing ring joint
- consumable insert joint

Joint Preparation

Joint preparation is the first step in creating an acceptable pipe weld. Many types of joints are used in the pipe welding industry, and each one requires a

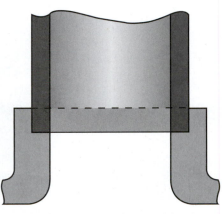

Figure 16-1. A typical socket joint used in pipe assemblies.

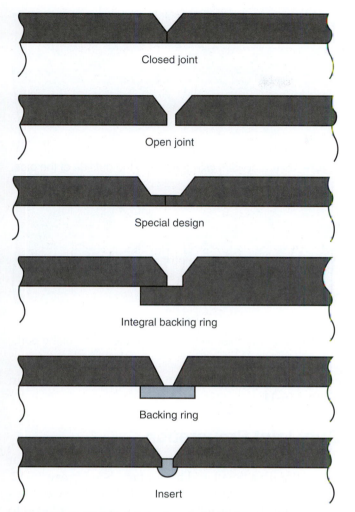

Figure 16-4. Weld joint designs used in butt welds.

Closed joint

Open joint

Special design

Integral backing ring

Backing ring

Insert

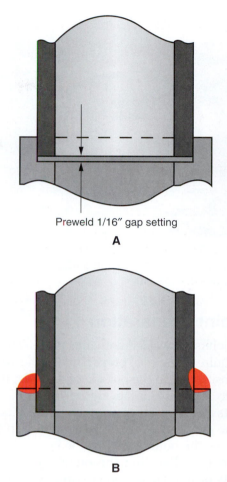

Preweld 1/16″ gap setting

A

B

Figure 16-5. A—A gap of 1/16″ (1.6 mm) is included between the end of the pipe and the bore seat. B—The gap allows weld shrinkage to prevent cracking in the root of the weld.

somewhat different preparation process. It is critical that joints be properly prepared and fit up before being tacked and welded. A good preparation process and practice fitup can reveal potential problems with the joint, such as inconsistent beveling, inconsistent root openings, or even pipe that might be out-of-round. The amount of allowable inconsistency depends on the application and the strictness of the code or specification being used.

Socket-Type Joint Preparation

Socket-type joints are welded with fillet welds, as shown in **Figure 16-5**. Because there is a small amount of weld shrinkage during welding, socket joints must be prepared in a special manner. They require a preweld minimum 1/16″ (1.6 mm) gap at the end of the pipe into the socket bore end, as shown in **Figure 16-5A**. If the gap is too small, cracks may occur at the weld root.

Establish the gap distance using one of the following methods:

- Insert the pipe into the socket until it seats at the bore end. Scribe a line at the flange end. Move the part out to a 1/16″ (1.6 mm) gap.
- Insert a spacer ring. Clean the ring and pipes before assembly. Insert the pipe until it seats against the ring.

Some spacer rings have manufactured kinks that can flatten out as the pipe expands, creating the space needed for thermal expansion. These kinks are shown in **Figure 16-6**. Other spacing rings are water soluble and are washed out with water after the weld is complete. Spacer rings provide a 1/16″ (1.6 mm) spacing in the socket to provide room for thermal expansion and contraction of the pipe. See **Figure 16-7**. Without the proper gap, the weld will eventually crack. Many pipe welding codes require spacer rings.

Figure 16-6. These metal spacing rings provide necessary spacing for expansion and contraction of the weld joint. (Mark Prosser)

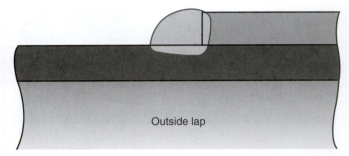

Outside lap

Figure 16-8. Lap joint with the lap on the outside of the pipe.

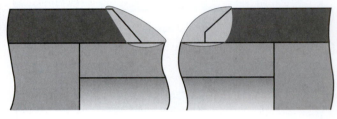

Inside lap

Figure 16-9. Lap joint with the lap on the inside of the part.

Lap Joint Preparation

Lap joints are welded with fillet welds. They usually do not require any special preparation. If the overlapping pipe is positioned on the outside of the joint, the joint is referred to as an outside lap, **Figure 16-8**.

Joints in which the overlapping pipe is positioned inside the joint are referred to as inside lap joints. In these joints, the outer pipe may be beveled, as shown in **Figure 16-9**. Beveling the outer pipe reduces the possibility of cold laps. Cold laps are areas of the weld that have not fused with the base material.

Lateral, T-Joint, and Y-Joint Preparation

These types of joints may require a butt weld, a fillet weld, or a combination of both types of welds. Fillet weld joints, as shown in **Figure 16-10**, usually do not require any special preparation. Butt welds or combination-type welds can be prepared as shown in **Figure 16-11**

Butt Joint Preparation

Standard pipes and fittings are often prepared at the mill with a bevel and a root face. See **Figure 16-12**. Different angles are used with different thicknesses of material. Thicker material requires a larger bevel to allow access to the root of the weld. For open or closed joints, no special joint preparation is required before welding. An *open joint* is a joint in which there is a large gap between the root faces of the joint. In other words, the joint is open at the root. A *closed joint*

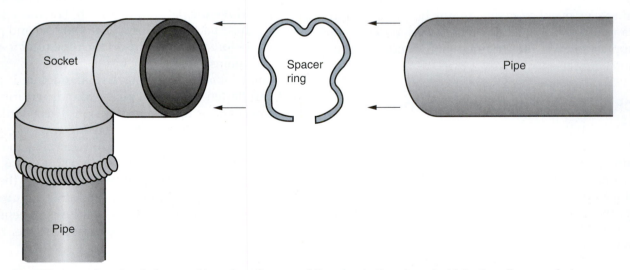

Figure 16-7. The spacing ring is inserted into the elbow, and the pipe is then inserted into the elbow socket.

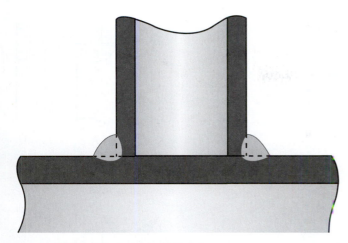

Figure 16-10. Fillet-welded T-joint.

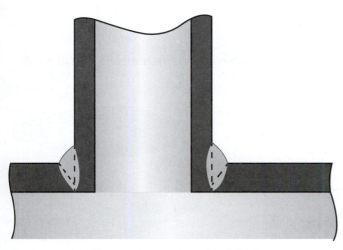

Figure 16-11. Fillet- and groove-welded T-joint.

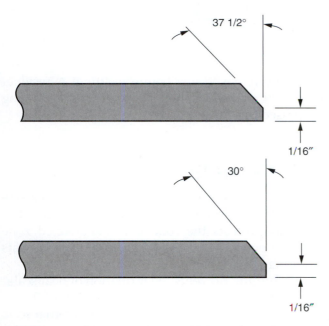

Figure 16-12. Standard pipe weld joint preparations, as prepared at the mill.

refers to a joint that is not open at the root, such as a butt joint with supportive backing.

Special machined joints are designed for a specific purpose and are more expensive to make. The designs shown in **Figure 16-13** are made from fittings, rings, or forgings. The ends of the pipes are machined as needed to make the root faces of the joint line up.

In a consumable insert joint, the end of each pipe is machined to match the profile of the insert installed between them. Various types of joints and inserts are shown in **Figure 16-14**. If a pipe has considerable *ovality*, it must be rounded prior to machining. There are many methods and tools used for removing ovality. If the ovality is not removed, the joint dimensions will not be true around the pipe face. This will cause numerous problems when the root pass (initial pass of a multipass weld) between the insert and the pipes is welded.

Backing ring joints require machining of the pipe's inner walls to fit the backing ring. Various types of backing rings are shown in **Figure 16-15**. As with the other types of joints, the ovality of the pipe must be removed prior to machining. If the ovality is not removed, the pipe wall thickness may vary, or the backing ring may not fit properly during assembly.

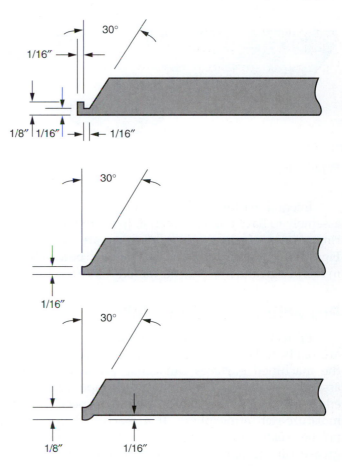

Figure 16-13. Special-purpose weld joints.

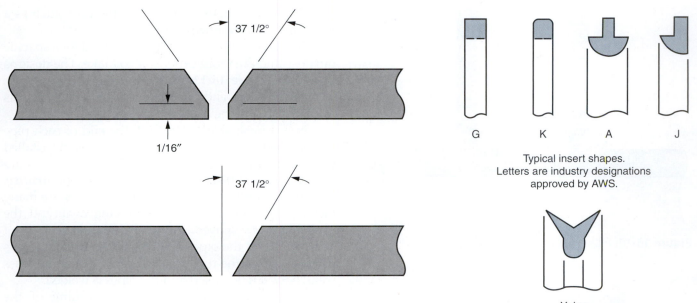

Figure 16-14. Consumable insert shapes and joint preparation. Specifications for consumable inserts are found in AWS A5.30.

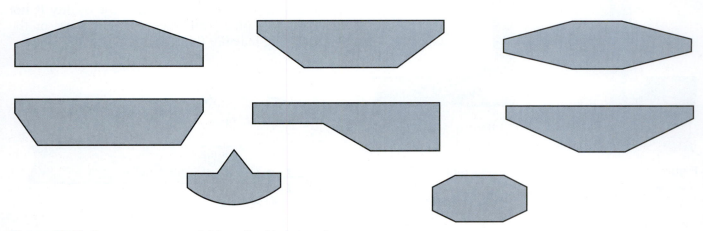

Figure 16-15. Common commercial-type backing ring shapes.

Integral backing joints require one part of the assembly to have a thicker section to make the backing ring. Some of the various designs are shown in **Figure 16-16**. Ovality of the mating pipe can be a problem in this design and must be considered before machining.

Protecting Machined Joints

Protect machined joints from damage until they are ready to be installed. Nicks, burrs, and dents in the machined surfaces can cause problems during assembly, welding, or both. Steel pipe machined edges are very susceptible to rusting, especially in a moisture-rich atmosphere. If a machined joint will not be welded immediately, a light coating of oil or grease can be applied to the machined surfaces to prevent them from rusting. The oil or grease must be thoroughly removed just before welding. A residue-free solvent, such as alcohol or acetone, and a clean rag can be used for this purpose.

Pipe Purging

When GTAW pipe welds are made, the back side of the weld should be purged with an inert gas. The purge gas protects the root and heat-affected zone inside the pipe. The following are some key benefits of purging:

- Purging prevents the formation of heat scale on the inside of steel pipe. If repairs of the root pass are required later, any internal scale could be absorbed into the repair weld. This, in turn, could require another repair.

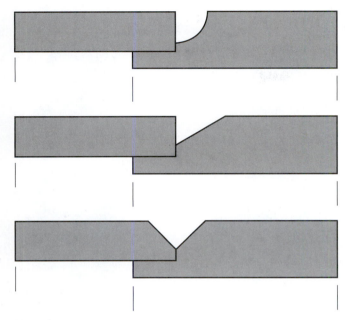

Figure 16-16. Integral backing-type joint designs.

- The inert atmosphere aids in the "wetting" of the penetration weld onto the inner pipe walls on butt joints that do not have backing rings.
- The inert atmosphere assists in the formation of the penetration weld contour on butt joints that do not have backing rings.

The basic gases used for purging oxygen from inside the pipe are argon, helium, and nitrogen. The gas selected depends on the following criteria:

- If the area to be purged is large enough, nitrogen gas can be used for initial purging to remove the majority of oxygen before a complete purge is performed with argon or helium.
- Different materials require different degrees of purging. For example, titanium requires a very pure, oxygen-free atmosphere to prevent contamination. Steel pipe can tolerate a slightly more reactive environment, and sometimes a simple purge with nitrogen is sufficient.
- Weld location is an important factor in selecting a purge gas. Because helium rises and argon settles, helium is a better purge gas for upper welds, and argon is better for lower welds.
- For weld joints with an integral backing bar or a backing ring, nitrogen can be used as a purge gas in some cases (carbon steel, stainless steels).

- Cost is a major factor when selecting purging gases. Nitrogen is less expensive than argon or helium. In many cases, nitrogen is all that is required for proper purging. In situations where it cannot be used for the final purge, it can still be used for initial purging. An initial purge with nitrogen greatly reduces the oxygen level, minimizing the quantity of more expensive gas required for the final purge.

Liquefied gases are often used if a large volume of gases is required. The use of liquefied gases greatly reduces the number of cylinders required and the time lost in replacing empty cylinders.

Purge Dams

Many types of dams or stoppers can be used to keep the purge area as small as possible in order to lower the purging cost. In situations where dams cannot be used, the cost rises accordingly. In each case, the design of the weldment, pipe, or joint should be considered in order to achieve the proper purge in the minimum amount of time.

If only a small number of welds are involved, the simplest method to use is shown in **Figure 16-17**. The ends of the pipe are sealed with masking tape and cardboard, purging is completed, welding is performed, and the caps are removed.

If several joints are to be welded, various types of commercial dams can be used, including soluble dams, thermally disposable dams, cone purge dams, and bladder dams. A hole is made in the dam, and a shielding gas hose is run into the pipe, as shown in **Figure 16-17**. A separate shielding gas cylinder, or the same cylinder with a dual regulator, provides the

Figure 16-17. Masking tape and cardboard are often used to make an effective seal for purging pipe. (Mark Prosser)

Figure 16-18. Special papers and adhesives are used where the dam is to be removed by water. (Junction Tool Supplies Pty. Ltd.)

shielding gas at a low pressure. Too much pressure can have negative effects on the weld joint during welding. Too much pressure can make the arc unstable.

Soluble dams are made of a paper that dissolves in a liquid. They are inserted into the pipe and held in place with a dissolvable adhesive. When welding is completed, the dams are flushed away. **Figure 16-18** shows the proper installation of a soluble dam. This type of dam cannot be used where the pipe is preheated, because the heat will destroy the dam. *Water-soluble films* are special films that are easily applied and form barriers for purging, **Figure 16-19**. These films are very effective and are easily removed by flushing the pipe with water.

Bladder dams, shown in **Figure 16-20**, are small bladders or balloons that are placed into the pipes and expanded to seal the pipe before purging begins. When the weld is complete, the bladders or balloons

Figure 16-19. Liquifilm is water soluble and easy to use. (Junction Tool Supplies Pty. Ltd.)

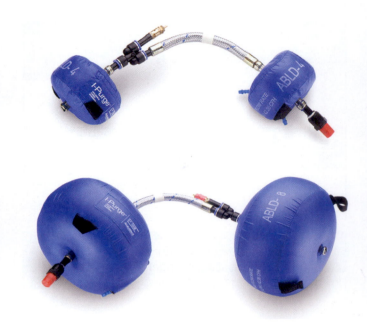

Figure 16-20. These purge dams are available in different sizes, are fast and easy to install, and provide an excellent barrier. (Aquasol Corporation)

are deflated and removed from the area. **Figure 16-21** shows a 12″ bladder inside a pipe.

Root Purging Operation

Purging of the weld root area requires admitting the purging gas and removing the atmosphere until the oxygen content is 1% or lower. To achieve this low oxygen level, purge gas must flow into and out of the purge volume until the total volume of gas in the purge zone has been completely replaced several times.

A high-volume flow at a low pressure uses the least amount of gas to achieve the desired oxygen level. A higher pressure increases the flow rate, but requires more gas to reach an acceptably low oxygen level. The chart in **Figure 16-22** shows typical gas flow rates and purge times for purging various size pipes.

Figure 16-21. A 12″ bladder inside a pipe. (Aquasol Corporation)

Pipe Size (inches)	Purge Inert Gas Flow Rate (cfh)	Purge Time (minutes)	Vent Size (inch)
3.0	20	3	1/16"
4.0	20	3	1/16"
5.0	20	5	1/8"
6.0	20	6	1/8"
8.0	25	8	1/8"
10.0	25	13	1/8"
12.0	30	13	1/8"
14.0	30	16	1/8"
20.0	35	25	1/8"

Figure 16-22. Typical purge times and flow rates for various pipe sizes. Rates are figured for 12" (30.5 cm) long pipe. For purge spaces longer than 12", the purge time should be increased accordingly.

If large-volume spaces are being purged, nitrogen gas is often used to reduce the initial atmosphere level. Argon is then used for the final purge. This method can result in significant cost savings due to the low cost of nitrogen compared to argon.

The amount of purge gas being delivered to the purge area should always be monitored by a flow-meter. Pressure gauges are not sufficiently accurate. Improper purging rates affect the number of gas volume exchanges necessary to reach the required reduced oxygen level.

Oxygen Level Test

Several methods and types of equipment can be used to determine the oxygen level of the purge area. Oxygen analyzers are very reliable and offer good response time in determining oxygen level. An analyzer can be used for spot or continuous checking on single or multiple stations. Oxygen analyzers are often used to make purge charts, similar to the chart shown in **Figure 16-22**. The purge gas flow rate and purge times for different sized pipes can be found on the chart. If the recommendations on the chart are followed closely, an analyzer is not required, except for measuring oxygen levels in very critical situations.

A *gas purge tester* is a simple instrument that uses a special lightbulb with a tungsten filament and extensions for attaching the purge gas hose from the purged area. The gas is introduced into the bulb, and an electrical current is used to heat the tungsten. The color of the tungsten indicates the amount of contamination. This unit does not require elaborate calibration and is considerably less expensive than an oxygen analyzer.

Fitup and Tack Welding

Aluminum and magnesium pipes are welded with ACHF current. Pipes made from all other metals are welded with DCEN current. Information on machine setup and electrode types can be found in the chapter covering the particular base metal. High-frequency arc starting and remote control of the welding current is not mandatory. However, thumb or foot controls for high-frequency start and current give the welder better control of the weld pool.

Fitup (Butt Joint Procedure)

The pipe weld joint should be fit up, if possible, to eliminate the following conditions:

- Out-of-round condition. If the pipe is not machined on the outside and inside, a tool can be used to compensate for the out-of-round by realigning the pipes. See **Figure 16-23**.
- Nonparallel root faces. The faces must be filed or ground to produce an even gap to accept the insert or to make an open-butt root joint.

The pipe insert, if used, should be fit up to eliminate the following:

- Uneven contact on the pipe. Rings must be straight and in contact with the joint to provide proper heat flow for penetration on the root pass. Twisted or bent rings should not be used.

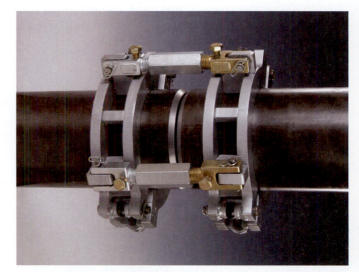

Figure 16-23. Alignment tools are used to remove out-of-round condition and to establish groove openings. (Walhonde Tools, Inc.)

- Improper end gap. Backing rings that do not fit on the inside diameter of the pipe must be split and fitted prior to welding. The ring should be tack welded to the proper diameter before installation. Inserts furnished in coils must always be cut and fit to the proper end gap before installation and tack welding.

Tack Welding Procedure

Prior to any tack welding, the pipe and rings (or inserts, if used) must be cleaned. After cleaning, the rings and purge dams are placed into position and the purge gas flow is started. A gas flowmeter, like the one shown in **Figure 16-24**, can be used to supply gas to both the torch and the purge dam at the same time.

Tack welding can start only after the pipe is purged to the proper atmosphere. Improper purging can cause contamination of the insert or ring. This contamination can result in incomplete fusion, melting, or wetting of the penetration onto the pipe wall.

Tack welds must be feathered at each end to reduce the amount of weld cross section before making the open joint root pass weld. The feathering must be done properly to ensure complete fusion into and out of the tack weld. Do *not* use grinders for this purpose. Grinding wheel residue causes porosity. High-speed mill cutters or tungsten carbide ball routers, like those shown in **Figure 16-25**, should be used.

Open Butt Weld Procedure

Figure 16-26A shows one method that can be used to align and space a butt joint in pipe. Since this joint design uses no inserts, the tack welds will shrink

Figure 16-25. Various types of cutters and routers used to feather tack welds. (Mark Prosser)

after welding. Always allow for this condition by using a larger spacer than the final gap desired.

The tack welds will become part of the root pass. Therefore, sufficient heat and filler metal must be added to form the underbead properly. Improperly made tack welds cause many problems when the root pass weld is made. Pipe joint spacing can also be done using a piece of bent wire, as shown in **Figure 16-26B**.

Procedure for Backing Rings

Figure 16-27 shows small fillet welds securely holding the backing ring in place. The amount of welding wire used should be only enough to hold the ring in place and should penetrate into the joint to prevent cold shuts. **Figure 16-28** shows a cold shut defect, which can result in a defective weld.

Procedure for Inserts

Tack welds are made to hold inserts in place. In most cases, welding wire is not used in tack welding. The required metal is obtained from the insert or the pipe root face. If welding wire is used, the wire composition must match the insert material.

Welding Techniques

Maintaining the torch and welding wire alignment during a pipe weld is very difficult under the best of circumstances. A method called "walking the cup" is commonly used to assist in maintaining the alignment. In this method, the edge of the torch gas nozzle is rounded off by grinding, and the cup is then kept in contact with the joint bevels during the welding. During welding, the torch is oscillated, and

Figure 16-24. A dual flowmeter allows the welder to monitor both the torch and purge gas at the welding station during the entire operation. (Mark Prosser)

A

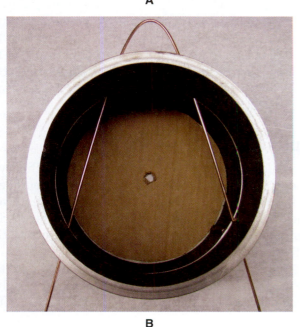

B

Figure 16-26. Aligning pipe joints. A—Assembly tools for aligning pipe joints. (Walhonde Tools, Inc.) B—Tack weld in the open V-groove, then remove the bent wire. Next, align space and tack weld 180° from first tack weld. Realign the joint and tack weld midway between the previous tack welds. (Mark Prosser)

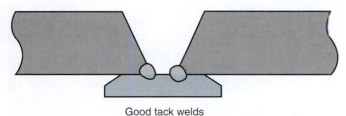

Good tack welds

Figure 16-27. Small fillet tack welds adequately hold backing rings in place.

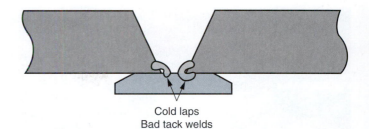

Cold laps
Bad tack welds

Figure 16-28. Large fillet tack welds may result in cold shuts between the ring and the pipe joint weld.

at each outward swing end, the gas nozzle is "walked" forward along the joint. The welder can increase or decrease the amount of welding current by raising or lowering the electrode from the weld. A dwell or hold point at the end of the oscillation also decreases the heat in the weld pool, which allows some freezing of the weld. **Figure 16-29** shows the preparation of the gas nozzle and the placement of the electrode tip relative to the nozzle.

The placement of the torch and welding wire for fixed uphill welding is shown in **Figure 16-30**. The placement of the torch and welding wire for fixed horizontal welding is shown in **Figure 16-31**.

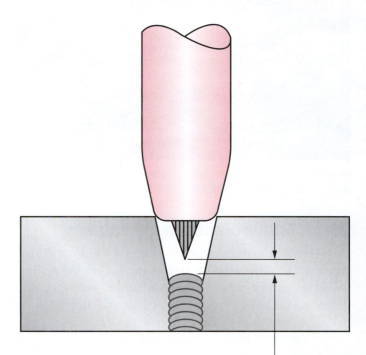

Approximately 3/32" arc length

Figure 16-29. Gas nozzle preparation and tungsten extension from the nozzle for "walking the cup" welding technique.

Figure 16-30. Torch and wire positions for fixed position, uphill welding. (Mark Prosser)

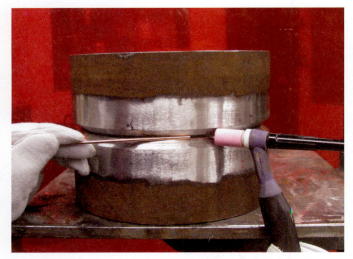

Figure 16-31. Torch and wire positions for fixed position, horizontal welding. (Mark Prosser)

Torch Manipulation with Pulsers

The need to manipulate the torch is greatly reduced when pulsers are used. A pulser raises and lowers the welding current automatically to preset levels, which decreases the need for manual oscillation of the torch.

The operation of the pulser includes:

- current high level (molten weld pool forms)
- high current level dwell time (welding wire added)
- low current level (molten weld pool freezes)
- low current dwell time (torch moved)

With this type of control, the weld is formed as a local spot weld. The torch is then moved the correct distance (about one-half the diameter of the spot weld) forward to form the weld again. The completed weld is a series of overlapping welds.

The sequencing of the pulser is established by testing to ensure that it produces adequate fusion into the root of the joint. Welder travel speed is then synchronized with the pulser operation.

Pulsers are built into many power sources. They can also be purchased separately and added to the system as an auxiliary item. Pulsers can be used to weld an entire pipe joint, regardless of the joint design. However, pulsers are generally used to their best advantage to make root passes.

Backing Ring Weld Root Passes

Welds of this type are made with the addition of welding wire. The addition of welding wire increases the cross section of the root pass to prevent cracking. The welding wire should always be added into the front of the weld pool to prevent cold shutting of the joint edges and the backing ring. A slight oscillation of the torch ensures adequate melting of the sidewalls into the backing ring.

Open Butt Weld Root Passes

Filler metal application is required in this type of joint to establish the root pass with sufficient metal to form the penetration contour and prevent weld cracking. Because metal thinning occurs during cooling of the weld metal, an insufficient cross section of the weld will often cause cracking.

Filler metal diameter is very important in this operation. The welding wire diameter must be slightly larger than the opening, as shown in **Figure 16-32**. The

Welding wire

Figure 16-32. The wire is slightly larger in diameter than the groove opening so the wire rests on the bottom of the V-groove walls. (Mark Prosser)

wire should be applied directly in the center of the root opening. Both sidewalls and wire are melted together with a slight oscillation movement. The wire should not be moved from the pool because this disturbs the proper flow of metal, and the cross section of the weld will vary. **Figure 16-33** shows a root pass weld with sufficient cross section and an undersized weld with a shrink crack.

Insert Ring Weld Root Passes

Weld joints with inserts rely on the melting of the insert to provide sufficient root pass cross section metal. The addition of welding wire disturbs the heat flow, with possible incorrect melting of the insert. In all cases, the insert ring must be melted completely to form the proper contour. A slight oscillation of the torch can be used to ensure adequate heating of the joint edges and proper fusion. As with any open butt joint, the weld is required to have complete penetration. The inspection criteria will determine allowable discontinuities.

Roll Welding Root Passes (1G Flat Position)

If a pipe is in a fixed location, a welder often needs to move around the pipe, welding in different positions to complete the weld joint. Sometimes a pipe can be rolled, continuously or by steps, to allow the welder to complete the weld in a single position. **Figure 16-34** shows a pipe that is being continuously rotated so the welder can complete the entire weld joint in the flat welding position.

The torch should be held at the angle shown in **Figure 16-34**, and the electrode should always be directed to the centerline of the pipe joint. The rotation speed should be maintained to keep the torch on the uphill part of the joint.

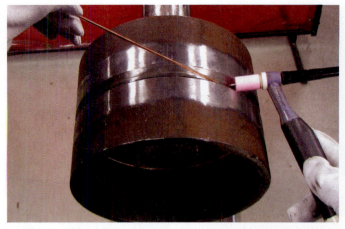

Figure 16-34. Torch and wire alignment for welding pipe, which is rolled during the operation. (Mark Prosser)

Horizontal Welding 2G (Pipe Vertical) Fixed Root Passes

Welding vertical pipe requires very close control of the torch angle and the addition of the proper amount of welding wire at the right time. Because of gravity, the molten metal will tend to sag, as shown in **Figure 16-35**. Therefore, the welds should be made as cool as possible, with the lowest possible amperage. Oscillation should be avoided. Sufficient filler metal should be added to freeze the molten pool each time wire is added. Stringer weld beads should be made.

Vertical Welding 5G (Pipe Horizontal) Fixed Root Passes

If the pipe is horizontal and the weld is being made in the vertical welding position, the weld should be started near the bottom of the pipe and made uphill. A slight oscillation movement of the torch ensures proper melting of the sidewalls for complete fusion.

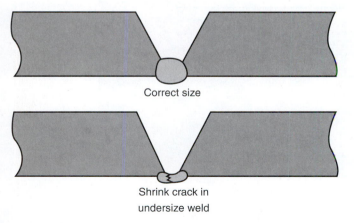

Figure 16-33. Weld root passes with insufficient cross section may crack during or after welding.

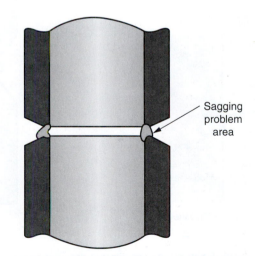

Figure 16-35. Horizontal in-position groove weld root pass problem areas are caused by gravity sagging the weld.

Two major problems encountered with open or insert joints welded in this position are a tendency to form a concave root surface on the bottom of the joint and excessive penetration at the top of the joint. These problem areas are shown in **Figure 16-36**. As the weld is made at the bottom, the molten weld pool must be chilled by adding filler metal or by adjusting the torch motion to the sidewall. This freezes the weld pool for a moment, preventing or reducing the concavity of the root surface.

As the welder works upward, the entire joint heats, which requires faster travel speed and faster application of the welding wire. The heat rising in the joint makes the weld pool very fluid, causing over-penetration. Oscillation of the torch to the joint sidewalls aids in cooling of the molten weld pool and diminishes the over-penetration.

Flat insert rings for this type of joint can be installed offset in order to reduce problem areas. **Figure 16-37** shows the flat ring installed so the bottom of the ring has more material inside of the pipe, and the top of the ring has less material inside the pipe at the top of the joint.

Multipass Welds

The initial pass of a multipass weld is referred to as a *root pass*. Intermediate passes are referred to as *hot passes*. The final passes made are referred to as *cap passes*. See **Figure 16-38**.

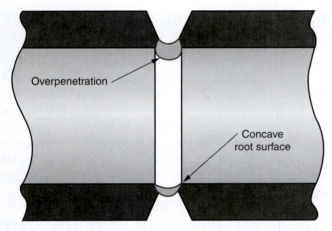

Figure 16-36. Vertical in-position groove weld root pass problem areas.

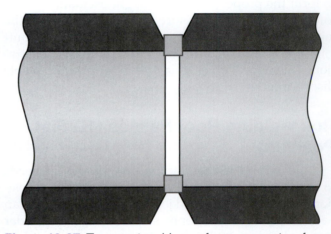

Figure 16-37. To correct problems of concave root surfaces and over-penetration, the flat insert ring is placed so there is more material inside the pipe at the bottom than at the top.

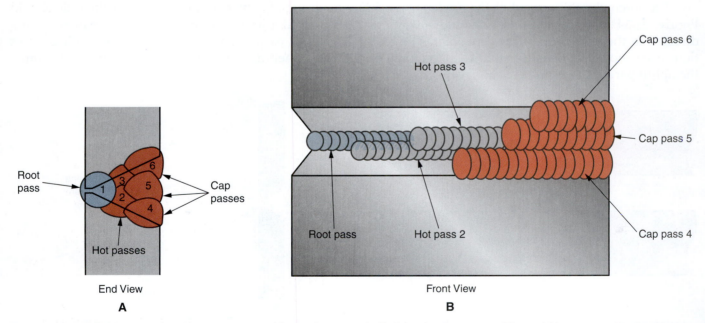

Figure 16-38. Each pass of multipass crown welds overlaps one-half of the previous pass. The weld layers must be kept level. When welding in the horizontal position, a new layer is always started at the bottom of the joint. A—End view. B—Front view.

When laying beads in the joint, the welder must make sure that each pass has proper fusion with the previous passes and with the sidewalls of the joint. Weld beads in the horizontal position should always be made with a slight upward work angle to prevent sagging. The first hot pass in a multipass weld should be placed at the bottom edge of the root pass. Each subsequent pass should be placed above the previous pass. Each weld bead needs the support of the previous weld bead. Adding too much welding wire will cause excessive crowning and possibly a lack of fusion at the joint interface. The pass placement for vertical and horizontal position welds is shown in **Figure 16-39**.

The total number of weld beads depends on the welding procedure and, more importantly, the thickness of the material being welded. Horizontal (2G) position welds are made with stringer beads, while vertical (5G) welds are made with weave beads. The weave bead technique usually requires fewer passes to fill the joint.

Crown welds should flow evenly into the pipe walls, without undercut or excessive height. For specific welding procedure information, refer to the chapter pertaining to the base material to be welded.

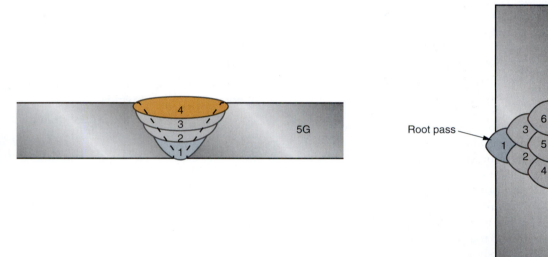

Figure 16-39. Sequence of passes and layers for filling groove joints and making crown welds. When welding in the horizontal position, always start a new layer at the bottom of the joint. Vertical position welds are done with the weave bead technique.

Summary

Joint designs include socket-type joints, lap joints, lateral joints, T-joints, Y-joints, and butt joints. The different types of joints used in pipe welding require somewhat different preparation processes.

It is critical that the joint be properly prepared and fit up before tacking and welding begins. Socket-type joints require a preweld minimum 1/16″ (1.6 mm) gap at the end of the pipe into the socket bore end. For butt joints, pipe ovality must be corrected prior to machining, or the backing ring may not fit properly during assembly. Machined joints must be protected from damage until they are ready to be installed.

The back side of GTAW pipe welds should be purged with an inert gas to prevent the formation of heat scale, aid in the wetting of the penetration, and assist in the formation of the penetration weld contour on butt joints without backing rings. Gases used for purging oxygen from inside a pipe are argon, helium, and nitrogen. Purge dams can be used to keep the purge area as small as possible in order to lower the purging cost. Types of purge dams include soluble dams, thermally disposable dams, cone purge dams, and bladder dams.

Purging gas is delivered to the weld root area and the atmosphere is removed until the oxygen content is 1% or lower. The amount of purge gas being delivered should be monitored by a flowmeter. Oxygen analyzers or gas purge testers can be used to determine the oxygen level of the purge area.

The pipe weld joint should be fit up, if possible, to eliminate out-of-round condition, nonparallel root faces, uneven contact of the pipe, and end gap. The pipe and rings (or inserts) must be cleaned prior to tack welding. Tack welding can start only after the pipe is purged to the proper atmosphere. Tack welds must be feathered at each end to reduce the amount of weld cross section before an open joint root pass weld is made.

Torch alignment when welding pipe is aided by a method called "walking the cup." The need for manual oscillation of the torch is decreased with the use of a pulser.

For backing ring weld root passes, welding wire is added to increase the cross section of the root pass and prevent cracking. For open butt weld root passes, the welding wire diameter must be slightly larger than the opening. Weld joints with inserts require the complete melting of the insert to provide adequate root pass cross section metal.

Pipes in fixed locations can sometimes be rotated to allow the welder to complete the weld in a single position. Welds made in the vertical position should be made as cool as possible, with the lowest possible amperage to help prevent sagging of molten metal.

For vertical welding on a horizontal pipe, the weld should be made uphill. The two main problems encountered with open or insert joints welded in this position are a concave root surface on the bottom of the joint and excessive penetration at the top of the joint. Flat insert rings can be installed offset in order to reduce problem areas.

In multipass welds, each weld bead needs the support of the previous weld bead. Weld beads in the horizontal position should be made with a slight upward work angle to prevent sagging. Horizontal (2G) position welds are made with stringer beads, while vertical (5G) welds are made with weave beads.

Review Questions

Write your answers on a separate sheet of paper. Do not write in this book.

1. List six factors involved in selecting pipe weld joint designs.
2. What type of joint is used to join two pipes with a coupling or sleeve?
3. What is a common cause of weld root cracking in socket-type joint welds?
4. The outer pipe of inside lap joints may be beveled to reduce the possibility of _____.
5. Standard pipes and fittings are prepared at the mill with a(n) _____ and a(n) _____.
6. Why must pipes with ovality be rounded?
7. Purging with an inert gas prevents the formation of _____ on the inside of steel pipe.
8. List the three gases used for purging oxygen from the inside of pipe.
9. For purging of large areas, which gas is the cheapest to use for the initial purging?
10. In what situation are liquefied gases often used?
11. What is the purpose of purge dams?
12. List four types of commercial purge dams.
13. Purging the weld root area requires reducing the oxygen content of the atmosphere to _____ or lower.
14. For purging the weld root area, a(n) _____-volume flow at a(n) _____ pressure uses the least amount of gas to achieve the desired oxygen level.
15. List two types of equipment used to determine the oxygen level of the purge area.
16. What operation must be done on open butt joints after tack welding but before root pass welding?
17. The use of a(n) _____ decreases the need for manual oscillation of the torch.
18. Is a filler metal used for welding root passes with inserts?
19. What are the two main problems encountered when welding root passes on a vertical weld?
20. List three things a welder can do to minimize sagging of molten metal when welding pipe in the vertical position.

Chapter 17

Semiautomatic and Automatic Welding

Objectives

After completing this chapter, you will be able to:

- ❑ Differentiate between semiautomatic and automatic welding.
- ❑ Recall the types of welds made and the materials that can be welded with a semiautomatic welding system.
- ❑ Explain the components of semiautomatic welding equipment, including the power source, arc voltage control, and travel control.
- ❑ Identify the various accessories that can be added to a semiautomatic welding system.
- ❑ Explain semiautomatic long seam welding systems, rotational welding systems, and spot welding systems.
- ❑ Recall welding techniques for tube-to-tube welding, tube-to-sheet welding, and pipe-to-pipe welding.
- ❑ Summarize the reasons for performing a weld test.
- ❑ Summarize the parameters that are recorded in welding schedules.
- ❑ Identify potential semiautomatic welding problems and explain how they can be corrected.
- ❑ Recall the equipment needed for automatic welding.
- ❑ Compare the benefits and costs of automatic welding.
- ❑ Explain the importance of joint tolerances and tooling in an automatic welding program.
- ❑ Explain how a welding schedule for an automatic welding operation is maintained.

Key Terms

automatic voltage control systems
automatic welding
cold wire feeder
full tolerance
hot wire feeder
mechanical voltage controls
oscillators
pulsers
semiautomatic welding
welding operator
welding schedule

Introduction

GTAW is a very versatile process that includes manual, semiautomatic, and automatic applications. In manual welding, the welder controls all aspects of the welding process, including positioning the torch, adjusting welding current, and adding filler metal.

In *automatic welding*, the welding is performed entirely by a machine of some type, usually a robot. The operator sets the parameters, and the machine tracks the torch, adds the filler metal, and maintains proper travel speeds, torch angles, and other variables. Automatic welding is most often used for high-production welding or where high-quality, repetitive welding is required. Advantages include improved quality, increased productivity, and reduced cost for most applications.

In *semiautomatic welding*, at least one aspect of the welding process is automated. However, the *welding operator* still maintains manual control of at least one aspect of the welding process. For example, filler metal may be added automatically, the torch may be mounted on a carriage or manipulator, but the welding operator may still directly control the welding speed and current. This would be an example of semiautomatic welding.

Semiautomatic Welding

Semiautomatic welding equipment can automatically adjust for various conditions or can be manually adjusted by the welding operator. Controls can be preset by the operator, thus allowing the welding sequence to take place without manual direction. It is the welding operator's responsibility to operate the equipment in a sequence and manner that produces satisfactory welds. Semiautomatic welding procedures are used to make individual or production-type welds.

Types of Welds

Two types of welds made semiautomatically are long seam welds and rotational welds. Seam welding is the welding of two faying surfaces of the same types of metal. This procedure allows for long straight welds with little down time and good deposition.

Rotational welding is done by rotating a tube while the torch mechanism remains stationary, usually in the flat position. Newer equipment carries the torch around a stationary tube or pipe.

Long Seam Welds

Several long seam welds that can be made using semiautomatic welding are shown in **Figure 17-1**. These joints are usually made in the flat position. Except for fillet welds, all the joints can be made without difficulty. Fillet welds require penetration at the root of the joint. However, it is difficult to properly align the weldment with the electrode tip to produce penetration at the root. **Figure 17-2** shows the effect of electrode misalignment to the joint centerline. Welding speed can be obtained by either the movement of the weldment or the movement of the welding torch, as shown in **Figure 17-3**.

Rotational Welds

As in long seam welding, the majority of rotational welding is done in the flat position. Fillet welds are difficult due to the joint alignment problem. The proper welding speed is attained by adjusting the rotational speed of the weldment on a fixture or a positioner.

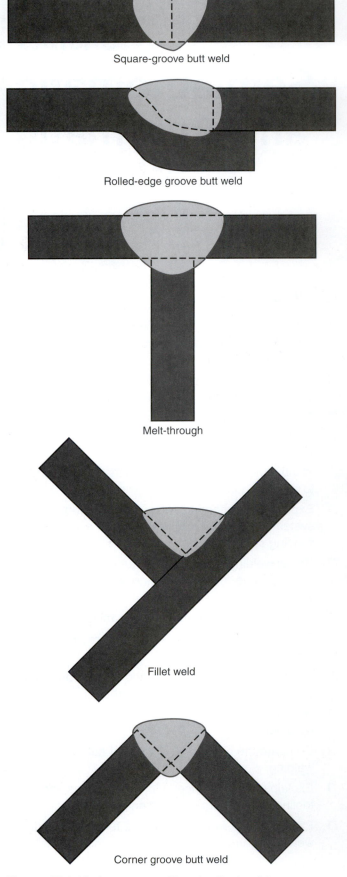

Square-groove butt weld

Rolled-edge groove butt weld

Melt-through

Fillet weld

Corner groove butt weld

Figure 17-1. Various types of longitudinal welds.

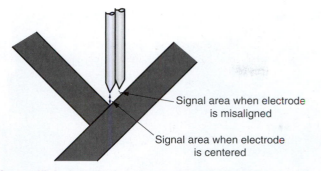

Figure 17-2. Automatic voltage-controlled welding heads will not operate properly unless the joint is properly aligned.

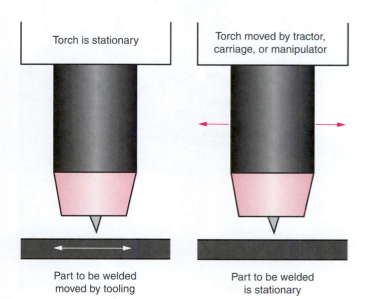
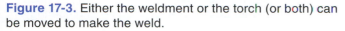

Figure 17-3. Either the weldment or the torch (or both) can be moved to make the weld.

Types of Materials

Virtually all materials that are welded manually can be welded with a semiautomatic system. In addition, aluminum, which is normally welded by ACHF, can be welded with DCEN polarity and helium gas if the arc length is automatically controlled. Thin-gauge materials are often welded by semiautomatic equipment using DCEP polarity. The welding equipment automatically adjusts travel speeds to prevent melt-through.

Equipment

The equipment required for semiautomatic welding, whether long seam or rotational, can range from simple to very complex. Major components include the power source, arc voltage control, and travel control.

Figure 17-4. Remote amperage controls on this manipulator allow adjustment by the operator. (Jetline Engineering)

Power Source

Power sources used for semiautomatic welding must be able to produce the desired welding current polarity. They must also produce the desired amperage for the weld and be capable of maintaining the current for the desired length of time (duty cycle). Lastly, semiautomatic welding power sources are usually equipped with remote controls that allow the welding operator to start the welding operation, adjust the current, and end the welding operation from outside the welding site. See **Figure 17-4**.

Arc Voltage Control

Mechanical voltage controls are used where the arc length is fixed and requires only occasional adjustment. The torch is moved up or down by adjusting screws or levers. The arc gap can be set using set blocks, dial indicators, etc., as shown in **Figure 17-5**.

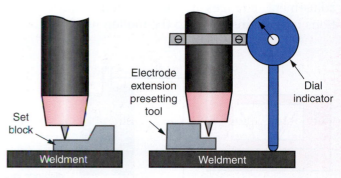

Figure 17-5. This drawing shows two methods of setting the electrode to a predetermined height from the weldment surface. A—A set block can be used to set the tip of the electrode and the bottom of the gas cup a predetermined distance from the workpiece. B—The electrode extension is set with a set block and the torch is moved to the desired height, as shown on the dial indicator. The dial indicator must be rotated out of welding zone before welding.

If continuous changes are required during welding, manual voltage controls can be used in conjunction with an arc voltmeter. The voltmeter is installed into the system across the arc gap, as shown in **Figure 17-6**. The operator then raises or lowers the torch as required to obtain the desired voltage.

Automatic voltage control systems, commonly called *AVC systems*, are used in automatic and semiautomatic welding operations to control the arc voltage (electrode to work gap) to a preset condition. AVC systems automatically and continuously adjust the distance between the workpiece and the electrode to maintain a preset gap. Where the arc gap may vary due to misalignment problems, AVC systems relieve the operator of the burden of constantly making adjustments. The equipment includes a control unit and a drive unit on which the machine welding torch is mounted. **Figure 17-7** shows an AVC head and controller.

An AVC head is a must for high-quality welds where arc voltage is a critical factor and tooling or driving mechanisms cannot be set with precision. The major advantages of AVC systems are as follows:

- Welding can be done at higher speeds.
- Welds can be made on uneven surfaces.
- Set-up time is minimized.
- Less operator skill is involved.
- Electrode deterioration is automatically compensated for.
- High-quality welds are produced.

Using an AVC system with cold wire feed is another advantage. (Cold wire feed is discussed later in the chapter.) Because the welding operator can set the wire feed manipulator to the desired height, the wire will stay in the proper relationship to the electrode and the molten weld pool. With manual voltage adjustment, changes in the arc length cause variations in the wire feed into the molten weld pool. This

Figure 17-7. Automatic-voltage-controlled welding head with sequencer for pulser lockout. (Jetline Engineering)

requires constant wire feed manipulator adjustments to compensate for the changes in arc length.

The major sequence of events in an AVC welding operation is as follows:

1. Welding operation starts.
2. Drive motor moves the welding head to the work at a preselected speed.
3. Electrode touches the workpiece and retracts to set the gap distance.
4. High-frequency voltage fires and starts the arc.
5. Voltage control sets the electrode-to-work gap and maintains preset voltage during the weld operation.
6. Welding operation stops.
7. Drive motor automatically retracts the welding head a set distance after the arc goes out.

The head operates by sending an electrical signal from the electrode to the workpiece. Receiving the signal, the head then responds to a preset voltage and drives the head up or down as required. The unit

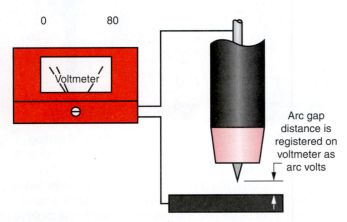

Figure 17-6. Voltmeters can be used to determine arc gap between the electrode tip and the weldment.

operates very well on butt welds. However, the signal may be deflected on fillet welds, as shown in **Figure 17-2**, if alignment is not perfect. Fillet weld joints designed as shown in **Figure 17-8** prevent this problem, since the point of signal contact is close to the electrode tip.

Travel Control

Control of welding speed is important in GTAW. If the speed is not consistent, the weld will have uneven penetration and uneven reinforcement along the length of the weld seam. Two methods of drive and control systems are in common use.

Alternating current motors and controller systems include ac motors operating at a single speed with a mechanical reduction to the required welding travel speed. The controls regulate the mechanical reduction section by changing the linkage between the electrical motor and the gear box.

Since utility power ac voltage input varies considerably, the electrical motor may not operate at a consistent rpm. This causes the weld travel speed to vary with each fluctuation in input voltage. This variation can affect the weld quality.

Direct current motors and controller drive systems do not rely on mechanical reduction because they use a variable-speed 12-volt or 24-volt electric motor. The motors are highly efficient and easily adjustable throughout a specific range by a potentiometer located in the controller. This controller is also known as a *governor*.

Accessories

Accessories are added to a basic system to improve the quality of the weld and to assist the welding operator during the operation. The accessories include automatic voltage controls, cold wire feeders and manipulators, hot wire feeders, oscillators, pulsers, and timers and slopers. Automatic voltage control systems were described earlier in this chapter. Other equipment was described in detail in Chapter 3, *Auxiliary Equipment and Systems*.

Cold Wire Feeder and Manipulator

A *cold wire feeder* supplies welding wire into a weld joint to maintain the weld profile. The feeder consists of a mount for the weld wire spool, speed governor, dc electrical motor, gear box, wire feed drive rollers, wire guide manipulator, conduit to the manipulator, and adjustment screws to move the manipulator. A complete cold wire feeder unit is shown in **Figure 17-9**. Power required for the unit is 110 volts ac, which is converted to 12-volt or 24-volt dc for operation of the motor in the drive box.

The manipulator is used to adjust the wire fed into the weld pool through the wire tip guide. The wire tip guide must be the proper size for the wire diameter used. When hard wires are used, these guides wear out quickly and need frequent replacement.

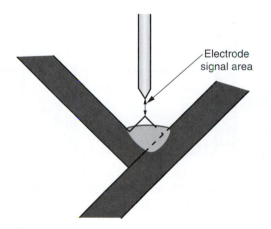

Figure 17-8. In this joint, the top workpiece has a small flange that melts to create the weld bead. Fillet welds designed in this manner have compatible base materials and do not need separate weld metal. This type of weld works well for semiautomatic welding.

Electrode signal area

Figure 17-9. Cold wire feeder system. (Jetline Engineering)

Cold wire feeders are used in manual, semi-automatic, and fully automatic welding operations where welding wire is required. The wire feed drive controller shown in **Figure 17-10** is powered by 110-volt ac. It can be used for manual start-and-stop operation by the welding operator or for automatic sequencing during the weld operation. Solid-state controllers with digital readouts control wire speed with reliability.

Another type of wire feeder, **Figure 17-11**, can be used for machine welding or adapted to manual welding. Drive rollers on these types of feeders are usually designed for one diameter of either a hard or soft wire. Hard wires require V-groove rollers, while soft wires require U-groove rollers. See **Figure 17-12**. The rollers must be changed every time the type or diameter of wire is changed. Conduits with liners are used to feed the wire from the feeder to the manipulator. The conduits must be replaced when a different diameter wire is used or when the conduits become excessively worn. Conduits should not be sharply bent. A sharp bend restricts the passage of wire and causes the liner to wear very rapidly.

Hot Wire Feeders

A *hot wire feeder* supplies preheated welding wire into a joint to maintain the weld profile. A basic *hot wire feeder system* is similar to a cold wire feeder system, except the wire is electrically resistance-heated to the melting point prior to entering the weld pool. The preheated wire has very few surface impurities because they are burned up during the heating.

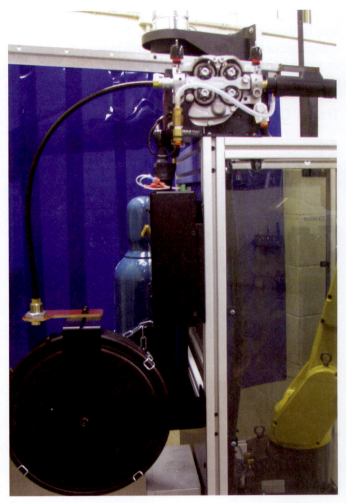

Figure 17-11. Wire feeders can be used for manual or automatic feeding of welding wire. All of the control and drive components are assembled in one unit. (Mark Prosser)

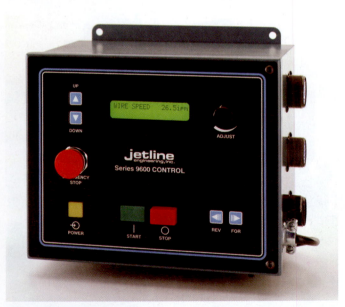

Figure 17-10. This wire feed drive control has a retract timer to draw the wire out of the molten weld pool after welding ceases. This prevents the wire from freezing into the weld metal. (Jetline Engineering)

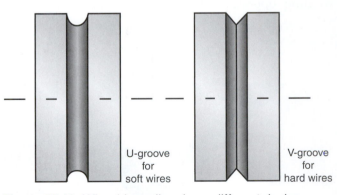

U-groove for soft wires

V-groove for hard wires

Figure 17-12. Wire drive rollers have different design grooves for hard or soft wires. Using incorrect grooves causes wire feeding problems.

Weld quality is very good, since impurities are not fed into the weld pool.

A hot wire feeder system is used only on semiautomatic and automatic operations. A hot wire feeder system has two major advantages over a cold wire feeder system:

- **High welding wire deposition rate.** The increase in welding wire deposition rates lowers production time and the cost of fabrication.
- **Excellent weld quality.** Because the welding wire is heated to its melting point before the wire is inserted into the molten weld pool, any surface contaminants, such as oil, grease and drawing lubricants, are driven off. As a result of the heating, the source of hydrogen porosity is removed, and the completed welds are virtually free of porosity.

A basic hot wire feeder system is shown in **Figure 17-13**. Equipment includes a high speed welding wire feeder, contact tube and holder, and an ac constant voltage power source for melting the wire. A direct current power source cannot be used for a hot wire feeder system.

Oscillators

Oscillators are used on semiautomatic and automatic torches to direct the arc back and forth laterally across the weld seam as the weld progresses. This oscillation controls and improves the weld profile and agitates the weld pool. The agitation allows gases to rise to the surface, thus reducing porosity in the weld.

Oscillation provides the following advantages:

- controls heat distribution and reduces weld shrinkage

- minimizes undercut
- reduces porosity
- improves penetration
- improves overall integrity of the weld
- improves weld contour

Oscillators can be used for all types of materials and can be either mechanical or magnetic. **Figure 17-14** shows some of the different patterns that can be achieved by adjusting the oscillator control settings.

Pulsers

Pulsers, either built into the power source or portable units added to a system, change welding current from one level to another for a set period of time. The pulsing operation develops enough heat for making the weld, and then drops to a lower setting, which allows the weld to solidify.

Since pulsers are set for high and low current levels, the addition of welding wire should also be pulsed and synchronized with the amperage pulse. This requires additional equipment. It is often difficult for a welding operator to properly start each sequence manually. For this reason, the joint design should not require welding wire when the joint is being welded semiautomatically using pulsers. Pulsing the wire feed and travel speed in synch with current is usually restricted to full automatic welding, where the sequencing can be electronically controlled. For more information about pulsers, review Chapter 3.

Timers and Slopers

Timer and sloper controls are often used to time weld current to and from one level to another level. An example is starting a spot weld at low amperage, increasing current to welding level, holding for a period of time, and finally decreasing the current to zero.

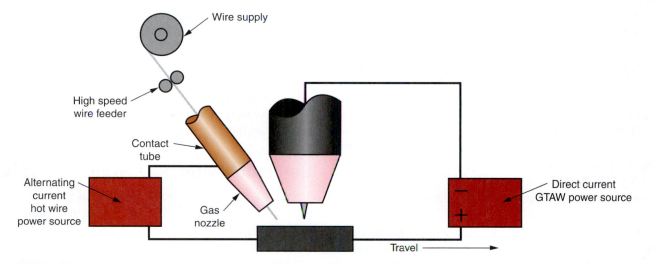

Figure 17-13. Hot wire feeder systems have a very high deposition rate. They are used only in semiautomatic and automatic welding systems.

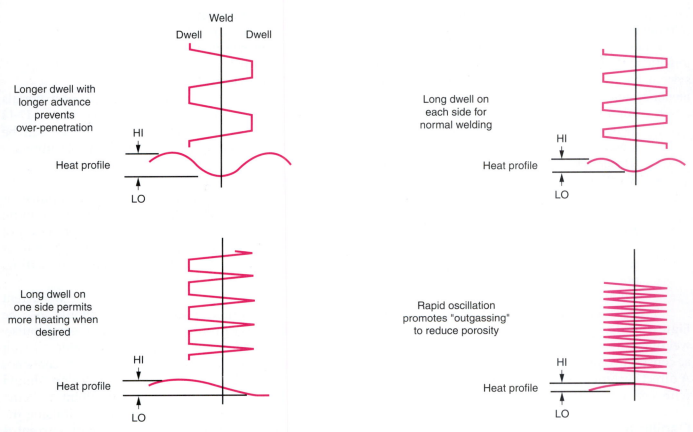

Figure 17-14. Oscillation patterns are designed to achieve specific results.

Operators of equipment without proper timers and slopers cannot be trained to produce welds with the level of accuracy obtained with these devices. For more information about timers and slopers, review Chapter 3.

Semiautomatic Long Seam Welding System

A semiautomatic long seam welding system, such as the one diagrammed in **Figure 17-15**, can be used on all types and thicknesses of metal. A tractor for longitudinal travel is added to move the welding torch along the weld joint. The arc gap is maintained along the joint by raising or lowering the torch with a gear-and-rack adjustment wheel.

A wire supply and feeder assembly can be used in this system. The addition of material is for filling groove weld joints and making a crown on the top of the weld. Longitudinal welds in cones, cylinders, channels, and tubes are welded using this type of system.

Long seams are an excellent application for semiautomatic welding systems because this type of system can drastically increase production rates and high-quality repeatability. It is very beneficial for

manufacturing firms to use semiautomatic long seam welding systems when applicable.

Semiautomatic Rotational Welding System

Semiautomatic rotational welding systems are used to make welds between round parts, such as rings and cylinders. See **Figure 17-16**. The machine welding torch is mounted above the part. Manual adjustments maintain the proper gap between the torch and the part.

A positioner rotates the part at a set speed under the torch during the weld. Tilt-table positioners are used where the weld joint must be angled to the welding torch. A wire supply and feeder assembly can also be added to this system.

Semiautomatic Spot Welding System

Spot welds can be used to hold mating parts together for assembly. Spot welding can also be the primary welding process in a fabrication operation. For example, many car bodies are put together almost entirely with spot welds. Spot welds are commonly made using resistance welding equipment, but the

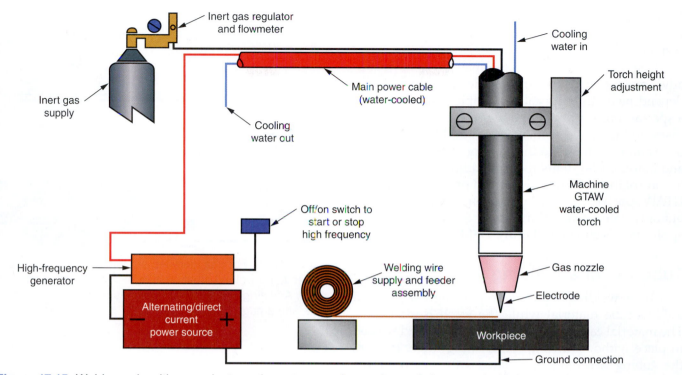

Figure 17-15. Welds made with a semiautomatic system can be made much faster and with fewer defects than manual welds. A wire supply and feeder assembly provides filler metal.

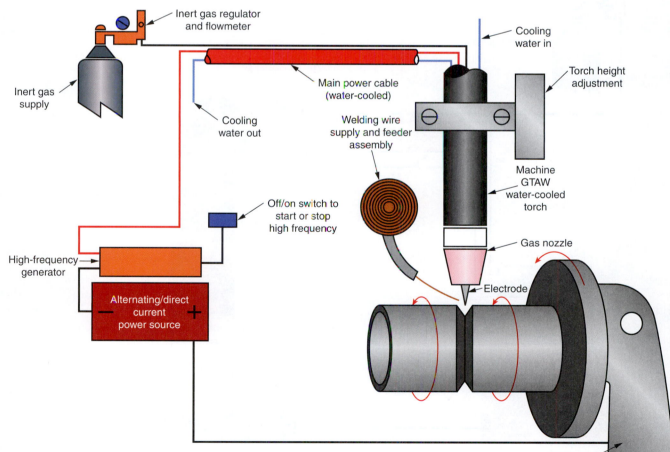

Figure 17-16. Positioners are used to position and rotate circular parts at an accurate speed for semiautomatic rotational welding.

GTAW process can be adapted to make spot welds as well. **Figure 17-17** shows a basic GTAW spot welding system.

The spot welding process may be done manually, semiautomatically, or completely automatically, depending on the application. Simple systems employ a special torch, a timer, and a power source. In some cases, additional equipment is required to start the arc or to add filler metal to the weld pool. Semiautomatic and automatic systems require additional equipment to control the welding current and program sequence. GTAW spot welding can be done on all types of metals, but only on relatively thin pieces. A GTAW spot weld application is shown in **Figure 17-18**.

Tube-to-Tube Welding

Tube welding equipment is designed to weld around tube material while the torch is stationary. The material can be prepared, assembled, and welded in place without turning the welding torch. Instead, the tubing assembly is rotated in relation to the

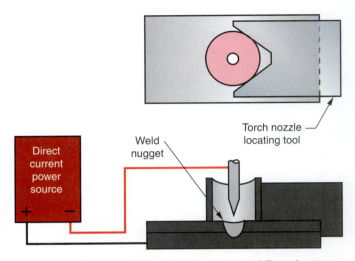

Figure 17-18. For gas tungsten arc spot welding, the two pieces to be welded must be in tight contact with each other. Any air gap between the pieces reduces the heat flow, and a nugget will not form.

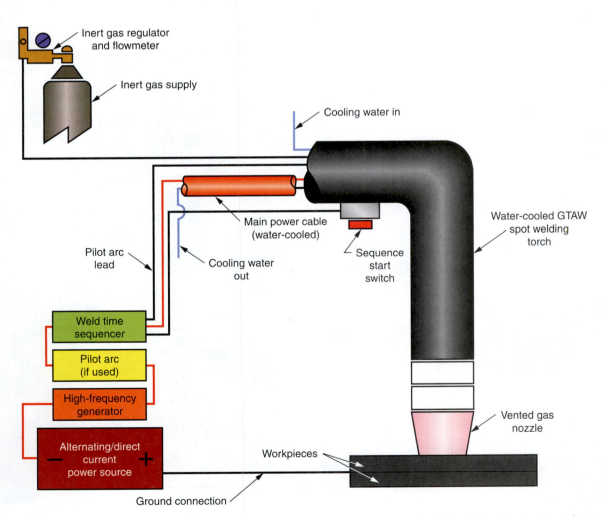

Figure 17-17. Overlapping materials are spot welded by heating the top plate and melting it into the bottom plate.

torch. Tube-to-tube intersections can be made in a variety of ways using different types of connecting configurations.

Material can be repaired in place by cutting out a section and replacing it with a new section. Sections can be replaced or components of new configurations added without completely removing the old system. In many areas of construction, mechanical connections, such as the bolting of tubes together with flanges, cannot be tolerated. These connections sometimes need to be removed and new sections of pipe welded in their place.

Tube welding systems operate on all types and thicknesses of tubing; however, each weld must be assembled to a specific tolerance. Welding parameters must be entered into the weld programmer before the welding operation starts. Typical joint designs for tubing welds are shown in **Figure 17-19**.

Major equipment components for tube welding include the power source, programmer, and the welding head. The following steps are required to complete a weld with a tube-welding system:

1. Weld parameters are entered into the programmer.
2. The torch is mounted over the tube weld joint.
3. Welding current levels are set in the power source.
4. Starting the cycle automatically sequences all of the events required for the weld.
5. The unit shuts down when the sequencer times out.

Tube-to-Sheet Welding

Boilers, condensers, and heat exchangers require the welding of tube to sheet as the major method of fabrication. The tubing, which is cylindrical, is welded to a flat sheet of metal. Because each tube weld is critical to the proper operation of the final product, each weld must be made with the highest quality possible. Weld joint design and machining tolerances must be established to accomplish this task. The welding equipment must be capable of making very high-quality welds with 100% repeatability. **Figure 17-20** shows some common tube-to-sheet weld designs.

Pipe-to-Pipe Welding

Because of the wide range of pipe thicknesses, the weld joint design and the equipment used for a pipe welding system are much more complex than those used for tube welding. Several pipe weld joint designs are shown in **Figure 17-21**.

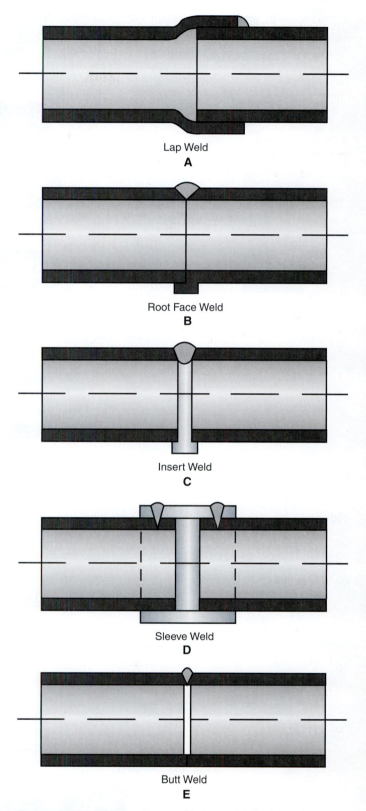

Figure 17-19. Various types of tube weld designs can be used with tube welding heads. Some or all of the filler metal for welds B–D come from the inserts.

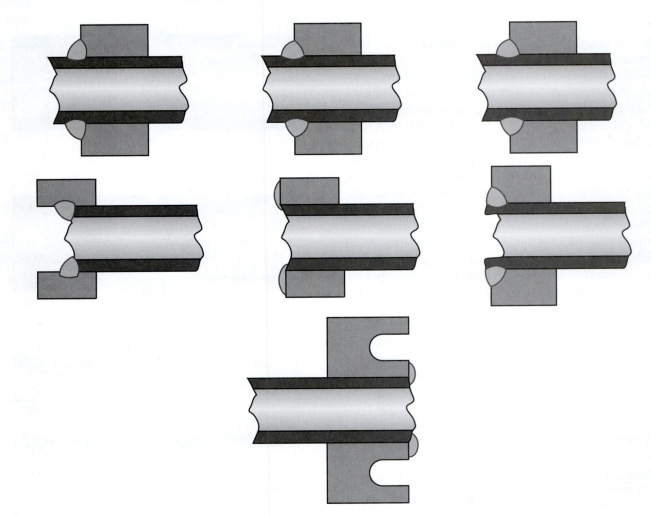

Figure 17-20. Tube-to-sheet weld joint designs.

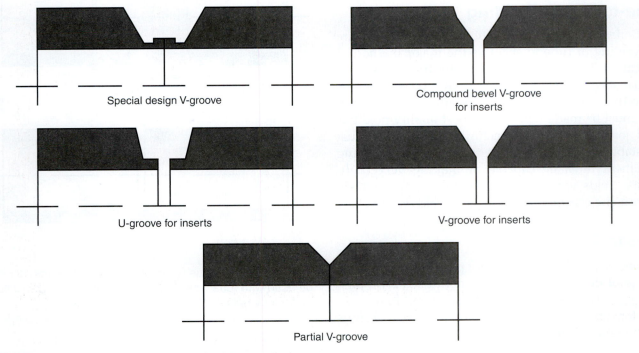

Figure 17-21. Standard pipe weld joint designs

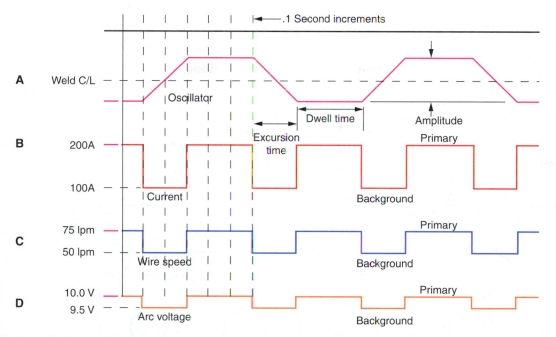

Figure 17-22. A welding schedule or program sequences each operation in proper relationship with other operations. A—This line graphs the oscillation of the torch back and forth across the weld centerline. B—This line graphs the pulsing of weld current. C—This line graphs the pulsing of the wire feed rate. D—This line graphs the changes in voltage.

Because the weld extends 360° around the pipe, the molten weld pool must be controlled to prevent a sagging, nonuniform contour. Automatic voltage controls, pulsers, and oscillators with dwell control are used to prevent sagging. These pieces of equipment can be used in any combination in the system. The pulsers can control the pulse of the current, the wire feed, and the travel speed.

Welding Schedules

Before the start of production welding, perform a weld test (make and examine a sample weld) to determine the welding parameters required for a particular joint design. During the weld test, record all of the actual parameters for use on the production weldment. These recorded parameters are called a *welding schedule*. Variables such as step pulsing of welding wire, welding current, and oscillation work together during the welding process, as shown in the graph in **Figure 17-22**. Sample forms used for recording welding schedules are shown in **Figures 17-23** and **17-24**.

A weld test is used to determine weld quality, tensile strength, and other requirements that are placed on the weld. The test can also be used as a training period for the operator and used for operator qualification. System operation is verified during weld testing, and required modifications are made prior to performing the final weld test.

The weld test used to develop the welding schedule should be performed on the same type of material, with the same thickness, the same joint design, and with the same type of tooling as will be used for the actual joint. The test weldment should duplicate the actual joint.

The following sections cover the main parameters that must be established in a welding schedule. Other parameters may be required to improve weld characteristics. As you study the following sections, think about how you would fill in the required information in forms like those shown in **Figures 17-23** and **17-24**.

Weld Joint Design

- Record the overall thickness of butt welds.
- Record the angles and dimensions of groove welds.
- Record the thicknesses of the tube walls when making fillet welds.
- Record the fitup criteria.

Variations in the base material, machining, and fitup can severely affect the quality of the welding performed on the production part. Proper tolerances must be established if the weld quality is to be the same on the weld test and the production part. If the tolerances vary considerably, a weld test should be performed for both extremes. When there is significant variation in the preassembled parts, a weld test may need to be made and a separate welding procedure developed for each half of the variation range. The parts would then be sorted into two groups, and the proper welding procedure would be used to weld each group.

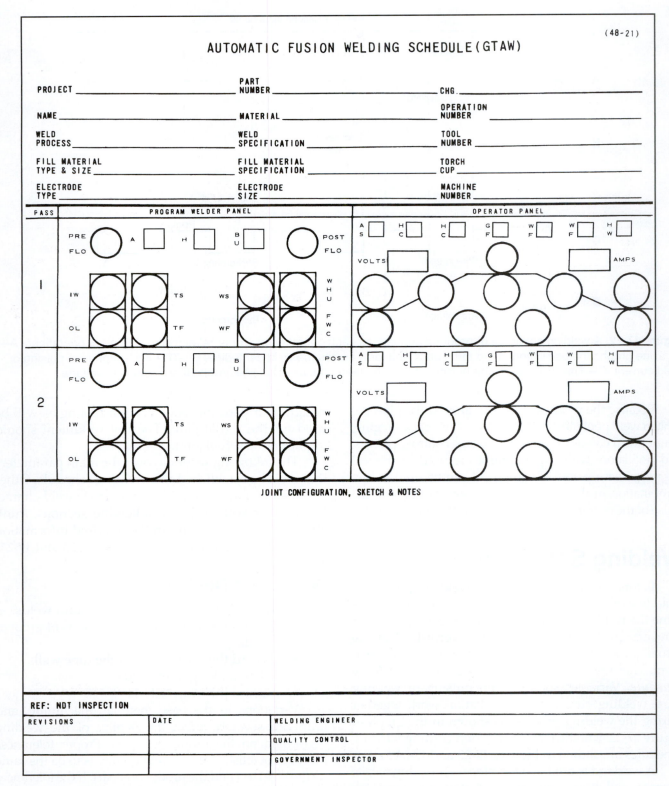

Figure 17-23. GTAW schedule made for a specific-type welding machine.

(48-21)

AUTOMATIC FUSION WELDING SCHEDULE (GTAW)

PROJECT _____ PART NUMBER _____ CHG. _____

NAME _____ MATERIAL _____ OPERATION NUMBER _____

WELD PROCESS _____ WELD SPECIFICATION _____ TOOL NUMBER _____

FILL MATERIAL TYPE & SIZE _____ FILL MATERIAL SPECIFICATION _____ TORCH CUP _____

ELECTRODE TYPE _____ ELECTRODE SIZE _____ MACHINE NUMBER _____

MACHINE SETTINGS

PASS NUMBER	AMPERAGE	ARC VOLTAGE	WELD SPEED IPM	WIREFEED MEASURED SPEED	SHIELDING TORCH		SHIELDING BACKUP OR PURGE		PRE-HEAT	INTER-HEAT	POST-HEAT	OSCILLATOR
					AR	HE	PURGE	TRAIL CUP				
1												
2												
3												
4												
5												
6												

JOINT CONFIGURATION, SKETCH & NOTES

REF: NDT INSPECTION

REVISIONS	DATE	
		WELDING ENGINEER
		QUALITY CONTROL
		GOVERNMENT INSPECTOR

Figure 17-24. GTAW schedule made for a general-type welding machine.

Weld Tooling

Weld tooling, such as heat sinks and weld backing, has a strong effect on the welding parameters when used in the immediate area of the weld. A test weld that will be used to develop a welding procedure must be made using tooling identical to the production tooling.

- Record the backup groove design and material type.
- Record the backing ring design, finger bar design, and material type.
- Record the backing ring and finger bar spacing.
- Record the backing bar temperature (if the weld is to be made at a specific temperature).
- Record the finger bar pressure on seamers.

Welding Current

- Select and record the current type.
- Record whether high-frequency voltage is used to start the arc (dc), used continuously (ac), or is not used.
- Select the amperage for the following operations:
  - initial start
  - upslope
  - weld current
  - downslope
- Record the type of amperage control—machine or manual.
- Select and record the type of arc start.
- Set and record the pulser parameters (if a pulser is used).
  - low current level and duration
  - high current level and duration

Welding Problem Areas

Several factors affect the quality of the finished weld made by semiautomatic welding. These factors include the following:

- **Welding operator training.** The welding operator must be thoroughly trained in the complete sequence of the operation. Since the welding operator enters all of the parameters and variables into the operation at the proper time, the controls must be available when needed. "Dry runs" are done often so the welding operator can become familiar with the locations of these controls. After the dry runs have been completed, several test components are welded to test

the welding schedule for reproducibility and to further train the operator prior to starting the production weld. Production starts only after the welding operator has made repeated acceptable welds on the test components.

- **Component part tolerances.** Repetitive quality welds cannot be made on parts that do not fit within the tolerances of the welding schedule. In manual welding, the welder can change many factors to correctly produce acceptable welds. However, with automatic and semiautomatic welding, the operator does not have this ability. The operator cannot change the required parameters while welding at a fixed rate of speed.

  To prevent inconsistent weld quality due to differences in part fitup, make additional test welds with the component parts at the *full tolerance* of the weld joint. Machine settings can then be readjusted for maximum gaps between parts. A weld test made on a weldment with maximum gaps also reveals if any additional auxiliary equipment is needed to weld the parts that make up the weldments. If the changes produce unsatisfactory welds, the tolerances must be reduced until acceptable quality welds can be made. If the parts range widely in size or shape, multiple welding schedules may need to be developed to ensure that acceptable welds can be made in all variations of the parts.

- **Welding equipment and tooling maintenance.** Worn equipment and tooling affect weld quality over a period of time. The weld quality can be affected to such an extent that the welded parts have to be scrapped. Old or even new equipment can malfunction due to the excessive demands placed on it because of the quantity of welds produced. To reduce and possibly prevent equipment and tooling problems, a systematic maintenance schedule should be established. The schedule should include each piece of equipment. Often the maintenance checks are performed at the start or at the end of the day, so production is not affected by a shutdown. These maintenance checks may indicate that major problems are developing. Major overhauls or maintenance can then be scheduled at a later period during nonproduction times.

Automatic Welding

Automatic welding is a welding operation using equipment that performs the entire welding sequence. The operator's role is to activate the machine and make sure the weld is performed properly. Prior to activating the machine, the operator must preset the parameters used during the automatic welding operation. The basic parameters to be set include amperage, voltage, wire feed, travel speed, shielding gas, and various timers and controls. Automatic welding is performed by robots in many industrial settings.

Equipment

The type of equipment used in automatic welding varies considerably. Simple welds can be made using several types of standard equipment assembled into a system and operated by timers, sequencers, relays, or special electronic circuitry.

Prior to operating the machine, the operator must manually preset each sequence of the operation for each individual type of weld to be made. If additional requirements are added to the welding operation (such as pulsing, slope up, or slope down), additional equipment can be added to the welding system to perform those functions. More complex welds that require directional or positional changes may require a complete computerized programmer.

Automatic and Robotic Welding

Robots, like the one shown in **Figure 17-25**, are used in many industrial applications and training situations. Robots are most useful when repetitive welds are required. Automatic welding and welding robots increase production and quality, reduce labor costs, and improve control of the welding operation. However, several factors must be taken into account when considering this type of welding:

- production quantity versus equipment cost
- welding program development and operator training cost
- component parts design and dimensional tolerances
- tooling design, development, manufacture, and maintenance
- welding equipment maintenance

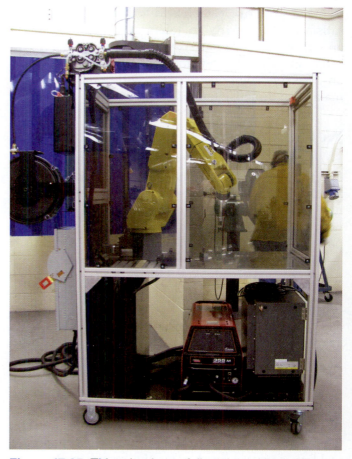

Figure 17-25. This robot has a full work envelope with precise movement in any longitudinal or circular direction. The specific weld parameters to be used are entered into the programmer. (Mark Prosser)

Creating a Welding Program

A welding program is a series of commands that directs the welding equipment to perform a weld according to the welding schedule. For an automatic welding machine to operate properly, each parameter must be established and then entered into the program. In semiautomatic welding, the welding operator can make some changes during welding. However, in an automatic welding system, every change must be programmed. Modifications to the welding program can be costly.

A major factor in the weld program is the dimension of the weld joint and the tooling. Joint tolerances and tooling have a pronounced effect on weld quality. Therefore, these tolerances must be considered in the total program.

The program established for the weld will be satisfactory for parts in tolerance. However, welds on out-of-tolerance parts may be unsatisfactory. The program should, therefore, be established based on tests on component parts with minimum and maximum dimension tolerances.

Welding Schedule Maintenance

The quality of the production weld can diminish over a period of time as the equipment and tooling is used. Therefore, periodic inspections must be scheduled to examine equipment areas that can deteriorate with use. A machine that has been inoperative for a period of time or exposed to degrading elements, such as moisture or heat, should also be checked. After any necessary adjustments or repairs are completed, a sample weld test should be made before production is resumed.

When a major change is made in equipment, such as installation of a different power source or holding fixture, a weld test should be performed to verify that the existing welding schedule is still usable. The weld test could reveal the need for modification of the existing parameters.

Since every application of automatic welding is customized for a particular setup or industry, a welding operator must enter the proper parameters when setting up the machine for a specific welding sequence. The welding operator must be able to operate many different types of equipment. Since so many different applications exist, a welding operator must be able to read the equipment manuals and closely follow the manufacturers' directions for setting up and directing each automatic welding sequence.

Designers and engineers continuously design welds that require special programming. The ability to program the parameters of these unique welding configurations into an automatic welding system enables a business to produce weldments in a timely manner and at a competitive price.

Summary

GTAW can be used for automatic and semiautomatic applications. Automatic welding is welding performed entirely by a machine of some type, usually a robot, and is most often used for high-production or repetitive welding. Semiautomatic welding is welding in which the welding operation is controlled by a welding operator. The equipment can automatically adjust itself, or it can be manually adjusted.

Long seam welds and rotational welds made using semiautomatic welding are usually made in the flat position. For fillet welds it is difficult to properly align the weldment with the electrode tip to achieve the required penetration at the root.

Semiautomatic welding can be performed on virtually all materials that can be welded manually. Aluminum can be semiautomatically welded with DCEN polarity and helium gas if the arc length is automatically controlled.

The major components of a semiautomatic welding system include the power source, arc voltage control, and travel control. Power sources are equipped with remote controls. Mechanical arc voltage controls are used where the arc length is fixed. Automatic voltage control (AVC) systems automatically and continuously regulating the distance between the workpiece and the electrode to a preset gap. Control of welding speed is important in GTAW. If the speed is not consistent, the weld will have uneven penetration and uneven reinforcement along the length of the weld seam. Two methods of drive and control systems—alternating current and direct current—are used to control welding speed.

Cold wire feeders and hot wire feeders are accessories that supply welding wire into a weld joint to maintain the weld profile. In a hot wire feeder system, wire is heated to the melting point prior to entering the weld pool. A hot wire feeder system lowers production time and produces welds that are virtually free of porosity.

Oscillators direct the arc back and forth laterally across the weld seam as the weld progresses. The oscillation agitates the weld pool, allowing gases to rise to the surface and reducing porosity.

Pulsers change welding current from one level to another for a set period of time. Pulsing the wire feed and travel speed in synch with current is usually restricted to full automatic welding, where the sequencing can be electronically controlled.

Semiautomatic welding systems include long seam, rotational, and spot welding systems. Tube-to-tube, tube-to-sheet, and pipe-to-pipe welding are also done semiautomatically.

During the weld test, all of the parameters for use on the production weldment are recorded into a welding schedule. The weld test used to develop the welding schedule should be performed on the same type of material, with the same thickness, the same joint design, and with the same type of tooling as will be used for the actual joint. After the welding operator completes several "dry runs," test components are used to test the welding schedule for reproducibility and to further train the welding operator. If there is a wide range in part size or shape, multiple welding schedules may need to be developed to ensure that acceptable welds can be made in all variations.

Equipment can malfunction due to excessive demands placed on it. A systematic maintenance schedule should be established for each piece of equipment.

In automatic welding, the operator presets the parameters to be used during the welding operation. These parameters include amperage, voltage, wire feed, travel speed, shielding gas, and various timers and controls. Robots are used in many industrial applications, particularly where repetitive welds are required. Automatic welding and welding robots increase production and quality, reduce labor costs, and improve control of the welding operation. However, the costs of equipment and training, as well as design considerations, must be taken into account.

Each parameter must be established and then entered into the welding program. Joint tolerances and tooling must be considered in the program. The program should be developed based on tests of component parts with minimum and maximum dimension tolerances.

Equipment areas that can deteriorate with use must be inspected periodically. A machine that has been inoperative or exposed to degrading elements should also be checked before being used in production. After necessary adjustments or repairs are made, a sample weld test should be performed before production is resumed. When a major change is made in equipment, such as installation of a different power source or holding fixture, a weld test should be performed to verify that the existing welding schedule is still usable.

Review Questions

Write your answers on a separate sheet of paper. Do not write in this book.

1. A person who controls semiautomatic welding equipment is called a(n) _____.

2. Semiautomatic long seam welds are generally made in the _____ position.

3. What is the problem encountered in making a fillet weld?

4. _____ allows aluminum to be welded using DCEN polarity and helium shielding gas.

5. A(n) _____ is installed into the arc circuit to determine arc gap.

6. Set blocks are used to _____.

7. What does AVC stand for? What is the purpose of AVC systems?

8. List three advantages of AVC systems over mechanical controls.

9. What two types of electrical motors and controllers are used to drive weld travel systems?

10. Why does the use of hot wire feeders produce welds that are virtually free of porosity?

11. What is the purpose of an oscillator?

12. What are the two types of oscillators?

13. A joint design should not require _____ if the joint is being welded semiautomatically using pulsers.

14. What is the purpose of a positioner in a semiautomatic rotational welding system?

15. Why must the two pieces to be welded in gas tungsten arc spot welding be in tight contact with each other?

16. Welding parameters are recorded on a(n) _____.

17. List four parameters recorded on a welding schedule for weld joint design.

18. To prevent inconsistent weld quality due to differences in part fitup, additional test welds should be made with the component parts at the _____ tolerance of the weld joint.

19. Robotic welding is most useful when _____ welds are required.

20. A major change to automatic welding equipment requires a(n) _____ to be performed to verify that the existing welding schedule is still applicable.

Chapter 18

Weld Inspection and Repair

Objectives

After completing this chapter, you will be able to:

- ❏ Identify areas that should be inspected by a GTAW welder.
- ❏ Recall the various types of nondestructive tests and their advantages and disadvantages.
- ❏ Recall the various types of destructive tests.
- ❏ Recall the tests used in weld integrity inspections.
- ❏ Identify the conditions that require dimensional repairs and the techniques used to make these repairs.
- ❏ Identify various types of surface defects and explain corrective actions for each type.
- ❏ Recall procedures for finding and repairing internal defects.
- ❏ Recall the preparation needed prior to making repairs.
- ❏ Summarize proper welding practices for repairs.
- ❏ Explain the need for postweld inspection and repair review.

Key Terms

bend tests
corrosion tests
cross section tests
defect
destructive tests
discontinuity
dye penetrant test
ferrite test
flaw
fluorescent penetrant test
gamma rays
hardness tests
hydrostatic test
image quality indicator (IQI)
macrotest
magnetic particle test (MT)
microhardness test
microtest
nick-break tests
nondestructive examination (NDE)
notch-toughness tests
penetrameters
penetrant test (PT)
pressure tests
quality control
radiographic test (RT)
temper beads
tensile tests
ultrasonic testing (UT)
visual test (VT)
X-rays

Introduction

Quality control is used throughout the welding industry to monitor the quality of the items produced. All manufactured items are made to specifications. Inspections must be made during and after the manufacturing cycle to ensure that parts meet the requirements of the specifications. The American Welding Society has increased the requirements for a CWI (certified welding inspector) certificate to ensure that certificate recipients are capable of performing quality control. Quality control is not only important to ensure quality parts but also to ensure that the correct procedures are performed as efficiently as possible. The use of proper quality control measures help make a company as strong as possible in an increasingly competitive industry.

Inspection Areas

GTAW welders need to ensure that all aspects of the operation are performed correctly. At a minimum, welders should check the following areas:

- Base material is as specified.
- Joint design is as specified and within required tolerances.
- Filler metal type and size are correct.
- Required welding equipment is available and operating satisfactorily.
- Tooling has been adequately tested to determine that it will properly support the operation.
- Parts have been properly cleaned.
- Welder training or certification is sufficient for the weld operation.
- Proper welding procedure is used and the welding equipment is set up properly for the operation.
- Inspections and tests required during the welding operation are performed as specified.
- Completed weld has been inspected to ensure that it will meet the visual requirements. Additional nondestructive examination is usually performed by other personnel.

Types of Inspection

Inspections are performed to determine whether a weld meets expectations. Depending on the final use of the weldment, several types of inspections may be required, ranging from simple visual inspection to rigorous testing.

Nondestructive Examination (NDE)

Inspections and tests of a weld that do not destroy any portion of the completed weld are called *nondestructive examination (NDE)*. Inspections and tests that destroy the completed weld, or samples of the completed weld, are called *destructive tests*.

Visual Test

A *visual test (VT)* is one of the most important methods of inspection and is widely used for acceptance of welds. VT is also used to identify bad welds before other more expensive or time-consuming forms of inspection are performed. Visual inspection is easy to apply, quick, and relatively inexpensive. Visual testing equipment includes rulers, fillet weld gauges, squares, magnifying glasses, and reference weld samples. Some of the various tools used in weld inspection are shown in **Figures 18-1**, **18-2**, and **18-3**. These tools include the following:

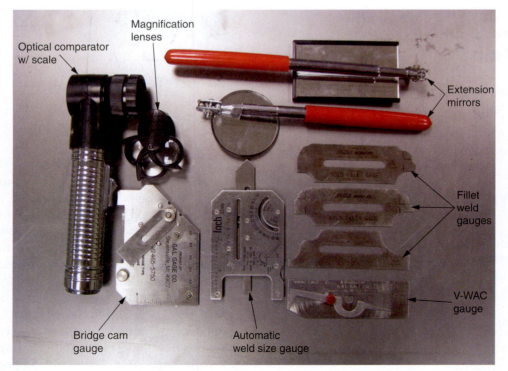

Figure 18-1. An inspection kit may include any of the tools shown here. These tools are used to inspect a variety of dimensions, including material thickness, bevel angle, crown height, undercut, mismatch, fillet weld leg length, and throat thickness. (Mark Prosser)

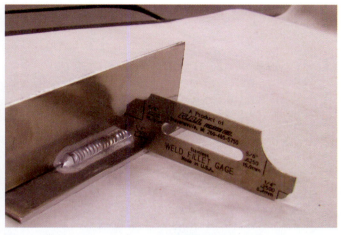

Figure 18-2. Fillet weld size and crown gauge. (Mark Prosser)

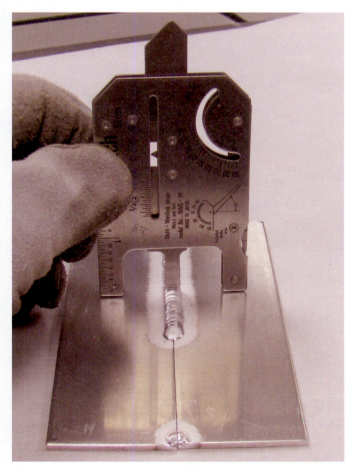

Figure 18-3. This adjustable fillet and crown gauge is being used to check the height of the weld bead. (Mark Prosser)

- **Optical Comparator.** Magnifies, illuminates, and precisely measures weld discontinuities.
- **Magnification lenses.** Pocket-sized magnification lenses.
- **Extension mirrors.** Used for root pass inspection of pipe welds

- **Fillet weld gauge.** Measures the size of fillet welds.
- **V-WAC gauge.** Used for measuring height and depth. The gauge checks undercut depth, porosity comparison, amount of porosity per linear inch and crown height.
- **Automatic weld size gauge.** Measures several aspects of a weld, including the height.
- **Bridge cam gauge.** Used for measuring several aspects of welds, such as depth of undercut, depth of pitting, and fillet weld throat size and leg length.

Visual tests provide very important information about a weld's general conformity to specifications. The following weld features are measured and compared to specifications to ensure that the weld meets expectations:

- crown height
- crown profile
- weld size
- weld length
- dimensional variation
- root side profile
- root side penetration
- surface color (titanium welds)

In addition, a visual test may reveal discontinuities in the weld. A *discontinuity* is any disruption in the consistency of a weld. A *flaw* in the weld is a discontinuity that is small enough that it does not render the weld unacceptable. A *defect* is a discontinuity that is serious enough to make the weld unacceptable. The following common problems can be detected by visual tests:

- underfill
- undercut
- overlap
- surface cracks
- crater cracks
- surface porosity
- joint mismatch
- warpage

Penetrant Test

A *penetrant test (PT)* is a sensitive method of detecting and locating minute discontinuities that are open to the surface of the weld. A penetrating liquid (dye) is applied over the surface of the weld. The fluid then enters the discontinuity. After a short period of time, the excess penetrant is removed from the surface. A developer is applied to the surface and allowed to dry. The penetrant in the discontinuity rises to the surface by capillary action, making the discontinuity easy to see. The penetrant test sequence is shown in **Figure 18-4.**

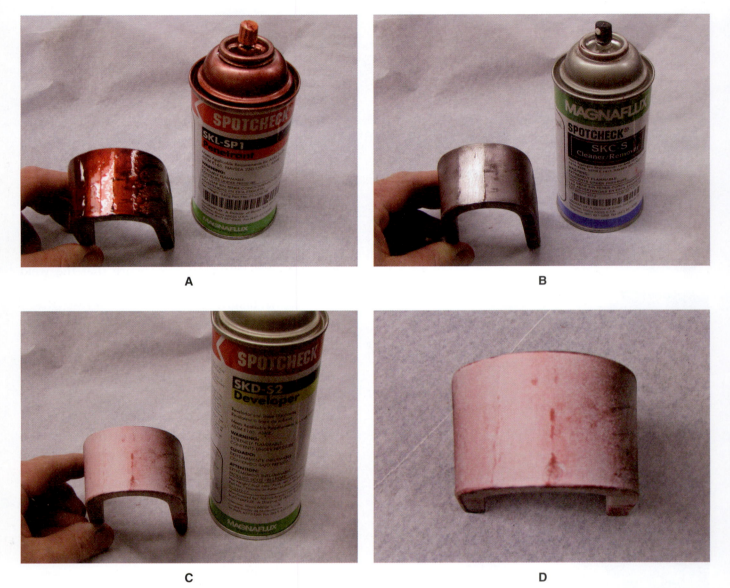

Figure 18-4. Dye penetrant test sequence. A—The penetrant is applied to the weldment. B—The penetrant is cleaned off the weldment. C—The developer is applied to the weldment. D—The weldment is inspected for discontinuities that appeared after the developer was applied. (Mark Prosser)

A penetrant test is particularly useful on nonmagnetic materials, where a magnetic particle test cannot be used. Penetrant tests are used extensively for exposing surface defects in welds on aluminum, titanium, magnesium, and austenitic stainless steel weldments. **Figure 18-5** shows how penetrant inspection can also be used to detect leaks in both open-top and sealed tanks or vessels. The penetrant is sprayed around all the weld areas. Penetrant works by capillary action and will identify any weld discontinuities or defects. When testing a sealed tank, the dye is sprayed (or brushed) on the external sides of the welds. The two types of penetrant tests are dye penetrant and fluorescent penetrant.

A *dye penetrant test* requires the surface of the weld to be sprayed generously with penetrant and allowed to soak for a specified time. Excessive penetrant is then removed with an aerosol cleaner. All of the penetrant is then wiped from the weld area. After the penetrant is removed, the developer is applied. The developer is a powdery white substance that is lightly applied from an aerosol can. Any imperfections in the weld will hold the dye and bleed through the white developer, identifying the problem. A dye penetrant test can be done anywhere because it is portable, and it can be done in any position. The results can be detected in normal light, without the use of special equipment.

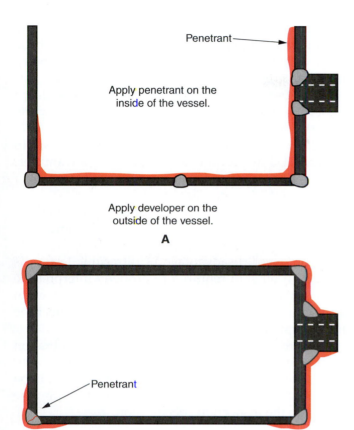

Apply penetrant on the inside of the vessel.

Apply developer on the outside of the vessel.

A

Penetrant

B

Figure 18-5. Minute leaks in tanks or vessels can be located using penetrant inspection. A—Dye penetrant inspection of an open-top tank. B—Dye penetrant inspection of a closed tank.

A *fluorescent penetrant test* requires an ultraviolet light (black light) to observe the test results. It may be necessary to enclose the viewing area in order to properly read the test results.

Magnetic Particle Test

A *magnetic particle test (MT)* is a nondestructive method of detecting cracks, seams, inclusions, segregations, porosity, or lack of fusion in magnetic materials. This test can detect surface defects that are too fine to be seen with the naked eye or that lie slightly below the surface.

When a magnetic field is established in a ferromagnetic material, minute poles are set up at any defects. These poles have a stronger attraction for magnetic particles than the surrounding material has. In a magnetic particle test, the ferromagnetic material is magnetized by an electric current, and iron particles or powder is applied to the magnetized area. If the magnetic field is interrupted by a defect, the iron particles form a pattern on the surface. The pattern is the approximate size of the defect. **Figure 18-6** shows how magnetic particle tests are performed.

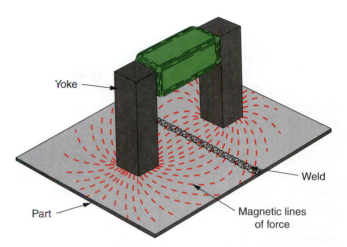

Figure 18-6. Magnetic particle tests. One pole of the yoke is placed on each side of the weld.

Small, portable, permanent magnets can be used for thin-gauge materials. Heavier material requires power from transformers, generators, or rectifiers. A typical magnetic particle unit is shown in **Figure 18-7**.

The magnetic particle test can be performed using either the wet or the dry method, depending on the individual application. The wet method, in which the particles are suspended in a fluid, is generally more sensitive than the dry method. Wet magnetic particle inspection allows for a more even distribution of particles over a large area and is better for detecting very small discontinuities on a smooth surface. The dry method, which uses finely divided dry particles that are dusted onto a magnetized surface, is better for rough surfaces. Either red or gray particles can be used in the test. The color selected should provide good contrast with the material being tested.

Figure 18-7. Magnetic particle units are very useful on small weldments and require only 110 volts for operation. (Magnaflux, A Division of Illinois Tool Works)

The test can be modified by adding fluorescent dye to the particles. In this method, an ultraviolet light is used to illuminate fluorescent dye on the iron particles, allowing the inspector to clearly see and interpret the formation of the particles at the defect. As with fluorescent penetrant testing, it may be necessary to examine the weldment in a darkened area.

Accurate interpretation of magnetic particle tests requires training. Discontinuities revealed by the test pattern can be misleading to the untrained eye and may have no consequence on the weld's acceptability. If the size of the discontinuity falls within allowable limits, the weld is still acceptable. If the size of the discontinuity is larger than the allowable limit, the weld is rejectable.

Ultrasonic Test

Ultrasonic testing (UT) is a nondestructive method of detecting the presence of internal cracks, inclusions, segregations, porosity, lack of fusion, and similar discontinuities in all types of metals. It can be used as the sole type of inspection, or it can be used with other types of testing. UT is often used in conjunction with radiographic testing because it determines the depth of the defect from the test surface.

In ultrasonic testing, very-high-frequency sound waves are transmitted through the part to be tested. The sound waves then return to the sender and are displayed as a graph on a monitoring screen for interpretation.

Since very-high-frequency sound waves travel only short distances in air, the test must be done with the part (signal sender) and the transducer (receiver) immersed in water or with the transducer coupled to the workpiece by a thin liquid film. These two methods are shown in **Figure 18-8**. UT inspection techniques include several different patterns and techniques. The technique used depends on the material, weld thickness, welding process, and inspection criteria being used. Where tests are required out-of-perpendicular with the transducer, a wedge or angle block is placed under the transducer at the desired angle to properly scan the material, as shown in **Figure 18-9**.

Ultrasonic testing is portable and nonhazardous. In addition, UT inspection has the following advantages:

- Great penetration power allows the testing of thick materials.
- High sensitivity allows detection of small discontinuities in a short period of time.
- Inspection can be done from one surface.

The major disadvantage of ultrasonic testing is the advanced skill required to properly interpret the results. Weld design, location of the defect, internal structure, and complexity of the weldment affect the interpretation of the ultrasonic signal.

In order to achieve the desired results, calibration blocks and reference weld samples are used to calibrate the equipment prior to making the test. With the proper calibration, the operator can then interpret the results to the inspection specification.

Radiographic Test

A *radiographic test (RT)* is a nondestructive method that reveals the presence and nature of discontinuities in the interior of welds. This test makes use

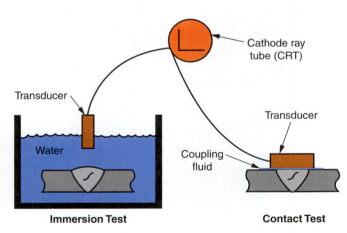

Figure 18-8. Ultrasonic tests are made with the part and the transducer submerged in water. If this is not practical, the transducer is coupled (connected) to the test area by a thin layer of liquid.

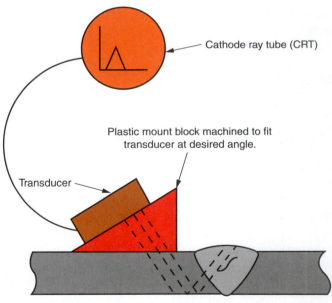

Figure 18-9. Wedges or angle blocks are used to send and receive ultrasonic signals in areas where signal transmission could be blocked in a straight-line plane.

of the ability of short wavelength radiations, such as X-rays or gamma rays, to penetrate material that is opaque to ordinary light.

X-rays are a form of electromagnetic radiation that penetrates most materials. An X-ray test is similar to a photograph. A machine in a fixed location transmits X-rays through the material being tested. A film or sensor on the other side of the material is exposed by the X-rays that pass through the test material. Any defects or inconsistencies in the metal change the amount of X-rays that are able to pass through. Because more or fewer X-rays pass through those locations, they look different in the developed film, or display.

The area surrounding the X-ray machine may be lead-shielded to prevent the escape of radioactivity. An X-ray testing station usually includes all of the support equipment, such as film-developing machines. The end result is a radiograph made in a minimum amount of time. Recently, more portable X-ray equipment has been developed and is becoming very common in the pipeline industry. This equipment consists of a small machine that is sent down the center of the pipe to X-ray each weld.

Gamma rays are electromagnetic waves that are similar to X-rays, but with a shorter wavelength. Gamma rays are produced from radioactive materials such as cobalt, cesium, iridium, and radium. These radioactive materials must be contained in a lead-shielded box and transported to the job site for in-place radiographs.

The film that is exposed by these rays is called a *radiograph*. Film is placed on one side of the weld, and the radiation source is placed on the other side of the weld. The radiation passes through the test material and exposes the film, revealing any inconsistencies in the weld. Different types of radiation sources are more or less powerful. The thickness of the material usually determines the type of radiation source used for the test.

A radiograph inspection of a fusion weld is shown in **Figure 18-10**. The film is developed for viewing on a special viewer. The radiograph must be compared by a skilled technician to a specification that defines discontinuities. The film must have sharp contrast for proper definition of the weld and identification of any defects. (Contrast is the degree of blackness of the darker areas compared with the degree of lightness of the brighter areas.)

To ensure sharp images on the film, *image quality indicators (IQI)*, also called *penetrameters*, are used to indicate the quality of the radiograph. A hole-type penetrameter consists of a thin shim of the base metal, usually with a thickness equal to 2% of the weld thickness. One, two, or more holes with various diameters are drilled into the metal shim. The shim is

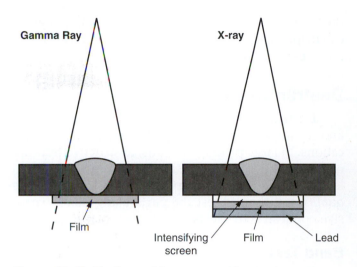

Figure 18-10. Radiographic test operation.

laid next to the weld before being x-rayed. The ability of the radiograph to show definite-sized holes in the penetrameter establishes the radiograph quality. The resolution of the X-ray is indicated by the smallest hole that is visible.

Another type of IQI is a wire type. A wire IQI is a series of wires embedded in plastic. The wires have decreasing diameters. The quality of the radiograph is determined by the thinnest diameter wire that can be seen on the image. **Figure 18-11** shows a wire-type IQI.

Radiographs are expensive; however, they provide a permanent record of the weld quality. It is often useful to compare a radiograph created when the weld was first made with later radiographs made after

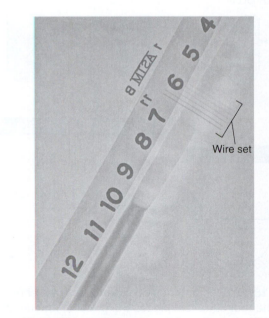

Figure 18-11. This radiograph used a wire-type IQI to monitor clarity in the development process. (VJ Technologies)

the weldment has been in service. This comparison can help engineers identify areas of the weldment that are stressed by service.

Destructive Tests

Destructive tests are used for welder qualification and certification, as well as welding procedure qualifications. In large production runs, destructive tests are often made by pulling apart sample units. It is often less expensive to scrap a part to make a destructive quality test than to test the parts using more expensive nondestructive tests.

Bend Test

Bend tests are used to determine internal weld quality. As shown in **Figure 18-12**, there are three different types of bend tests:
- face bend (face of the weld is tested)
- root bend (root of the weld is tested)
- side bend (sides of the weld are tested)

In bend tests, a weldment is sliced into test strips, called *coupons*. The weld is then bent around a die of a specific size, creating a horseshoe of the coupon. This process stretches the weld to test the weld's integrity.

Figure 18-13 shows a radius bend testing machine. This machine bends the prepared test coupon into a U form over a specified radius, which is dependent on the thickness and strength of the material. After bending, the outer surface and the inner surface of

the U are checked for cracks and other indications as required by the weld inspection criteria. The outer face of the bend may be examined by a visual, penetrant, or magnetic particle test to detect defects such as cracks, lack of fusion, and lack of penetration.

Tensile Test

Tensile tests are used to compare the weldment to the base metal mechanical values and specification requirements. The weldment is sliced into coupons, and then each end of the coupon is pulled in opposite directions until the coupon fails (breaks). A tensile test machine is shown in **Figure 18-14**.

Tensile tests are made to determine the following:
- **Ultimate strength of the weld.** This is the point at which the weld fails under tension.
- **Yield strength of the weld.** This is the point at which the weld yields or stretches under tension and will not return to its original dimensions.
- **Elongation.** This is the amount of stretch that occurs during the tensile test. It is measured by placing gauge marks on the sample or coupon before testing and comparing the after-break distance with the original gauge marks.

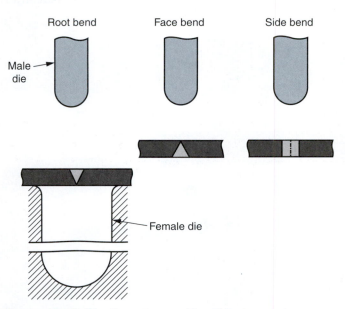

Figure 18-12. The three types of bend tests are shown here. The root bend test places the greatest amount of stress on the weld root. The face bend test places the greatest amount of stress on the weld face. The side bend places the greatest amount of stress along the weld axis.

Figure 18-13. A radius bend testing machine. (Mark Prosser)

Notch-Toughness Test

Notch-toughness tests are used to define the ability of welds to resist cracking or crack propagation at low temperatures under loads. These tests are used on welds that are intended for use in low temperature environments with pulsating or vibrating loading. The weldment is cut into test coupons, which are then notched, cooled to a low temperature, and put under pressure until they fail.

The test coupons are cut from the test weld. They are prepared for either a Charpy or an Izod impact test, **Figure 18-15**. The test bars are cooled to the test temperature and then placed into the test machine and broken, **Figure 18-16**. The results are measured in the energy required to make the coupon break and are expressed in foot-pounds. Comparisons are then made with the original material and specification requirements.

Cross Section Test

Cross section tests are used to define the internal quality and structure of the weld. The weldment is cut

Figure 18-14. This tensile test machine has a recorder mounted on the side to record the test operation parameters and results. A tensile test machine can be equipped with fixtures for holding weld coupons for testing. (Photo courtesy of Tinius Olsen)

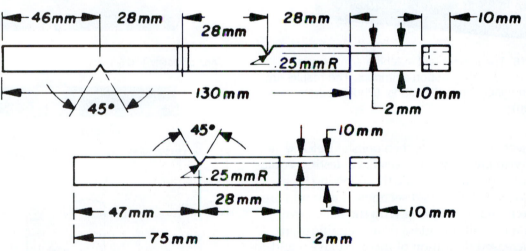

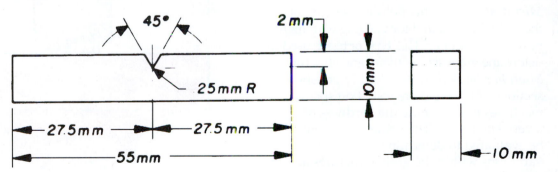

Figure 18-15. Charpy and Izod test bar dimensions.

Figure 18-16. The impact test machine arm swings downward to break the coupon at impact. The results are shown on the gauge in foot-pounds. (Photo courtesy of Tinius Olsen)

into cross sections, which are then polished, etched, and examined visually or with specialized testing equipment. Cross section tests include the following types:

- *Macrotest.* Polished sections of the weld are prepared for viewing with the naked eye or a magnifying glass (low magnification). Increased definition of the weld layers and passes can often be obtained by etching the sample with a suitable etchant.
- *Microtest.* Very highly polished sections of the weld are prepared for viewing with high-power microscopes. This type of test is used to determine grain size, content, and structure.
- *Microhardness test.* Very highly polished sections of the weld are tested on special machines to determine the hardness of a very small area. The results can then be evaluated to determine the hardness variations within the grain structures and the weld zones.

Nick-Break Test

Nick-break tests are destructive tests that are very simple to make. They are used to determine the internal quality of a weld with regard to porosity, lack of fusion, and slag. Notches are cut in the sides of a weld coupon in the weld area. The coupons are then laid across a support on each end and force is applied with a hammer to try to break the weld sideways for a simple internal inspection. A nick or groove cut into the weld helps the specimen break when force is applied.

A section of the weld to be tested is removed from the weld and prepared as shown in **Figure 18-17**. The coupon is then placed into a vise, **Figure 18-18**, and broken with a sharp blow to the upper section. Defects can then be observed in the broken areas.

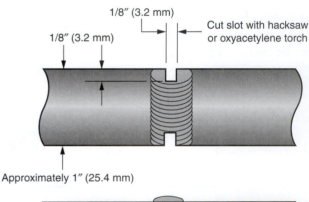

Figure 18-17. Nick-break test dimensions.

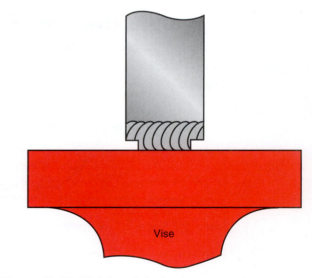

Figure 18-18. Nick-break test setup.

Weld Integrity Inspections

Many weldments require destructive tests in addition to nondestructive examination to verify the weld quality. These tests may be imposed as part of the qualification test program, or an additional test may be conducted during or after the manufacturing cycle. The test requirements are usually a part of the fabrication specification.

Pressure Test

Pressure tests subject a vessel, tank, piping, or tubing to internal pressure. Pressure tests can use either air or fluid. If a fluid is used, the test is called a *hydrostatic test*.

The test program may require a number of cycles to be performed, simulating the use of the part in actual service. During the test, the part will expand. This expansion should not be restricted with tools, or undue stresses will build within the part.

When conducting a pressure test, be alert to the possible failure of the unit. Before beginning the test, make sure the test procedure ensures the safety of everyone in the testing area.

Hardness Test

Hardness tests involve pressing a test probe into the surface of the weld. The amount of pressure required to deform the surface of the weld metal is an indication of the hardness of the weld or weldment. If the weldment has been heat-treated (annealed, hardened, tempered, etc.) after fabrication, the hardness test will determine the effect of the heat treatment. A hardness tester is shown in **Figure 18-19**.

Hardness test results can also be used to determine the ultimate tensile strength of the material. The test results can be compared with standard tables. (Refer to the hardness conversion table included in the reference section of this book.) Weldment designs that include areas such as flanges and extensions can be tested by portable hardness testers as shown in **Figure 18-20**.

Weldment designs that do not have clear areas for testing must have test bars of the same material (welds if required) submitted with the weldment during the heat-treating cycle. The bars are then tested on a stationary tensile machine, as shown in **Figure 18-14**, for the hardness results. Round tensile test bars are shown in front of the tester in **Figure 18-21**. A flat tensile test bar is being tested in **Figure 18-22**.

Corrosion Test

Corrosion tests measure the ability of a weld to restrict corrosion by a specific material. These tests are usually performed on test weldments. The test

Figure 18-19. Rockwell hardness testing machine for bench testing. (Mark Prosser)

Figure 18-20. Portable Rockwell hardness tester for flange testing. (Qualitest International LC, www.WorldOfTest.com, info@qualitest-inc.com)

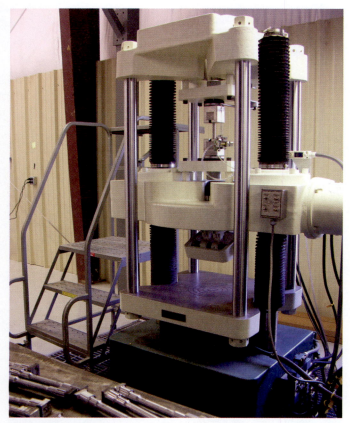

Figure 18-21. Round-type tensile bars are shown in the foreground of this image. (Photo courtesy of Tinius Olsen)

Figure 18-22. A flat tensile test bar is being tested. (Photo courtesy of Tinius Olsen)

parts are sprayed with a salt mixture and allowed to corrode, simulating years of exposure to the weather. Such tests are usually required by the specification used by the fabricator. Weld test parameters made on the qualification test weldments must be duplicated on the production part without deviation to maintain the proper corrosion resistance.

Ferrite Test

A *ferrite test* on completed stainless steel welds determines the amount of magnetic ferrite in an austenitic (nonmagnetic) weld. Insufficient ferrite in a weld made under high restraint is prone to cracking at red heat. Limits of ferrite will depend on the use of the final weldment. These amounts are usually specified in the fabrication specification.

Weld test parameters and filler metals used on qualification test weldments must be duplicated on the production part without deviation to maintain the proper ferrite content.

Weld Repair

If a weldment fails inspection, the welding inspector will review it in order to determine the extent of damage that may be caused by repairing the weld and whether the weldment can fulfill its function if the defect is allowed to remain in place. If the function of the weldment is affected by the defect, the weldment must be discarded and replaced. In some cases, the defect may not affect the functionality of the weldment, in which case it can be left. These determinations are made on a case-by-case basis.

If a part requires rework, a thorough welding procedure should be established to minimize the effect of the repair on the remaining portion of the weld. This procedure must consider the procedure used to create the original weld. It must also consider the following:

- the condition of the base metal and weld
- the type of filler metal to be used in the repair
- the welding sequence
- any in-process inspection required during the repair
- tooling required for the repair
- the final weld's mechanical properties

Incomplete consideration of any of these factors may result in further rejection of the weld repair and possible failure of the weld when placed into service.

Dimensional Repairs

Dimensional repairs are repairs that are required because the weld is too small for the material and joint type. These repairs involve the addition of material to increase weld size and are usually necessary due to the insufficient addition of filler metal during the welding operation. Conditions that require dimensional repairs are as follows:

- **Crown height is too low, Figure 18-23.** A low crown does not provide adequate reinforcement of the weld. Repair this type of defect with stringer beads to minimize weld shrinkage. Add only enough new filler metal to build the crown to height requirements. Do not overweld.
- **Surfacing or overlay type weld height is too low, Figure 18-24.** Surfacing or overlay that is not high enough reduces the durability and service life of the surfaced material. Repair this type of defect with stringer beads to minimize dilution and distortion.
- **Fillet weld size is too small, Figure 18-25.** A small fillet weld does not provide adequate strength in the joint. The weld is repaired by removing the inadequate weld and rewelding to create the proper size fillet weld.

Regardless of the repair technique, these repairs require close control to avoid overwelding and adding additional stress to the original weld.

Surface Defect Repairs

Defects seen on the surface of the weld can extend deep into the weld. For this reason, the defect must be removed. After the defect has been removed, the area must be reinspected before a repair can be attempted. Common defects and factors to be considered when repairing them are as follows:

- **Longitudinal, transverse, or crater cracks, Figure 18-26.** On steel and steel alloys, use a small grinding wheel like the one shown in **Figure 18-27** to remove cracks. Remove only the amount of metal required to eliminate the crack.

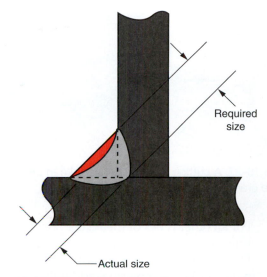

Figure 18-25. Fillet weld leg length is satisfactory; however, the weld is concave, which reduces the actual fillet weld size.

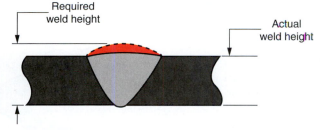

Figure 18-23. A groove weld crown with insufficient height.

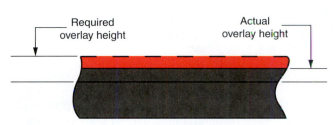

Figure 18-24. An overlay weld deposit with insufficient height.

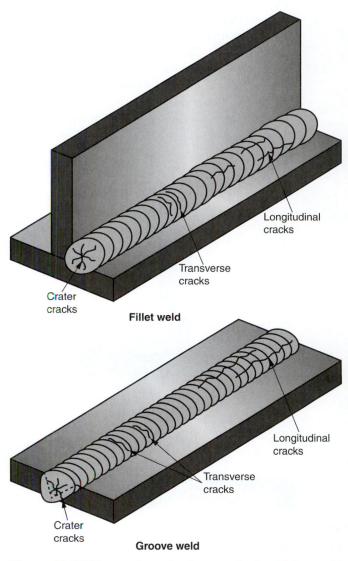

Fillet weld

Groove weld

Figure 18-26. Types of cracks that may be found on groove and fillet welds.

Figure 18-27. A cutting or grinding wheel can be used to remove cracks from the weld. (Mark Prosser)

For all other types of metal, use small rotary tungsten carbide tools to remove the crack. Do *not* use grinding wheels on nonferrous material. When repairing, add only sufficient filler metal to match the adjacent weld contour.

- **Undercut** at the edges of the weld, **Figure 18-28**. Dirt, scale, and oxides may be present in the undercut area. These impurities can cause further defects if not removed prior to weld repair. Remove these impurities by grinding or routing as previously described. Be careful not to remove base metal adjacent to the undercut. Since the repair will widen the original crown size, use lower currents and sufficient wire to prevent additional undercuts and underfill.

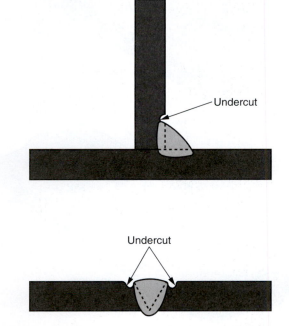

Figure 18-28. Undercut on flat groove welds can occur on either edge of the weld. Groove and fillet welds made in the horizontal position will usually have undercut on the upper part of the weld.

- **Porosity**, or pores in the weld, **Figure 18-29**. Remove isolated or single pores with a rotary tool for weld repair. Remove multiple and linear (aligned in a row) pores by grinding or machining. Then reinspect the weld by radiographic or ultrasonic inspection to ensure that the porosity has been completely removed before repairs are started. For weld repair, always fill the deepest part of the recessed area first. Keep each layer of weld level until the area is filled.

- **Cold laps**, **Figure 18-30**, are areas of the weld that have not fused with the base metal. Cold laps can occur on fillet welds or butt welds, usually as a result of a travel speed that is too slow. Since the extent of the overlap cannot be determined by NDT, remove the entire

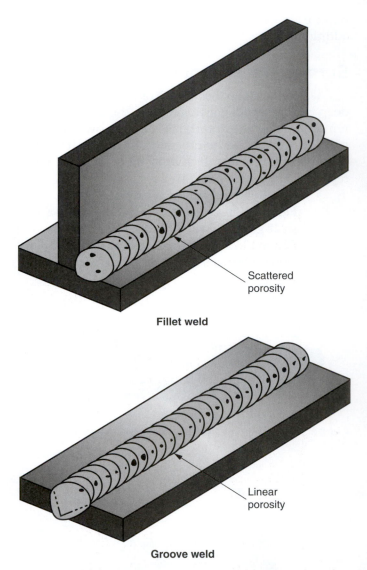

Fillet weld

Groove weld

Figure 18-29. Isolated pores can occur in any portion of the weld. Linear (aligned) pores usually are found near the bottom, along the sidewall, or at weld intersections.

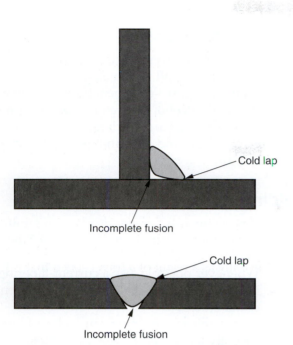

Figure 18-30. Fillet weld cold laps are usually located on the bottom side of the weld. Butt weld cold laps can occur on either side of the weld crown.

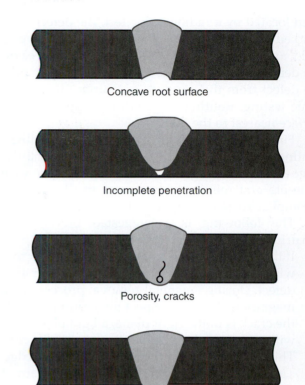

Figure 18-31. The defects shown can occur on the root side of the weld.

area by grinding or routing. Use extreme care when grinding into lap joints to prevent grinding into adjacent metal and creating more problems. When the overlap material has been removed, perform a penetrant test to determine if the defect is entirely gone. Continue removing material until the penetrant test is satisfactory. If weld repair is required to satisfy crown height requirements, use low currents and sufficient wire to match the crown with adjoining material.

- **Incomplete penetration** on the root side of butt welds. Other types of defects can also occur on the root side of the weld, such as concave root surface, cracks, porosity, melt-through, etc. See **Figure 18-31**. Remove all of these areas by grinding or routing. To ensure complete removal of all defects, perform a penetrant test before repairing the weld. Since oxides form in this area during welding, clean the repair area to bright metal before rewelding. Use stringer beads and add only enough wire to build a small crown.

Internal Defect Repairs

Internal defects, **Figure 18-32**, may or may not extend to the surface and might not be found by surface inspection. They are generally found by radiographic and ultrasonic testing. Once the defect has

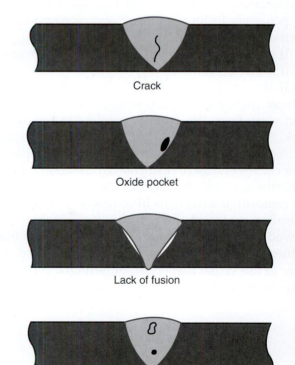

Figure 18-32. The defects shown can occur in the interior of the weld.

been located by a radiographic test, the defect can be marked on the X-ray film. The film is aligned over the weld, and a punch can be used to indent the area over the defect. The next step is to determine the depth of the defect from the top and root surface using ultrasonic testing. Routing or grinding is done from the surface nearest to the defect.

Finding the defect in the weld by grinding or routing requires both skill and patience. Porosity and large areas exhibiting lack of fusion are generally easy to locate and remove. Cracks and small areas with incomplete fusion are more difficult to locate.

The following is a suggested procedure for repairing an internal defect in a groove weld:

1. If the defect depth is known, remove metal to within approximately 1/16″ (1.6 mm) from the defect. During the metal-removal process, use a magnifying glass to inspect the ground area. If the crack is in the right plane, a light blue surface will sometimes be found at the edge of the crack. This is caused by overheating of the crack edge. This is also a good situation for a dye penetrant test.
2. Perform a penetrant test on the grooved area. If no indication of a crack or defect is seen, remove the penetrant.
3. Grind .010″–.015″ (.25 mm–.38 mm) deeper.
4. Penetrant test the grooved area again. Continue penetrant testing, grinding, and retesting until the defect is found.
5. If the defect is still not found, X-ray to determine if the defect remains.

If the crack is not found after removing metal halfway through the part, reweld the ground area. Then work from the opposite surface to remove the crack.

Never grind a slot or a hole through the part. Repairing a slot causes excessive distortion in the adjacent areas or shrinkage and the possibility of more defects. The removal of defects in fillet welds is difficult due to limited access of the grinder. If the penetrant is allowed to penetrate into the joint, it can cause many problems during weld repair. For a fillet weld, it is easier to use visual inspection or radiographic tests to ensure that the defect has been removed.

Preparing for Repair

After the defect is removed, prepare the area for welding by removing all rough edges on the ground area. Any oil, grease, scale, or penetrant residue must be removed with alcohol or acetone. Do *not* use grit blasting in the grooved area. Grit material can become embedded in the ground area and become trapped in the weld repair.

Welding for Repairs

- If possible, use stringer beads with minimum amperage for minimum shrinkage of the joint.
- Whenever possible, use a current-tapering (crater fill) control on amperage to prevent crater cracks.
- Clean scale and oxides from each weld pass.
- Visually inspect each weld pass after cleaning.
- If the weld repair is deep, have an X-ray made after two or three completed passes to confirm that new cracks have not formed. This should also be done if there is any doubt about removal of the original crack.
- Always use a backing gas if the root of the weld may be exposed to air.
- Where possible, use the original parameters for preheat, interpass temperature, and postheat.
- Do *not* build the repair crown any higher than is required. Each pass that is made stresses the base of the weld due to shrinkage.
- When the grain size must be controlled throughout the repair, weld *temper beads* on top of the weld, as shown in **Figure 18-33**. These beads reduce the surface grain size and are removed after welding.

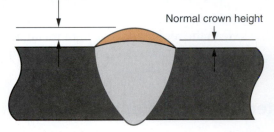

Figure 18-33. Temper beads are used to obtain an even structure throughout the top area of the weld. Since they add significant height to the crown, they are usually removed after welding.

Postweld Inspection

All of the nondestructive testing (NDT) required for final acceptance of the weld must be completed after the weld repair is done. This means that even if several inspections were satisfactory before the weld was rejected, all of the inspections must be redone. Repairs can cause new problems in a weld. After a repair is made, the entire weldment must be reinspected.

Repair Review

Repairs are expensive and often detract from the appearance of the final weld. Everything within reason should be done to eliminate defects that require costly repairs. Review every flaw and defect in the weld, regardless of its severity, to determine its cause. Plan the possible corrective action that can be taken in the future to eliminate similar problems.

Repair review should include the following areas:
- base material
- tooling
- preparation for welding
- joint preparation
- process application (welding variables)
- welder training and skill

Summary

Inspections are made during and after the manufacturing cycle to ensure that parts meet specifications. Types of inspection include nondestructive examination (NDE), in which no portion of the completed weld is destroyed, and destructive tests, in which the weld is destroyed. Weld integrity inspections may also be required in addition to the nondestructive test requirements to verify the weld quality. NDE includes visual tests, penetrant tests, magnetic particle tests, ultrasonic tests, and radiographic tests. Destructive tests include bend tests, tensile tests, notch-toughness tests, cross section tests, and nick-break tests. Weld integrity inspections include pressure tests, hardness tests, corrosion tests, and ferrite tests.

Intended weld repairs must be evaluated as to whether the discontinuity should be removed. Repairing a completed weld could cause buckling, distortion, or misalignment of the reworked area. In some cases, a discontinuity can be left in a weld without affecting the weld's function.

Repairs can be made to correct dimensional problems, surface defects, and internal defects. Common surface defects include longitudinal, transverse, or crater cracks; undercut at the edges of the weld; porosity; cold laps; and incomplete penetration. Internal defects are generally found by radiographic and ultrasonic testing.

Stringer beads should be used for welding repairs in order to minimize shrinkage of the joint. If the weld repair is deep, an X-ray should be made after two or three completed passes to confirm that the original crack has been fully repaired and new cracks have not formed. Where the grain size must be controlled throughout the repair, temper beads should be used to reduce the surface grain size.

All nondestructive testing required for final acceptance of the weld must be completed after the weld repair is done. Repairs can cause new problems in a weld.

Review Questions

Write your answers on a separate sheet of paper. Do not write in this book.

1. Inspections and tests made on a weld that do not destroy any portion of the completed weld are called _____.
2. A(n) _____ test is used to identify bad welds before other more expensive and time-consuming types of inspection are performed.
3. What is the difference between a flaw and a defect?
4. A penetrant test (PT) will only reveal discontinuities that are located _____.
5. In a magnetic particle test, _____ particles form a pattern on the surface of the material where a defect is located.
6. What does an ultrasonic test of a weld reveal?
7. List four factors that can cause misinterpretation of an ultrasonic test.
8. Radiographic tests use the ability of short wavelength radiations, such as _____ or _____, to penetrate opaque material.
9. What is the purpose of a penetrameter?
10. List the three types of bend tests.
11. List the three mechanical values of a weld determined by tensile tests.
12. _____ tests are used to define the ability of welds to resist cracking or crack propagation at low temperatures under loads.
13. A pressure test that uses fluid is called a(n) _____ test.
14. In a corrosion test, test weldments are sprayed with a(n) _____ mixture and allowed to corrode.
15. Why should all rejected welds be thoroughly reviewed before attempting a repair?
16. List three conditions that require dimensional repair.
17. What should be avoided when grinding out undercut defects?
18. What type of defects located in the body of a weld are very hard to locate using grinding or routing?
19. What should be done when making a deep weld repair to confirm that new cracks have not formed?
20. _____ beads reduce the surface grain size and are removed after welding.

Chapter

Qualification, Certification, and Employment

Objectives

After completing this chapter, you will be able to:

❑ Explain how specifications and codes are used in the welding industry.

❑ Recall the information contained in a welding procedure specification (WPS).

❑ Summarize the process of procedure qualification.

❑ Explain how welder/operator certification is obtained and recall the limitations of certification.

❑ Identify specifications used for welding.

❑ Recall the various careers available in the welding field.

❑ Give examples of soft skills needed in the workplace.

❑ Recall the major factors involved in estimating the cost of making a weld.

❑ Recall the tasks involved in workplace maintenance and equipment inspection.

❑ Explain the requirements for and the advantages of becoming a SENSE-certified welder.

Key Terms

arc hours
automatic operator certification
clock hours
codes
consumables
hourly flat rate cost
in-process inspections
labor
procedure qualification
soft skills
specifications
welder certification
welding procedure qualification record (WPQR)
welding procedure specification (WPS)

Introduction

All welding is done to specifications. The specification may be only a verbal order to the welder, or it may be a very detailed written document, with control over every aspect of the job. Control is necessary to assure the client that the weld will perform the job intended. The amount of assurance is relative to the job requirements. For example, a weld on a steel cabinet base requires very little assurance. However, a weld on a nuclear reactor part must have very high assurance of weld quality.

To achieve this assurance of weld quality, rules are made to cover the procedures, methods, materials, qualification, and testing that will be used to produce and test the weld. The rules, in the form of orders, codes, specifications, or instructions, are generally a part of the purchase order or contracts for the weldment.

Specifications and Codes

Specifications are documents that detail types of materials, welding processes, preparation of joints, qualification and welding requirements, and testing that will be used to produce the weldment. Specifications may be specific or general in nature and can be for one specific weld or several welds. Since specifications can be revised, welders must ensure that the latest revision is always used. In cases where a specific revision is required, the specified revision should always be used.

Codes are rules that specify the way a weld must be performed. Code compliance ensures that a weld or a component is done to a specific standard, which may be dictated by local, state, or federal law. AWS D1.1 *Structural Welding Code—Steel* is one of the most widely used and most respected code books.

It is strongly recommended that the referenced specification or code be studied thoroughly before work begins. Having a thorough knowledge of the specification or code requirements enables a welder to create a reliable weldment.

Welding Procedure Specification

A *welding procedure specification (WPS)* is a document that details all of the requirements and instructions for welding a specific joint. The WPS is created by the fabrication company to develop the proper technique of fabrication. A typical WPS includes the following information, as well as additional variables as needed:

- material type and condition
- tooling
- filler metals
- joint design, weld type, and position
- weld process and related parameters with tolerances
- final weld configuration
- inspection and/or test criteria

A WPS is developed by making test welds of the various joints required by the weldment drawings. If the test welds produce acceptable results, the data is recorded in a welding schedule for the welder's use on the actual weldment. The welding schedule denotes the WPS, and it is imperative that the test weld conditions parallel the actual weldment conditions. If the test and actual weldment conditions do not coincide, the welds on the production part may be of substandard quality.

Procedure Qualification

Procedure qualification is the process of performing nondestructive testing (NDT) and destructive testing (DT) on completed test welds. This testing determines whether the settings and variables in the welding procedure specification produce acceptable results. The settings used in the WPS and the results of the destructive and nondestructive testing are recorded in a document called a *welding procedure qualification record (WPQR)*.

The nondestructive testing includes one or more tests as required by the specification or code. The destructive testing may include tensile, bend, Charpy impact, and macro and microhardness tests to determine the mechanical values of the weld. Together, the destructive and nondestructive tests determine the quality of the weld. Acceptance or rejection of the weld quality is usually based on a comparison of the testing results to the standards or limits in the fabrication specification. After the NDT and DT are completed and the results are deemed satisfactory, the welding procedure is considered qualified and ready to be used for the production weld.

Requalification is required when a major change is made to the joint design, filler metal, type of weld current, tooling, or any other essential variable that may affect the quality or mechanical properties of the weld.

Welding Schedule

The welding schedule is created from the WPS data recorded during the welding of the test samples. A welder uses the welding schedule to set up the joint, set up the power source and related equipment, and make the weld.

With the many tolerances involved in material thickness, joint preparation, and tooling conditions, it is suggested that a tolerance be included in the welding parameters. Accommodations can then be made for these tolerance areas.

Welder/Operator Certification

Welder certification or *automatic operator certification* is the verification that a welder or operator is able to use the welding schedule to make an acceptable weld. In order to become certified for a weld, a welder must successfully complete a welder performance qualification test. During this test, the welder must create a sample weldment using the welding procedure specification. The sample weldment is then inspected and tested. If the results meet specifications, the welder is certified to perform the weld described in the WPS.

The welder/automatic operator certification may only allow a specific weld type, or it may include several different joint designs. This will depend on the fabrication specification requirements.

The length of time a certification is effective is variable. Some certifications are for a 6-month period, or only for welding a specific part or weld. Some certifications may last as long as the quality of the production welds is satisfactory. Certification can be rescinded if poor-quality welds are produced. Certification can also be rescinded if a welder does not use the process within three months, or an even shorter period of time if prescribed in the fabrication contract or specification.

Various types of welder certification test positions are shown in **Figures 19-1** through **19-3**. Welder/operator qualification tests are very specific to the weldment being produced or the job the welder will be working on. There are many variations of tests, and multiple processes are often required to complete a test. For example, a welder may have to use the GTAW process for a root pass on pipe and then use the SMAW process to finish the weld.

Qualification (certification) of the welder remains in effect while the welder is employed using the qualified procedure. Termination of employment automatically nullifies the welder certification. Under the rules of these codes, welders cannot have a certification unless they are employed, and certifications cannot be transferred from one employer to another.

Specification and Testing Organizations

There is no general specification in current use because there are so many variables and requirements for GTAW. Large companies that use the GTAW process usually have a company specification that states the rules for qualification, certifications, and the limits of the process within the company's structure.

The following are specifications used to define the limits of qualification, certification, and use of the process:

- ASME Boiler Code Section IX
- AWS 2.1 *Specification for Welding Procedure and Performance Qualification*
- MIL-STD-1595

The specifications stipulate that each weld type and material type must be qualified by test procedures. Welders use the qualified test procedure parameters and variables to achieve welder certification. Qualification of the weld procedure remains in effect until an essential variable is changed. These types of changes require a new certification test and possibly a new test for the welder.

Careers

A good understanding of welding processes, techniques, and print reading, in addition to the necessary skills to perform welds, can lead to many opportunities. Many careers are available in the welding field.

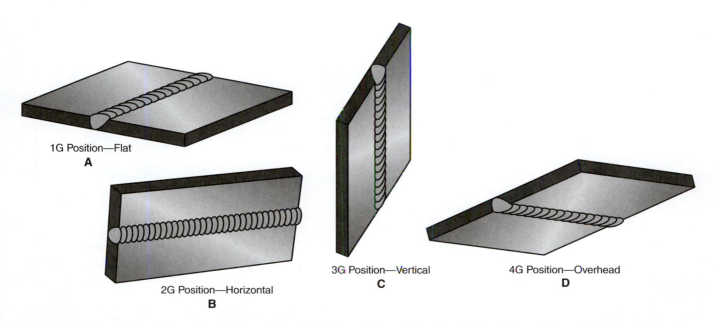

Figure 19-1. Four test positions for groove welds. A—Flat, or 1G, position. B—Horizontal, or 2G, position. C—Vertical, or 3G, position. D—Overhead, or 4G, position.

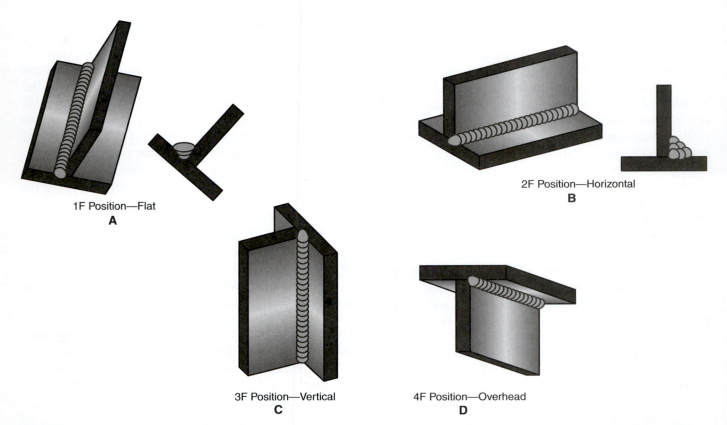

Figure 19-2. Four test positions for fillet welds. A—Flat, or 1F, position. Note that the multipass weld is completed using weave beads. B—Horizontal, or 2F, position. Note that the multipass weld is completed using stringer beads. C—Vertical, or 3F, position. D—Overhead, or 4F, position.

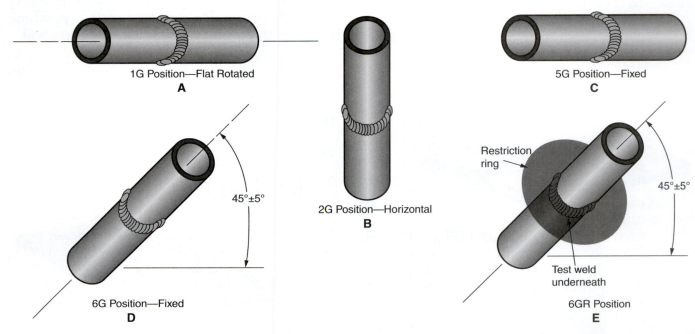

Figure 19-3. Five test positions for welding pipe. A—Flat rotated, or 1G, position. The pipe is rotated so that all welding is done in the flat position. B—Horizontal, or 2G, position. C—Multiple fixed, or 5G, position. The pipe is horizontal and remains stationary, so the welding must be performed in a variety of positions. D—Multiple fixed, or 6G, position. The pipe is at a 45° angle and remains stationary, so the welding must be performed in a variety of positions. E—Multiple fixed and restricted, or 6GR, position. The pipe is at a 45° angle and remains stationary. A restrictor ring limits access to the joint.

Careers in the welding field include the following:

- **Master welder**—A master welder performs many different welds with different processes in all positions.
- **Welding supervisor**—A welding supervisor monitors other welders to ensure production and quality.
- **Weld inspector**—A weld inspector performs quality assurance by inspecting welds, materials, and procedures.
- **Shop foreman**—A shop foreman orders materials, monitors welders, and organizes teams.
- **Sales**—Welding salespeople sell equipment, consumables, and gases. They also frequently provide help in setting up new equipment.
- **Troubleshooter**—A troubleshooter fixes and repairs equipment, solves problems, and maintains equipment.
- **Instructor**—An instructor teaches welding procedures, welding processes, and shop safety.
- **Independent shop owner**—An independent shop owner is a decision maker who oversees all of the shop business.

Other careers in the welding area require further education. The following careers generally require advanced knowledge, including knowledge of math, physics, and metallurgy:

- **Welding engineer**—Welding engineers design parts, select welding processes, determine joint types, select materials, and oversee the process.
- **Welding research engineer**—Research engineers research new welding processes, materials, and strength-to-weight ratios. They also play an important role in the design process.
- **Welding professor**—Professors train the future of the welding industry. They teach students about welding theory and demonstrate proper welding procedures.
- **Certified Welding Inspector (CWI)**—A CWI's duties vary depending on the job they are hired for. These duties include inspection of welds, evaluation of materials and welding procedures, and testing of welders' abilities.

- **Consultant**—Consultants are typically engineers or welders with advanced training or expertise in a specialized area within the welding field. Companies often hire consultants for advanced problem solving or to provide specialized training to the company's employees.

All of these career possibilities start with basic welding skills and understanding of the welding processes. Hard work, determination, and dedication to the trade can lead to a successful career in the welding and manufacturing industry.

Workplace Skills

Employers are very focused on the skills they want their employees to have. There has been a recent focus on the soft skills needed to maintain a job. *Soft skills* include the ability to begin work on time, work well with others, and follow directions with a positive attitude. It is as important for an employee to have good soft skills as it is to have good technical skills.

Good communication skills, both speaking and listening, are essential for a welder. In a fast-paced shop setting, verbal directions are sometimes the preferred way of conveying information to workers. Employers expect employees to accurately follow directions. Mistakes made due to lack of communication can be extremely costly. Welders must be able to communicate effectively with supervisors, coworkers, and customers.

Written communication is also very important in manufacturing and welding industries. For example, welders are required to read and be able to understand welding procedure specification sheets. They must also correctly fill out paperwork, such as time cards or reports.

Estimating Costs

How much will it cost? Give me a ballpark figure! How long will it take? Is the welder certified? What kind of inspection is required on the completed weld?

These and many more questions arise when the cost of making a weld is considered. The true cost of making a weld is probably never known. However, by considering the major factors, a reasonable estimated cost can be determined.

Labor and Consumables

Labor is all of the work required to complete the job. This should include welding setup time, actual welding time, and the time used after welding for checking the weld, removal of the part from the tool, etc. The term *clock hours* refers to the number of hours a welder is at work. The term *arc hours* refers to the number of hours he or she is actually welding. The number of arc hours spent on a project is usually small in comparison to the number of clock hours spent because the weld needs to be prepped and finished.

When labor cost is computed, a welder's efficiency and operating factors must also be considered. Actual arc hours versus clock hours for GTAW vary considerably depending on many factors, including the following:
- position of the weld
- accessibility of the weld
- tungsten preparation and maintenance
- welding sequence/welding equipment setup
- proper gas shielding (changing bottles)
- preheat, interpass, and postheat operations
- interpass cleaning
- interpass weld inspection
- variance in weld procedure for incorrect weld joint dimensions
- addition of or disassembly of tooling during the welding sequence
- application of NDT during the welding sequence
- welder fatigue
- change of wire type or a diameter during the welding sequence
- equipment duty cycle
- welder training or experience for the type of weld being made and the equipment being used

Having a skilled welder clean parts, sharpen electrodes, change gas bottles, keep records, and move parts to and from the area is not a wise use of time or money. Using a helper to perform these duties increases the number of arc hours that a welder can work in a day. As the welder is welding, the helper is prepping the next weld. The helper's wage, which is lower than the welder's, is included in the actual labor cost. The decreased labor cost makes the job more cost-effective, and the increased amount of time the welder spends welding boosts overall productivity.

Consumables are the materials or parts expended during the welding operation. They include the following:
- filler metal
- consumable inserts
- backing rings (welded into the joint)
- gases
- fluxes

Welding power is the energy consumed during the operation. The cost of this energy should be considered, as well as energy consumed by all auxiliary equipment, including the following:
- electrical heaters
- water coolers
- wire feeders
- oscillators
- positioners and manipulators
- grinders, brushes, weld shavers

Tooling and Inspection Cost

For simple shop welds, tooling and inspection costs can be considered part of the general overhead cost. However, if special equipment, tooling, and inspections are required, these costs should be calculated separately.

Tooling cost will depend on the complexity of the weldment, the quantity of the items to be produced, and the production schedule. The total cost must also include tool design time, tool tryout, tool modifications (if required), and maintenance.

Inspections made during the fabrication cycle are called *in-process inspections*. They are often used to verify the quality of root passes in multilayer welds. They are also used as a preliminary inspection for possible cracking on welds with high restraint. The results of these inspections are not usually considered in the final NDT required for the welds.

Some specifications require the use of an official inspector for in-process and final inspection. In most cases, the in-process inspections are called *hold-points*, and the part cannot be moved beyond this point until the inspection is complete and the part is found to be free from defects. The inspector's time is billed to the company doing the work.

Procedure Qualification and Welder Certification

In some cases, procedure qualification and welder certification may not be required for simple jobs around the shop. In other cases, they are required, but the process is relatively simple and quick and may be considered part of the shop's normal operating expenses, or overhead. However, for some jobs, procedure qualification and welder certification is a very involved process that can add considerable cost to the project.

Many specifications limit changes in the procedure qualification to a few areas. Therefore, if major changes are to be made during production, new welding procedures must be developed and qualified.

Welder training and welder qualification tests must be thoroughly considered when calculating the cost of a weldment. While expense is associated with testing and certification, savings result from the superior productivity of qualified welders.

Welder recertification is usually required after a set period of time. In some cases, a part is removed from production and tested until destruction in order to verify that the welder should remain certified. Changes in the welding procedure specification may nullify a welder's certification. In these cases, a welder must again certify to the new procedure. Welders who do not pass the certification test may be required to have additional training. The specification may then require a double test to satisfy the requirements. All of these factors impact cost.

Nondestructive Testing

Nondestructive testing costs relate to the contract or specification. All types of inspections take time and add nothing to the quality of the part. However, inspection does promote quality in the process and production of higher quality parts. Inspection costs may be included in overhead cost, or they may be applied as a direct cost to the job.

If exact costs are not known for each type of inspection required, it is advisable to seek expert help. Inspection labs experienced in these areas can compare the proposed testing to tests they have performed in the past and give an accurate cost analysis. When figuring cost, always remember that welds rejected by NDT must be completely reinspected before they can be accepted.

Material Processing

The cost of material processing varies considerably, since all metals are not processed the same. Some metals require a chemical etch before welding, while others require only a wire brushing before welding. Specifications can be quite explicit regarding how the material must be processed before, during, and after welding.

A commonly used method for estimating material processing is to prepare a job routing sheet listing each material processing operation required to complete the job. Then, the estimated costs of each operation are added together to calculate the estimated cost of material processing for the entire job.

Repair

It is often difficult to accurately measure repair costs. Welds made under very tight inspection restrictions often require special test procedures before weld repair can be done. In areas where dimensions are held to close tolerances, repairs often ruin the part and the unit must be scrapped.

Researching previous welds that are similar to the proposed weld aids in making the best possible estimate of the extent and difficulty of repairs that can be expected. Where repairs were required, what was done to eliminate the defects, and what was the cost of the repair? Records and experience are the best tools for estimating repair costs.

To minimize the number of weld repairs, it is suggested that all operating sequences be closely controlled. In-process inspections should be added as required to maintain weld quality.

Overhead

Overhead cost usually includes the day-to-day expenses of doing business, such as the following items:
- rent
- telephone and office supplies
- insurance
- salaries or commissions of supervisors and officials
- utilities
- licenses, permits, donations, advertising
- transportation
- services, such as engineering, tool design, and estimating
- shop equipment, maintenance
- Social Security
- unemployment compensation
- paid time off
- incentive pay, overtime
- welder certifications, training

Final Cost

To compute the final cost of the project, all the preceding costs are figured individually and added together. To this figure, a profit margin is added. The profit margin is variable and can range from 10%–100% or more of the computed cost. The lower percentage figure may be allowed for simple jobs. A higher figure is allowed for work that is delicate, complicated, or requires considerable development prior to or during production.

Flat Rate Cost

Some welding facilities figure all of the cost in one rate and then consider the total hours to be expended to complete the job. This is called the *hourly flat rate cost*. Many smaller shops use this system, to which material, tooling, and other special areas needed for the job may be added.

Maintaining the Workplace

Maintaining the workplace is the job of everyone that uses the shop. A clean workplace has many benefits: reduced hazards, increased production, and a more positive atmosphere. A well-maintained workplace promotes safety in the shop and is much more efficient. A tremendous amount of time can be wasted looking for tools or equipment in a cluttered workspace. A cluttered shop also presents safety hazards and even the possibility of equipment damage. A good rule of thumb is "everything has a place, and every place has a thing."

Many companies require employees to spend a given amount of time each day cleaning the work area, returning tools to their place, properly rolling up cables and cords to eliminate trip hazards, and making sure all equipment is in proper working order for the next shift or the next day. Floors need to be swept clean each day to avoid slipping hazards and even possibly a fire hazard.

Equipment Inspection

The welder using the equipment should perform an equipment inspection on a regular basis. Preventive maintenance of the welding machines keeps cost down and increases the longevity of the equipment.

On a daily basis, the welder should check that hose fittings are tight with no leaks, all cable ends are tight with good connections, and hoses and feed lines are not kinked or damaged and are properly rolled up. It is also a good idea to check the connections on both ends of the ground cable. Most importantly, the welder should make sure the machine's cooling system is working properly.

On a weekly basis, machines should be moved, if possible, and the area underneath them swept to prevent dirt or foreign objects from entering the cooling fans. All equipment vents should be blown out with compressed air, and the machines should be blown off and wiped down. Dust and dirt can wreak havoc on a machine's switches and dials.

Many machine malfunctions can be avoided by proper periodic maintenance. The cost of a downed machine is something that can usually be avoided. When a welding machine has a larger problem, it should be immediately removed from service and repaired by a qualified technician only. Working on the inside of a welding machine can result in damage to the machine and even possible electrocution.

AWS (SENSE) Welder Certification Program

SENSE stands for *Schools Excelling through National Skill Standards Education* and was introduced by the American Welding Society to help establish a national standard for students and welding programs. The SENSE program has many components that help ensure a proper professional training program. Many welding programs have adopted SENSE guidelines, and some have become certified SENSE programs, which offer their students SENSE certification upon completion of the program.

SENSE guidelines leave room for an individualized curriculum, as long as the students learn certain key topics. When a student passes tests proving knowledge of these key topics and the basic skills required by the SENSE program, he or she can become a SENSE-certified welder. In order to be certified, a student must demonstrate general knowledge about welding symbol interpretation, weld testing and inspection, welding safety, and thermal cutting. In addition, the student must demonstrate proficiency using one or more of the following welding processes: GTAW, GMAW, FCAW, and SMAW.

The advantage of a SENSE certification is that it shows prospective employers that the student has demonstrated the necessary skills to be competitive right from the start. Also, SENSE-certified students become registered in the SENSE certification national database. This gives SENSE-certified welders an advantage when seeking employment.

The SENSE program was developed and based around standards set by the industry on a national level. It is recognized throughout the welding industry as an excellent training outline, providing students with the skills and knowledge they need to be successful.

Summary

To ensure that a weld will perform the job intended, rules are made to cover the procedures, methods, materials, qualification, and testing that will be used to produce and test the weld. The rules may be in the form of orders, codes, specifications, or instructions.

A welding procedure specification (WPS) is a document created by the fabrication company. The WPS details all of the requirements and instructions for welding a specific joint. The WPS is developed by making test welds of the various joints required by the weldment drawings. The data from acceptable test welds is recorded in a welding schedule, which is then used by the welder to set up the joint, set up the power source and related equipment, and make the weld.

Nondestructive testing and destructive testing are performed on completed test welds to confirm that the settings and variables in the WPS produce acceptable results. The settings used in the WPS and the results of the DT and NDT are recorded in a welding procedure qualification record (WPQR).

In order to become certified for a weld, a welder must successfully complete a welder performance qualification test. The length of time a certification is effective is variable. Certification can be rescinded if poor-quality welds are produced or if a welder does not use the process within a specified period of time.

There is no general specification in current use for GTAW. Large companies that use GTAW usually have a company specification. Other specifications are found in Section IX of the ASME Boiler Code, the AWS B 2.14 specification, and the MIL-STD 1595 specification.

Each weld type and material type is qualified by test procedures. Welders use qualified test procedure parameters and variables to achieve certification. Qualification of the weld procedure remains in effect until an essential variable is changed. A welder's certification remains in effect while the welder is employed using the qualified procedure.

Many different careers are available in the welding industry. Employers want employees with good soft skills, including the ability to work cooperatively with others, communicate effectively, and accurately follow directions. Welders must be able to read and understand documents such as welding procedure specification sheets. They must also be able to correctly complete paperwork.

Estimating costs includes a consideration of the cost of labor, consumables, tooling, inspections, procedure qualification and welder certification, material processing, and weld repair. Overhead cost includes all items associated with the expense of doing business, such as rent, utilities, and insurance. Adding all costs together and adding a profit margin results in the final cost of the project. Figuring all the costs in one rate and then considering the total hours needed to complete the job is called an hourly flat rate cost.

The benefits of a clean and uncluttered workplace include increased safety and increased production. Many companies require employees to spend time each day cleaning and organizing the work area and checking that all equipment is in proper working order.

Welders are required to regularly inspect the equipment they use. Basic maintenance of welding machines performed on a daily and weekly basis keeps cost down and increases the longevity of the equipment.

AWS SENSE (Schools Excelling through National Skill Standards Education) program guidelines have been adopted by many welding programs. When a student passes tests proving knowledge of key topics and the basic skills required by the SENSE program, he or she can become a SENSE-certified welder.

Review Questions

Write your answers on a separate sheet of paper. Do not write in this book.

1. List three variables specified in a welding procedure specification.
2. Rules that specify the way a weld must be performed are called _____.
3. What is a welding procedure specification (WPS)?
4. The settings used in the WPS and the results of the destructive and nondestructive testing are recorded in a document called a(n) _____.
5. Performing nondestructive testing and destructive testing on completed test welds to verify that the welding procedure produces acceptable welds is called procedure _____.
6. What does a welder refer to the welding schedule for?
7. What happens to a welder's certification if his or her employment is terminated?
8. What AWS specification is used to define the limits of qualification, certification, and use of the welding process?
9. Qualification of a weld procedure remains in effect until a(n) _____ is changed.
10. List four job responsibilities of welding engineers.
11. The ability to get along with others, follow directions, and act professionally are considered _____ skills.
12. What are *consumables*?
13. The day-to-day expenses of doing business are referred to as _____ cost.
14. What should a welder look for when checking hoses and feed lines?
15. In order to become SENSE-certified, a student must demonstrate proficiency in one or more of which four welding processes?

Reference Section

The following pages contain 22 charts that are useful references in a variety of welding-related areas. To make locating the information easier, the charts are listed below by title and page number.

Welding Safety Checklist		
Hazard	**Factors to Consider**	**Precautionary Summary**
Electric shock can kill	• Wetness • Welder in or on workpiece • Confined space • Electrode holder and cable insulation	• Insulate welder from workpiece and ground using dry insulation, rubber mat, or dry wood. • Wear dry, hole-free gloves. (Change as necessary to keep dry.) • Do not touch electrically "hot" part or electrode with bare skin or wet clothing. • If wet area and welder cannot be insulated from workpiece with dry insulation, use a semiautomatic, constant-voltage welding machine or stick welding machine with voltage-reducing device. • Keep electrode holder and cable insulation in good condition. Do not use if insulation is damaged or missing.
Fumes and gases can be dangerous	• Confined area • Positioning of welder's head • Lack of general ventilation • Electrode types, i.e., manganese, chromium, etc. • Base metal coatings, galvanizing, paint	• Use ventilation or exhaust to keep air-breathing zone clear and comfortable. • Use helmet and position of head to minimize fume breathing zone. • Read warnings on electrode container and material safety data sheet for electrode. • Provide additional ventilation/exhaust where special ventilation requirements exist. • Use special care when welding in a confined area. • Do not weld unless ventilation is adequate.
Welding sparks can cause fire or explosion	• Containers that have held combustibles • Flammable materials	• Do not weld on containers that have held combustible materials (unless strict AWS F4.1 explosion procedures are followed). Check before welding. • Remove flammable materials from welding area or shield from sparks, heat. • Keep a fire watch in area during and after welding. • Keep a fire extinguisher in the welding area. • Wear fire-retardant clothing and hat. Use earplugs when welding overhead.
Arc rays can burn eyes and skin	• Process: gas-shielded arc most severe	• Select a filter lens that is comfortable for you while welding. • Always use helmet when welding. • Provide nonflammable shielding to protect others. • Wear clothing that protects skin while welding.
Confined space	• Metal enclosure • Wetness • Restricted entry • Heavier-than-air gas • Welder inside or on workpiece	• Carefully evaluate adequacy of ventilation, especially where electrode requires special ventilation or where gas may displace breathing air. • If basic electric shock precautions cannot be followed to insulate welder from work and electrode, use semiautomatic, constant-voltage equipment with cold electrode or stick welding machine with voltage-reducing device. • Provide welder helper and method of welder retrieval from outside enclosure.
General work area hazards	• Cluttered area	• Keep cables, materials, tools neatly organized.
	• Indirect work (welding ground) connection	• Connect work cable as close as possible to area where welding is being performed. Do not allow alternate circuits through scaffold cables, hoist chains, ground leads.
	• Electrical equipment	• Use only double insulated or properly grounded equipment. • Always disconnect power to equipment before servicing.
	• Engine-driven equipment	• Use only in open, well-ventilated areas. • Keep enclosure complete and guards in place. • Refuel with engine off. • If using auxiliary power, OSHA may require GFCI protection or assured grounding program (or isolated windings if less than 5 KW).
	• Gas cylinders	• Never touch cylinder with the electrode. • Never lift a machine with cylinder attached. • Keep cylinder upright and chained to support.

Chart R-1.   Welding safety checklist. (Lincoln Electric Company)

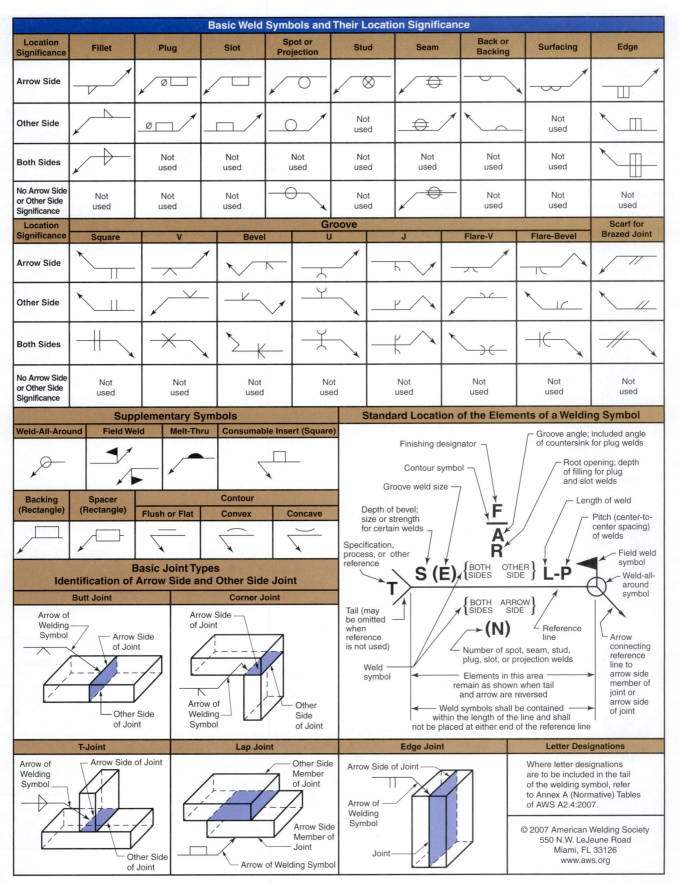

Chart R-2.   Basic welding symbols and their location significance. (AWS A2.4-2007, Standard Symbols for Welding, Brazing, and Nondestructive Examination, reproduced with permission from the American Welding Society (AWS), Miami, FL USA)

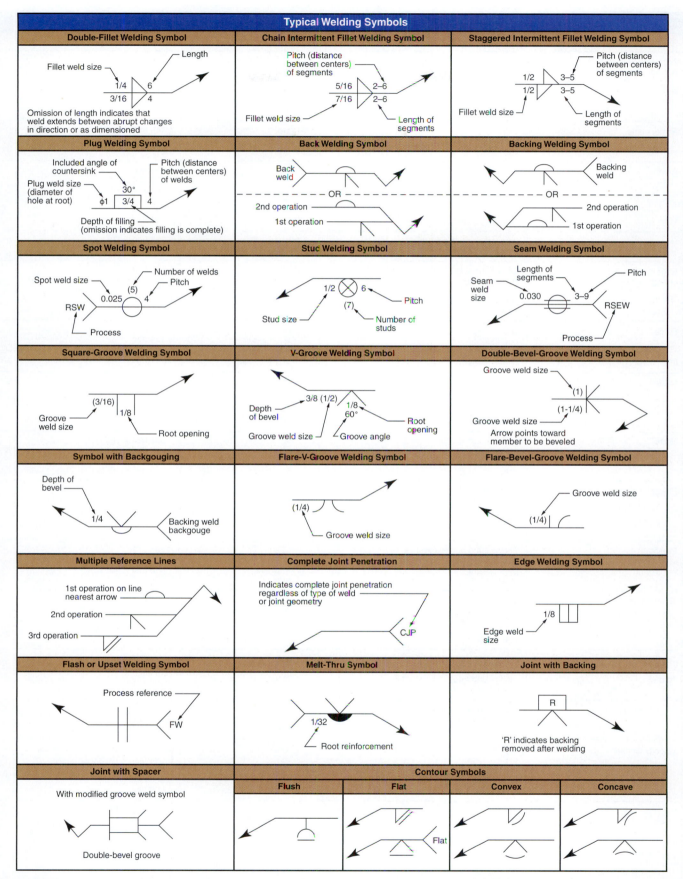

General Metric/U.S. Customary Conversions			
Property	**To Convert from**	**To**	**Multiply by**
Acceleration (angular) Acceleration (linear)	revolution per minute squared	rad/s^2	$1.745\,329 \times 10^{-3}$
	in/min^2	m/s^2	$7.055\,556 \times 10^{-6}$
	ft/min^2	m/s^2	$8.466\,667 \times 10^{-5}$
	in/min^2	mm/s^{2}mm/s^2	$7.055\,556 \times 10^{-3}$
	ft/min^2	m/s^2	$8.466\,667 \times 10^{-2}$
	ft/s^2	m/s^2	$3.048\,000 \times 10^{-1}$
Angle, plane	deg	rad	$1.745\,329 \times 10^{-2}$
	minute	rad	$2.908\,882 \times 10^{-4}$
	second	rad	$4.848\,137 \times 10^{-6}$
Area	in^2	m^2	$6.451\,600 \times 10^{-4}$
	ft^2	m^2	$9.290\,304 \times 10^{-2}$
	yd^2	m^2	$8.361\,274 \times 10^{-1}$
	in^2	mm^2	$6.451\,600 \times 10$
	ft^2	mm^2	$9.290\,304 \times 10^4$
	acre (U.S. Survey)	m^2	$4.046\,873 \times 10^3$
Density	pound mass per cubic inch	kg/m^2	$2.767\,990 \times 10^4$
	pound mass per cubic foot	kg/m^2	$1.601\,846 \times 10$
Energy, work, heat, and impact energy	foot pound force	J	$1.355\,818$
	foot poundal	J	$4.214\,011 \times 10^{-2}$
	Btu*	J	$1.054\,350 \times 10^3$
	calorie*	J	$4.184\,000$
	watt hour	J	$3.600\,000 \times 10^3$
Force	kilogram-force	N	$9.806\,650$
	pound-force	N	$4.448\,222$
Impact strength	(see Energy)		
Length	in	m	$2.540\,000 \times 10^{-2}$
	ft	m	$3.048\,000 \times 10^{-1}$
	yd	m	$9.144\,000 \times 10^{-1}$
	rod (U.S. Survey)	m	$5.029\,210$
	mile (U.S. Survey)	km	$1.609\,347$
Mass	pound mass (avdp)	kg	$4.535\,924 \times 10^{-1}$
	metric ton	kg	$1.000\,000 \times 10^3$
	ton (short, 2000 lb/m)	kg	$9.071\,847 \times 10^2$
	slug	kg	$1.459\,390 \times 10$
Power	horsepower (550 ft lbs/s)	W	$7.456\,999 \times 10^2$
	horsepower (electric)	W	$7.460\,000 \times 10^2$
	Btu/min*	W	$1.757\,250 \times 10$
	calorie per minute*	W	$6.973\,333 \times 10^{-2}$
	foot pound-force per minute	W	$2.259\,697 \times 10^{-2}$
Pressure	pound force per square inch	kPa	$6.894\,757$
	bar	kPa	$1.000\,000 \times 10^2$
	atmosphere	kPa	$1.013\,250 \times 10^2$
	kip/in^2	kPa	$6.894\,757 \times 10^3$
Temperature	degree Celsius, t°C	K	tK = t°C + 273.15
	degree Fahrenheit, t°F	K	tK = (t°F + 459.67)/1.8
	degree Rankine, t°R	°R	tK = t°R/1.8
	degree Fahrenheit, t°F	°C	t°C = (t°F − 32)/1.8
	kelvin, tK	°C	t°C = tK − 273.15
Tensile strength (stress)	ksi	MPa	$6.894\,757$
Torque	inch pound force	N·m	$1.129\,848 \times 10^{-1}$
	foot pound force	N·m	$1.355\,818$
Velocity (angular)	revolution per minute	rad/s	$1.047\,198 \times 10^{-1}$
	degree per minute	rad/s	$2.908\,882 \times 10^{-4}$
	revolution per minute	deg/min	$3.600\,000 \times 102$
Velocity (linear)	in/min	m/s	$4.233\,333 \times 10^{-4}$
	ft/min	m/s	$5.080\,000 \times 10^{-3}$
	in/min	mm/s	$4.233\,333 \times 10^{-1}$
	ft/min	mm/s	$5.080\,000$
	mile/hour	km/h	$1.609\,344$
Volume	in^3	m^3	$1.638\,706 \times 10^{-5}$
	ft^3	m^3	$2.831\,685 \times 10^{-2}$
	yd^3	m^3	$7.645\,549 \times 10^{-1}$
	in^3	mm^3	$1.638\,706 \times 10^4$
	ft^3	mm^3	$2.831\,685 \times 10^7$
	in^3	L	$1.638\,706 \times 10^{-2}$
	ft^3	L	$2.831\,685 \times 10$
	gallon	L	$3.785\,412$

*Thermochemical

Chart R-4.   General metric/U.S. Customary conversions.

Metric – Inch Equivalents

Inches Fractions	Decimals	Millimeters	Inches Fractions	Decimals	Millimeters
	.00394	.1	15/32	.46875	11.9063
	.00787	.2		.47244	12.00
	.01181	.3	31/64	.484375	12.3031
1/64	.015625	.3969	1/2	.5000	12.70
	.01575	.4		.51181	13.00
	.01969	.5	33/64	.515625	13.0969
	.02362	.6	17/32	.53125	13.4938
	.02756	.7	35/64	.546875	13.8907
1/32	.03125	.7938		.55118	14.00
	.0315	.8	9/16	.5625	14.2875
	.03543	.9	37/64	.578125	14.6844
	.03937	1.00		.59055	15.00
3/64	.046875	1.1906	19/32	.59375	15.0813
1/16	.0625	1.5875	39/64	.609375	15.4782
5/64	.078125	1.9844	5/8	.625	15.875
	.07874	2.00		.62992	16.00
3/32	.09375	2.3813	41/64	.640625	16.2719
7/64	.109375	2.7781	21/32	.65625	16.6688
	.11811	3.00		.66929	17.00
1/8	.125	3.175	43/64	.671875	17.0657
9/64	.140625	3.5719	11/16	.6875	17.4625
5/32	.15625	3.9688	45/64	.703125	17.8594
	.15748	4.00		.70866	18.00
11/64	.171875	4.3656	23/32	.71875	18.2563
3/16	.1875	4.7625	47/64	.734375	18.6532
	.19685	5.00		.74803	19.00
13/64	.203125	5.1594	3/4	.7500	19.05
7/32	.21875	5.5563	49/64	.765625	19.4469
15/64	.234375	5.9531	25/32	.78125	19.8438
	.23622	6.00		.7874	20.00
1/4	.2500	6.35	51/64	.796875	20.2407
17/64	.265625	6.7469	13/16	.8125	20.6375
	.27559	7.00		.82677	21.00
9/32	.28125	7.1438	53/64	.828125	21.0344
19/64	.296875	7.5406	27/32	.84375	21.4313
5/16	.3125	7.9375	55/64	.859375	21.8282
	.31496	8.00		.86614	22.00
21/64	.328125	8.3344	7/8	.875	22.225
11/32	.34375	8.7313	57/64	.890625	22.6219
	.35433	9.00		.90551	23.00
23/64	.359375	9.1231	29/32	.90625	23.0188
3/8	.375	9.525	59/64	.921875	23.4157
25/64	.390625	9.9219	15/16	.9375	23.8125
	.3937	10.00		.94488	24.00
13/32	.40625	10.3188	61/64	.953125	24.2094
27/64	.421875	10.7156	31/32	.96875	24.6063
	.43307	11.00		.98425	25.00
7/16	.4375	11.1125	63/64	.984375	25.0032
29/64	.453125	11.5094	1	1.0000	25.4000

Chart R-5. Metric–inch equivalents.

Converting Measurements for Common Welding Properties			
Property	**To convert from**	**To**	**Multiply by**
Area dimensions (mm²)	in² mm²	mm² in²	$6.451\ 600 \times 10^2$ $1.550\ 003 \times 10^{-3}$
Current density (A/mm²)	A/in² a/mm²	A/mm² A/in²	$1.550\ 003 \times 10^{-3}$ $6.451\ 600 \times 10^2$
Deposition rate** (kg/h)	lb/h kg/h	kg/h lb/h	0.045** 2.2**
Electrical resistivity (Ω·m)	Ω·cm Ω·m	Ω·m Ω·cm	$1.000\ 000 \times 10^{-2}$ $1.000\ 000 \times 10^2$
Electrode force (N)	pound-force kilogram-force N	N N ibf	4.448 222 9.806 650 $2.248\ 089 \times 10^{-1}$
Flow rate (L/min)	ft³/h gallon per hour gallon per minute cm³/min L/min cm³/min	L/min L/min L/min L/min ft³/min ft³/min	$4.719\ 475 \times 10^{-1}$ $6.309\ 020 \times 10^{-2}$ 3.785 412 $1.000\ 000 \times 10^{-3}$ 2.118 880 $2.118\ 880 \times 10^{-3}$
Fracture toughness (MN·m⁻³ᐟ²)	ksi·in¹ᐟ² MN·m⁻³ᐟ²	MN·m⁻³ᐟ² ksi·in¹ᐟ²	1.098 855 0.910 038
Heat input (J/m)	J/in J/m	J/m J/in	$3.937\ 008 \times 10$ $2.540\ 000 \times 10^{-2}$
Impact energy	foot pound force	J	1.355 818
Linear measurements (mm)	in ft mm mm	mm mm in ft	$2.540\ 000 \times 10$ $3.048\ 000 \times 10^2$ $3.937\ 008 \times 10^{-2}$ $3.280\ 840 \times 10^{-3}$
Power density (W/m²)	W/in² W/m²	W/m² W/in²	$1.550\ 003 \times 103$ $6.451\ 600 \times 10^{-4}$
Pressure (gas and liquid) (kPa)	psi lb/ft² N/mm² kPa kPa kPa torr (mm Hg at 0°C) micron (μm Hg at 0°C) kPa kPa	Pa Pa Pa psi il/ft² N/mm² kPa kPa torr micron	$6.894\ 757 \times 10^3$ $4.788\ 026 \times 10$ $1.000\ 000 \times 10^6$ $1.450\ 377 \times 10^{-1}$ $2.088\ 543 \times 10$ $1.000\ 000 \times 10^{-3}$ $1.333\ 22 \times 10^{-1}$ $1.333\ 22 \times 10^{-4}$ $7.500\ 64 \times 10$ $7.500\ 64 \times 10^3$
Tensile strength (MPa)	psi lb/ft² N/mm² MPa MPa MPa	kPa kPa MPa psi lb/ft² N/mm²	6.894 757 $4.788\ 026 \times 10^{-2}$ 1.000 000 $1.450\ 377 \times 10^2$ $2.088\ 543 \times 10^4$ 1.000 000
Thermal conductivity (W[m·K])	cal/(cmÃs·°c)	W/(m·K)	$4.184\ 000 \times 10^2$
Travel speed, electrode feed speed (mm/s)	in/min mm/s	mm/s in/min	$4.233\ 333 \times 10^{-2}$ 2.362 205

*Preferred units are given in parentheses.
**Approximate conversion.

Chart R-6.   Converting measurements for common welding properties.

Metric Units for Welding		
Property	**Unit**	**Symbol**
Area dimensions	Square millimeter	mm^2
Current density	Ampere per square millimeter	A/mm^2
Deposition rate	Kilogram per hour	kg/h
Electrical resistivity	Ohmmeter	$\Omega \cdot m$
Electrode force (upset, squeeze, hold)	Newton	N
Flow rate (gas and liquid)	Liter per minute	L/min
Fracture toughness	Meganewton meter$^{-3/2}$	$MN \cdot m^{-3/2}$
Impact strength	Joule	$J = N \cdot m$
Linear dimensions	Millimeter	mm
Power density	Watt per square meter	W/m^2
Pressure (gas and liquid)	Kilopascal	$kPa = 1000\ N/m^2$
Tensile strength	Megapascal	$MPa = 1\,000\,000\ N/m^2$
Thermal conductivity	Watt per meter Kelvin	$W/(m \cdot K)$
Travel speed	Millimeter per second	mm/s
Volume dimensions	Cubic millimeter	mm^3
Electrode feed rate	Millimeter	mm/s

Chart R-7.   Metric units for welding.

Wire Size	
Inch	**mm**
0.030	0.8
0.035	0.9
0.040	1.0
0.045	1.2
1/16	1.6
5/64	2.0
3/32	2.4
1/8	3.25
5/32	4.0
3/16	5.0
1/4	6.0

Chart R-8.   Wire size conversion to metric.

Hardness Conversion Table						
Brinell	Hardness No.	Vickers or Firth Hardness No.	Rockwell		Scleroscope	Tensile Strength 1000 psi
Dia. in mm, 3000 kg Load 10 mm Ball			C 150 kg Load 120° Diamond Cone	B 100 kg Load 1/16″ dia. Ball		
2.05	898					440
2.10	857					420
2.15	817					401
2.20	780	1150	70		106	384
2.25	745	1050	68		100	368
2.30	712	960	66		95	352
2.35	682	885	64		91	337
2.40	653	820	62		87	324
2.45	627	765	60		84	311
2.50	601	717	58		81	298
2.55	578	675	57		78	287
2.60	555	633	55	120	75	276
2.65	534	598	53	119	72	266
2.70	514	567	52	119	70	256
2.75	495	540	50	117	67	247
2.80	477	515	49	117	65	238
2.85	461	494	47	116	63	229
2.90	444	472	46	115	61	220
2.95	429	454	45	115	59	212
3.00	415	437	44	114	57	204
3.05	401	420	42	113	55	196
3.10	388	404	41	112	54	189
3.15	375	389	40	112	52	182
3.20	363	375	38	110	51	176
3.25	352	363	37	110	49	170
3.30	341	350	36	109	48	165
3.35	331	339	35	109	46	160
3.40	321	327	34	108	45	155
3.45	311	316	33	108	44	150
3.50	302	305	32	107	43	146
3.55	293	296	31	106	42	142
3.60	285	287	30	105	40	138
3.65	277	279	29	104	39	134
3.70	269	270	28	104	38	131
3.75	262	263	26	103	37	128
3.80	255	256	25	102	37	125
3.85	248	248	24	102	36	122
3.90	241	241	23	100	35	119
3.95	235	235	22	99	34	116
4.00	229	229	21	98	33	113
4.05	223	223	20	97	32	110
4.10	217	217	18	96	31	107
4.15	212	212	17	96	31	104
4.20	207	207	16	95	30	101
4.25	202	202	15	94	30	99
4.30	197	197	13	93	29	97
4.35	192	192	12	92	28	95
4.40	187	187	10	91	28	93
4.45	183	183	9	90	27	91
4.50	179	179	8	89	27	89
4.55	174	174	7	88	26	87
4.60	170	170	6	87	26	85
4.65	166	166	4	86	25	83
4.70	163	163	3	85	25	82
4.75	159	159	2	84	24	80
4.80	156	156	1	83	24	78
4.85	153	153		82	23	76
4.90	149	149		81	23	75
4.95	146	146		80	22	74
5.00	143	143		79	22	72
5.05	140	140		78	21	71
5.10	137	137		77	21	70
5.15	134	134		76	21	68
5.20	131	131		74	20	66
5.25	128	128		73	20	65
5.30	126	126		72		64
5.35	124	124		71		63
5.40	121	121		70		62
5.45	118	118		69		61
5.50	116	116		68		60
5.55	114	114		67		59
5.60	112	112		66		58
5.65	109	109		65		56
5.70	107	107		64		56
5.75	105	105		62		54
5.80	103	103		61		53
5.85	101	101		60		52
5.90	99	99		59		51
5.95	97	97		57		50
6.00	95	95		56		49

Chart R-9.   Hardness conversion table.

Metal	Melting Point	
	(°C)	(°F)
Admiralty brass	900–940	1650–1720
Aluminum	660	1220
Aluminum bronze	600–665	1190–1215
Brass	930	1710
Carbon steel	1425–1540	2600–2800
Cast iron, gray	1175–1290	2150–2360
Chromium	1860	3380
Cobalt	1495	2723
Copper	1084	1983
Cupronickel	1170–1240	2140–2260
Hastelloy C	1320–1350	2410–2460
Incoloy	1390–1425	2540–2600
Inconel	1390–1425	2540–2600
Iron	1536	2797
Lead	327.5	621
Magnesium	650	1200
Manganese	1244	2271
Manganese bronze	865–890	1590–1630
Molybdenum	2620	4750
Monel	1300–1350	2370–2460
Nickel	1453	2647
Red brass	990–1025	1810–1880
Silicon	1411	2572
Stainless steel	1510	2750
Tin	232	449.4
Titanium	1670	3040
Tungsten	3400	6150
Yellow brass	905–932	1660–1710

Chart R-10.   Melting points of metals.

Etching Procedures			
Reagents	**Composition**	**Procedure**	**Uses**
Solutions for Aluminum			
Sodium hydroxide	NaOH . . . . 1 g Acetic acid . . . . .20 ml	Swab 10 seconds	General microscopic
Tucker's etch	HF. . . . . .15 ml HCl . . . . .45 ml HNO_3 . . . . .15 ml H_2O. . . . . .25 ml	Etch by immersion	Macroscopic
Solutions for Stainless Steel			
Nitric and acetic acids	HNO_3 . . . . .30 ml Acetic acid . . . . .20 ml	Apply by swabbing	Stainless alloys and others high in nickel and cobalt
Cupric sulphate	$CuSO_4$. . . . . 4 gms HCl . . . . .20 ml H_2O . . . . .20 ml	Etch by immersion	Structure of stainless steels
Cupric chloride and hydrochloric acid	$CuCl_2$ . . . . 5 gms HCl . . . . .100 ml Ethyl alcohol .100 ml H_2O . . . . .100 ml	Use cold immersion or swabbing	Austenitic and ferritic steels
Solutions for Copper and Brass			
Ammonium hydroxide and ammonium persulphate	NH_4OH . . . . 1 part H_2O. . . . . 1 part $(NH_4)_2S_2O_8$ (2 1/2%) . . . . 2 parts	Immersion	Polish attack of copper and some alloys
Chromic acid	Saturated aqueous solution (CrO_3)	Immersion or swabbing	Copper, brass, bronze, nickel, silver (plain etch)
Ferric chloride	$FeCl_3$. . . . . 5 parts HCl . . . . 10 parts H_2O . . . . 100 parts	Immersion or swabbing (etch lightly)	Copper, brass, bronze, aluminum bronze
Solutions for Iron and Steel			
Macro Examination			
Nitric acid	HNO_3 . . . . .5 ml H_2O . . . . .95 ml	Immerse 30 to 60 seconds	Shows structure of welds
Ammonium persulphate	$(NH_4)_2S_2O_3$ . . . . 10 gms H_2O. . . . . .90 ml	Surface should be rubbed with cotton during etching	Brings out grain structure, recrystallization at welds
Nital	HNO_3 . . . . .5 ml Ethyl alcohol . . . . .95 ml	Etch 5 min. followed by 1 sec. in HCl (10%)	Shows cleanness, depth of hardening, carburized or decarburized surfaces, etc.
Micro Examination			
Picric acid (picral)	Picric acid. . . . . 4 gms Ethyl or methyl alcohol (95%) . . . . . . .100 ml	Etching time a few seconds to a minute or more	For all grades of carbon steels

Chart R-11.   Etching procedures.

Etching Reagents for Microscopic Examination of Iron and Steel			
Application	**Etching**	**Composition**	**Remarks**
Carbon low-alloy and medium-alloy steels	Nital	Nitric acid (sp gr 1.42) . . . . . 1–5 ml Ethyl or methyl alcohol. . . 95–99 ml	Darkens perlite and gives contrast between adjacent colonies; reveals ferrite boundaries; differentiates ferrite from martensite; shows case depth of nitrided steel. Etching time: 5–60 secs.
	Picral	Picric acid . . . . 4 g Methyl alcohol . . . . 100 ml	Used for annealed and quench-hardened carbon and alloy steel. Not as effective as No. 1 for revealing ferrite grain boundaries. Etching time: 5–120 secs.
	Hydrochloric and picral acids	Hydrochloric acid . . . . 5 ml Picric acid . . . . 1 g Methyl alcohol . . . . 100 ml	Reveals austenitic grain size in both quenched and quenched-tempered-steels.
Alloy and stainless steel	Mixed acids	Nitric acid . . . . 10 ml Hydrochloric acid. . . . . 20 ml Glycerol . . . . 20 ml Hydrogen peroxide . . . . 10 ml	Iron-chromium-nickel-manganese alloy steel. Etching: Use fresh acid.
	Ferric chloride	Ferric chloride . . . . 5 g Hydrochloric acid . . . . 20 ml Water, distilled . . . . 100 ml	Reveals structure of stainless and austenitic nickel steels.
	Marble's reagent	Cupric sulfate . . . . 4 g Hydrochloric acid . . . . 20 ml Water, distilled . . . . 20 ml	Reveals structure of various stainless steels.
High-speed steels	Snyder-Graff	Hydrochloric acid . . . . 9 ml Nitric acid . . . . 9 ml Methyl alcohol . . . . 100 ml	Reveals grain size of quenched and tempered high-speed steels. Etching time: 15 secs. to 5 min.

Chart R-12.   Etching reagents for microscopic examination of iron and steel.

Filter Plate Lenses		
Welding Process	**Approximate Welding Range (In amps)**	**Federal Specification Filter Shade Required**
Metal arc welding (coated electrodes). Continuous covered electrode welding. Carbon dioxide shield continuous covered electrode welding.	100 100–300 Over 300	8 or 9 10 or 11 12 or 14
Metal arc welding (bare wire). Carbon arc welding. Inert gas metal arc welding. Atomic hydrogen welding.	200 Over 200	10 or 11 12 or 14
Automatic carbon dioxide shield. Metal arc welding (bare wire).	Over 500	15 or 16
Inert gas tungsten arc welding.	15	8
	15–75	9
	75–100	10
	100–200	11
	200–250	12
	250–300	14

Chart R-13.   Filter plate lenses are designed to protect the welder against harmful infrared and ultraviolet rays. The lenses are heat-treated and manufactured to meet or exceed current standards and federal specifications. (Thermacote-Welco Company)

Alloy Wire Weights and Measures						
Brown & Sharpe Gauge No.	Decimal	Phos. Bronze Ft. per lb.	18% Nickel Ft. per lb.	Aluminum Ft. per lb.	Copper Ft. per lb.	Brass Ft. per lb.
4/0	0.4600	1.559	1.599	5.207	1.561	1.640
3/0	0.4096	1.966	2.016	6.567	1.968	1.854
2/0	0.3648	2.480	2.543	8.279	2.482	2.068
1/0	0.3249	3.127	3.206	10.44	3.130	2.608
1	0.2893	3.943	4.043	13.16	3.947	3.289
2	0.2576	4.972	5.098	16.60	4.977	4.147
3	0.2294	6.269	6.428	20.94	6.276	5.229
4	0.2430	7.906	8.106	26.40	7.914	6.594
5	0.1819	9.969	10.22	22.20	9.980	8.315
6	0.1620	12.57	12.89	41.98	12.58	10.49
7	0.1443	15.85	16.25	52.91	15.87	13.22
8	0.1285	19.99	20.49	66.73	20.01	16.67
9	0.1144	25.20	25.84	84.19	25.23	21.02
10	0.1019	31.78	32.59	106.1	31.82	26.51
11	0.09074	40.08	41.09	133.9	40.12	33.43
12	0.08081	50.53	51.82	168.8	50.59	42.15
13	0.07196	63.72	65.39	212.5	63.80	53.15
14	0.06408	80.35	82.34	268.2	80.44	66.88
15	0.05707	101.3	103.9	337.9	101.4	84.68
16	0.05082	127.8	131.0	426.9	127.9	106.6
17	0.04526	161.1	165.2	536.9	161.3	134.4
18	0.04030	203.2	208.3	678.4	203.4	169.7
19	0.03589	256.2	262.7	854.9	256.5	213.7
20	0.03196	323.0	331.2	1076.0	323.4	269.5
21	0.02846	407.3	417.7	1356.0	407.8	339.8
22	0.02535	513.6	526.7	1721.0	514.2	428.5
23	0.02257	647.7	664.1	2157.0	648.4	540.2
24	0.02010	816.7	837.4	2727.0	817.7	681.3
25	0.01790	1030.0	1056.0	3439.0	1031.0	859.0
26	0.01594	1299.0	1332.0	4358.0	1300.0	1083.0
27	0.0142	1638.0	1679.0	5464.0	1639.0	1366.0
28	0.01264	2065.0	2117.0	6940.0	2067.0	1723.0

Chart R-14.   Alloy wire weights and measures.

Inches per Pound of Wire												
Wire Diameter												
Decimal	Fraction	Mag.	Alum.	Alum. Bronze (10%)	Stain-less Steel	Mild Steel	Stain-less Steel	Si. Bronze	Copper Nickel	Nickel	De-ox. Copper	Ti
.020		50500	32400	11600	11350	11100	10950	10300	9950	9900	9800	19524
.025		34700	22300	7960	7820	7680	7550	7100	6850	6820	6750	12492
.030		22400	14420	5150	5050	4960	4880	4600	4430	4400	4360	8776
.035		16500	10600	3780	3720	3650	3590	3380	3260	3240	3200	6372
.040		12600	8120	2900	2840	2790	2750	2580	2490	2480	2450	4884
.045	3/64	9990	6410	2290	2240	2210	2210	2040	1970	1960	1940	3852
.062	1/16	5270	3382	1220	1180	1160	1140	1070	1040	1030	1020	2028
.078	5/64	3300	2120	756	742	730	718	675	650	647	640	
.093	3/32	2350	1510	538	528	519	510	480	462	460	455	964.8
.125	1/8	1280	825	295	289	284	279	263	253	252	249	499.92

Chart R-15.   Inches per pound of wire.

| GTAW Amperage Ranges and Argon Gas Flow | | | | | | | | | |
| Electrode Diameter (inches) | Cup Size | Welding Current (amperage) | | | | Argon Flow (cfh) Ferrous Metals | | Argon Flow (cfh) Aluminum | |
		AC Pure	AC Thoriated	DCEN Pure	DCEN Thoriated	Standard Collet Body	Gas Lens Collet Body	Standard Collet Body	Gas Lens Collet Body
.020	4-5	5–15	5–20	5–15	5–20	5–8	5–8	5–8	5–8
.040	4-5	10–60	15–80	15–70	20–80	5–10	5–8	5–12	5–10
1/16	4-5-6	50–100	70–150	70–130	80–150	7–12	5–10	8–15	7–12
3/32	6-7-8	100–160	140–235	150–220	150–250	10–15	8–10	10–20	10–15
1/8	7-8-10	150–210	220–325	220–330	240–350	10–18	8–12	12–25	10–20
5/32	8-10	200–275	300–425	375–475	400–500	15–25	10–15	15–30	12–25
3/16	8-10	250–350	400–525	475–800	475–800	20–35	12–25	25–40	15–30
1/4	10	325–700	500–700	750–1000	700–1100	25–50	20–35	30–55	25–45

Chart R-16. GTAW amperage ranges and argon gas flow.

AWS Specifications for GTAW Filler Metals	
A5.7	Copper and Copper-Alloy Bare Welding Rods and Electrodes
A5.9	Bare Stainless Steel Welding Electrodes and Rods
A5.10	Bare Aluminum and Aluminum-Alloy Welding Electrodes and Rods
A5.13	Surfacing Electrodes for Shielded Metal Arc Welding
A5.14	Nickel and Nickel-Alloy Bare Welding Electrodes and Rods
A5.16	Titanium and Titanium-Alloy Welding Electrodes and Rods
A5.18	Carbon Steel Electrodes and Rods for Shielded Metal Arc Welding
A5.19	Magnesium Alloy Welding Electrodes and Rods
A5.21	Bare Electrodes and Rods for Surfacing
A5.24	Zirconium and Zirconium-Alloy Welding Electrodes and Rods
A5.28	Low-Alloy Steel Electrodes and Rods for Gas Shielded Metal Arc Welding
A5.30	Consumable Inserts

Chart R-17. AWS specifications for GTAW filler metals.

| GTAW Amperage Ranges and Argon Gas Flow | | | | | | | | | |
| Electrode Diameter (inches) | Cup Size | Welding Current (amperage) | | | | Argon Flow (cfh) Ferrous Metals | | Argon Flow (cfh) Aluminum | |
		AC Pure	AC Thoriated	DCEN Pure	DCEN Thoriated	Standard Collet Body	Gas Lens Collet Body	Standard Collet Body	Gas Lens Collet Body
.020	4-5	5–15	5–20	5–15	5–20	5–8	5–8	5–8	5–8
.040	4-5	10–60	15–80	15–70	20–80	5–10	5–8	5–12	5–10
1/16	4-5-6	50–100	70–150	70–130	80–150	7–12	5–10	8–15	7–12
3/32	6-7-8	100–160	140–235	150–220	150–250	10–15	8–10	10–20	10–15
1/8	7-8-10	150–210	220–325	220–330	240–350	10–18	8–12	12–25	10–20
5/32	8-10	200–275	300–425	375–475	400–500	15–25	10–15	15–30	12–25
3/16	8-10	250–350	400–525	475–800	475–800	20–35	12–25	25–40	15–30
1/4	10	325–700	500–700	750–1000	700–1100	25–50	20–35	30–55	25–45

Chart R-18. AWS designations for solid steel wires for GTAW.

Liquid to Gas Equivalents				
		Standard Cubic Feet of Gas (at 70°F, sea level pressure)		
Gallons of Liquid (at sea level pressure)	Approximate Number of 244 SCF Oxygen Cylinders	Oxygen	Nitrogen	Argon
1	1/2	115.6	92.9	113.2
10	5	1156	929	1132
20	9	2312	1858	2264
30	14	3468	2787	3396
40	19	4624	3716	4528
50	24	5780	4645	5660
60	28	6936	5574	6792
70	33	8092	6503	7924
80	38	9248	7432	9056
90	43	10,404	8361	10,188
100	47	11,560	9290	11,320
150	71	17,340	13,935	16,980
200	95	23,120	18,580	22,640
300	142	34,680	27,870	33,960
400	190	46,240	37,160	45,280
500	237	57,800	46,450	56,600
600	284	69,360	55,740	67,920
700	332	80,920	65,030	79,240
800	379	92,480	74,320	90,560
900	426	104,040	83,610	101,880
1000	474	115,600	92,900	113,200
1200	569	138,720	111,400	135,840
1500	711	173,400	139,350	169,800
2000	948	231,200	185,800	226,400
2500	1184	289,000	232,250	283,000
3000	1421	346,800	278,700	339,600
5000	2369	578,000	464,500	566,000

Boiling Points				
	°C	°F	°K	°R
Oxygen	−183.0	−297.4	90.1	162.2
Nitrogen	−195.8	−320.4	77.3	139.1
Argon	−185.7	−302.3	87.4	157.3

$$°C = 5/9 \ (°F - 32) \qquad °F = 9/5 \ (°C) + 32$$
$$°K = °C + 273.19 \qquad °R = 9/5 \ (°K)$$

Chart R-19.   Liquid to gas equivalents.

Chemical Treatments for Removal of Oxide Films from Aluminum Surfaces			
Solution	**Concentration**	**Procedure**	**Purpose**
Nitric Acid	50% water, 50% nitric acid, technical grade.	Immersion 15 min. Rinse in cold water, then in hot water. Dry.	Removal of thin oxide film for fusion welding.
Sodium hydroxide (caustic soda) followed by	5% sodium hydroxide in water.	Immersion 10-60 seconds. Rinse in cold water.	Removal of thick oxide film for all welding processes.
Nitric acid	Concentrated	Immerse for 30 seconds. Rinse in cold water, then hot water. Dry.	Removal of thick oxide film for all welding processes.
Sulfuric-chromic	H_2SO_4 1 gal. CrO_3 45 oz. Water 9 gal.	Dip for 2-3 min. Rinse in cold water, then hot water. Dry.	Removal of films and stains from heat treating, and oxide coatings.
Phosphoric-chromic	H_3PO_3 (75%) 3.5 gal. CrO_3 1.75 lbs. Water 10 gal.	Dip for 5-10 min. Rinse in cold water. Rinse in hot water. Dry.	Removal of anodic coatings.

Chart R-20.   Chemical treatments for removal of oxide films from aluminum surfaces.

Weld Metal Requirements for Fillet Welds				
Size of Fillet	**45° Fillets**		**30–60° Fillets**	
Inches	**Pounds of Metal per Foot**	**Pounds of Rod per Foot**	**Pounds of Metal per Foot**	**Pounds of Rod per Foot**
1/8	.027	.039	.054	.078
3/16	.063	.090	.126	.180
1/4	.106	.151	.212	.302
5/16	.166	.237	.332	.474
3/8	.239	.342	.478	.684
7/16	.325	.465	.650	.930
1/2	.425	.607	.850	1.214
5/8	.663	.948	1.226	1.896
3/4	.955	1.364	1.910	2.728
7/8	1.300	1.857	2.600	3.714
1	1.698	2.425	3.396	4.850

Chart R-21.   Weld metal requirements for fillet welds.

Welding Codes and Specifications	
AIA Aerospace Industries Association www.aia-aerospace.org	ASME American Society of Mechanical Engineers www.asme.org
AISI American Iron and Steel Institute www.steel.org	AWS American Welding Society www.aws.org
ANSI American National Standards Institute www.ansi.org	MIL Department of Defense www.defense.gov
ASTM International www.astm.org	

Chart R-22.   Organizations that publish welding codes and specifications.

Glossary of Welding Terms

A

Abrasive pads. Mineral-impregnated fiber pads used to remove oxide film.

Air-cooled. Torches that are cooled by the flow of the inert shielding gas through a passage in the combination cable. Also called *gas-cooled torches*.

Annealing. Heating a material to a temperature that redefines the crystalline structure and reduces internal stress of the material, making the material softer and more ductile.

Anode. In a DCEN setup, the anode is the positively charged workpiece. In a DCEP setup, the anode is the positively charged electrode.

Arc hours. The number of hours a welder is actually welding.

Arc plasma. Gaseous particle ionization.

Arc wandering. The arc moves erratically around the end of the tungsten and will not focus in one concentrated area.

Arrow. The part of the welding symbol that indicates and marks the point at which the weld is to be made.

Artificially aged. Refers to a material that has been held at an elevated temperature for a period of time for the purpose of increasing its hardness and strength.

Atmosphere chambers. Enclosures that provide shielding for the weld by surrounding the weld with a protective atmosphere.

Austenitic stainless. Stainless steels that consist of 18% chromium steel to which at least 8% nickel is added.

Autogenous weld. Weld which uses no additional filler metal.

Automatic operator certification. The verification that an operator is able to use the weld schedule to make an acceptable weld.

Automatic voltage control systems. Systems used in automatic and semiautomatic welding operations to automatically and continuously regulate the distance between the workpiece and the electrode to a preset gap. Commonly called *AVC systems*.

Automatic welding. Welding performed entirely by a machine of some type, usually a robot.

Auxiliary controls. Pieces of equipment that can be built into or added onto a power source to give the welder greater control of the welding process than is possible with the amperage control alone.

B

Backing rings. Inserts that provide quick and easy alignment for pipe and tubing by providing small tabs that establish the root opening.

Backing weld. Weld made from the back side of the weldment to fill in the root of the weld before the front side of the joint is welded.

Basic weld symbols. Symbols that direct the welder to select the proper type of weld joint.

Bend tests. Tests in which a weldment is sliced into coupons. The weld is then bent around a die of a specific size, stretching the weld to test its integrity.

Beta alloys. Nonweldable titanium alloys.

Bladder dams. Small bladders or balloons that are placed into the pipes and expanded to seal the pipe before purging begins. After welding is done, they are deflated and removed from the area.

Brass. An alloy of copper and zinc.

Bright metal. Bare metal without oxides and scale.

Bronze. An alloy of copper and tin.

Buildup. A process in which a welder applies layers of weld beads to a base metal in order to give the surfaces the physical properties of the filler metal.

Butt joint. Joint created when two pieces of metal are aligned edge-to-edge in the same plane.

Buttering. The process of laying stringer beads to the faces of a weld joint before actually welding the joint.

Buttering materials. Materials that are used to join metals that are very different.

C

Carbide precipitation. Chromium moves out of the grains and into the grain boundary to combine with carbon that has precipitated from the solid solution, forming chromium-rich carbides.

Cast. The diameter of one complete circle of spool wire as it lies on a flat surface.

Casting. Pouring molten metal into mold that has the desired shape, keeping the metal in the mold until it solidifies, then removing the mold.

Cathode. In a DCEN setup, the cathode is the negatively charged electrode. In a DCEP setup, the cathode is the negatively charged workpiece.

Ceriated tungsten. Type of electrode that has very good starting characteristics, excellent performance in low amperage ranges, and can be used continuously for extended periods with either ac or dc current.

Certificate of conformance. A statement that the filler metal meets all the requirements of the material specifications.

Certified chemical analysis report. A report that lists the result of a chemical analysis of an individual heat or lot of filler metal.

Chill bars. Bars that draw excess heat out of the weld zone.

Chromium-molybdenum steels. A class of extremely strong and hard steels.

Cladding. A series of overlaid welds on the surface of a part to form a weld joint or to protect the base material. Also called *surfacing*.

Clock hours. The number of hours a welder is at work.

Closed joint. A term referring to a joint that is not open at the root, such as a fillet weld made to weld the pipe into a socket.

Closed-circuit voltage. The voltage that is present while the weld is being made.

Cobalt. A metallic element acquired as a by-product of mining copper and nickel.

Codes. Laws that require that a weld or a component is made to a specific standard that may be enforceable by state or federal law.

Cold wire feeder. Device that supplies welding wire into a weld joint to maintain the weld profile.

Cold work. An operation of mechanically working material without heat.

Collet (collet body). Sets that are designed to grip one diameter size of tungsten and cannot be used with any other size.

Constant-current power source. Type of power source in which the welding current will remain the same if the welder maintains a constant arc length. The movement of the torch raises or lowers the welding power supply output current. Also called *drooper* or *variable-voltage power source*.

Consumables. The materials or parts expended during the welding operation.

Contactor. Switch that turns the flow of welding current from the power supply to the torch on and off.

Contaminate. To make unfit for use because undesirable material has entered.

Corner joint. Joint created when two pieces of material are matched up, edge-to-edge, at a 90° angle or nearly 90° angle, creating a corner.

Corrosion tests. Tests that measure the ability of a weld to restrict corrosion by a specific material.

Cracking. Partially opening and then immediately closing the cylinder valve before attaching regulator to blow out any particles or dirt in the valve opening.

Crater. A depression in a weld.

Cross section tests. Tests in which the weldment is cut into cross sections, which are then polished, etched, and examined visually or with specialized testing equipment. Used to define the internal quality and structure of the weld.

D

Defect. A discontinuity that is serious enough to make the weld unacceptable.

Destructive tests. Inspections and tests made on the completed weld or samples of the completed weld that destroy the weld.

Dewar flasks. Liquefied gas cylinders, which are basically vacuum bottles. The shielding gas in a Dewar flask is changed from liquid form to gas form as it passes through heat exchangers within or on the system. Also called *micro bulk systems*.

Dilution. The mixing of filler alloy with base alloy, which changes the chemical composition and physical or mechanical properties of the metal.

Direct current electrode negative (DCEN). Current that flows from the welding machine negative terminal to the electrode holder, across the arc gap to the workpiece, and back to the welding machine.

Direct current electrode positive (DCEP). Current that flows from the welding machine to the workpiece, across the arc gap to the electrode, then back through the lead to the welding machine.

Direct current reverse polarity (DCRP). Another term for direct current electrode positive (DCEP).

Direct current straight polarity (DCSP). Another term for direct current electrode negative (DCEN).

Discontinuity. Any disruption in the consistency of a weld.

Downslope timer. Device that determines the period of elapsed time from weld current to final current.

Droopers. Type of power supply in which the welding current will remain the same if the welder maintains a constant arc length. The movement of the torch raises or lowers the welding power supply output current. Also called *constant-current power sources* or *variable-voltage power sources*.

Dross. Oxidized metal or impurities produced during the cutting process.

Ductility. A property of a material to deform permanently, or to exhibit plasticity without breaking while under tension (strain).

Duty cycle. The length of time the machine can be continuously operated at a given amperage.

Dye penetrant test. Test that requires the weld surface to be sprayed with penetrant and allowed to soak. The excessive penetrant is then removed and developer is applied. Weld imperfections will hold the dye and bleed through the white developer, identifying the problem.

E

Edge joint. Joint created when two pieces of material are fit face-to-face. One piece of material is stacked perfectly on top of another piece of material so that the edges align.

Emissivity. The relative power of a surface to emit heat by radiating electrons.

End caps. Turning the end cap into the torch head forces the collet into the collet body. This grips and locates the tungsten in the position desired.

Engine-driven welding generators. Portable units used for welding in the field when utility power is not available.

F

Ferrite test. Test on completed stainless steel welds that determines the amount of magnetic ferrite in an austenitic (nonmagnetic) weld.

Ferritic stainless. Group of iron-chromium and carbon alloys that are nonhardenable by heat treatment.

Filler metal. Additional metal added to build up the weld.

Filter lens. Optical material that protects the welder's eyes against excessive ultraviolet, infrared, and visible radiation. Also called *filter plate*.

Final current. Current that is set at a low amperage to prevent craters at the end of the weld.

First degree burns. Burns that are identified by red, painful, and tender skin. There is no sign of any broken skin.

Flash burn. A burn that results from exposure to the UV light generated by a welding arc.

Flaw. A discontinuity that is small enough that it does not render the weld unacceptable.

Flowmeters. Devices that control the amount of shielding gas that flows to the weld area.

Fluorescent penetrant test. Penetrant test that requires an ultraviolet light (black light) to observe the test results.

Forced ventilation. Ventilation that uses electric fans and ducts to remove fumes from the work area.

Forging. Pressing a shaped die into hot metal under high pressure. The metal, which is heated to a plastic state before forging, takes on the shape of the die that is pressed into it.

Full tolerance. The far extreme of the tolerance range.

Fusion. Melting together of filler metal and base metal or of base metal only.

G

Gamma rays. Electromagnetic waves similar to X-rays, but with a shorter wavelength.

Gas lens. A device that uses a series of stainless steel wire mesh screens to make the column of shielding gas flowing out of the nozzle less turbulent and more streamlined.

Gas nozzle. Device at the end of the torch that directs the shielding gas. Also called *cup*.

Gas purge tester. A simple instrument that uses a special lightbulb with a tungsten filament and extensions for attaching the purge gas hose from the purged area. The gas is introduced into the bulb, and an electrical current is used to heat the tungsten. The color of the tungsten indicates the amount of contamination.

Gas tungsten arc welding (GTAW). Welding process that fuses metals by heating them between a nonconsumable tungsten electrode and a workpiece. The heat necessary for fusion is provided by an arcing electric current between the tungsten electrode and the base metal.

Gas-cooled torches. Torches that are cooled by the flow of the inert shielding gas through a passage in the combination cable. Commonly called *air-cooled torches*.

General use wire. Filler wire that meets a specification requirement, but no record of chemical composition or strength level is submitted to the user.

Grain structures. The arrangement of individual crystals in a metal or alloy.

H

Hard tooling. Major tools designed for assembly of component parts and welding.

Hardfacing. Base metal is overlaid with weld beads intended to protect the base metal from wear or corrosion.

Hardness tests. Tests in which a test probe is pressed into the surface of the weld. The amount of pressure required to deform the weld metal surface indicates the hardness of the weld or weldment.

Heat treating. A process in which the material is heated and cooled to specific temperatures for specific amounts of time in order to obtain desirable qualities or to reduce undesirable qualities.

Helix. The maximum height of any point of the circle of spool wire above the flat surface.

High-frequency arc stabilizer box. Add-on box that superimposes a high frequency above the electrical current that establishes the arc without touching the electrode to the base metal. The high-frequency arc stabilizer box also maintains arc stability when welding with ac. Also called *high-frequency arc starter*.

High-frequency arc starter. Add-on box that superimposes a high frequency above the electrical current that establishes the arc without touching the electrode to the base metal. The high-frequency arc stabilizer box also maintains arc stability when welding with ac. Also called *high-frequency arc stabilizer box*.

High-frequency generator. Provides the spark necessary to ignite the arc during reverse polarity part of the cycle and to maintain and stabilize the arc. The high voltage of the current generated by the high-frequency generator causes the arc gap to become ionized, thus becoming a path for the electrons to flow.

Hot short. Has low strength at high temperatures.

Hot wire feeder. Device that supplies preheated welding wire into a joint to maintain the weld profile.

Hourly flat rate cost. All of the cost is figured into one rate, and then the total hours to be expended to complete the job is considered.

Hydrostatic test. A pressure test that uses a fluid.

I

Image quality indicator (IQI). An indicator of the quality of the radiographic image. Either a wire-type or hole type IQI is placed on the test and its image is recorded on the radiograph. Also called *penetrameter*.

Inert gas. Gas that does not normally combine chemically with base metal or filler metal.

Infrared (IR) light. Light that has a wavelength which is longer than the wavelengths of visible light. Infrared light is emitted from any hot object.

Initial current. The current after starting the welding current, but before it is established.

Initial time period. The period of time the initial current is maintained.

In-process inspections. Inspections made during the fabrication cycle.

Interference fit. The outer part inside diameter (ID) dimension is made smaller than the inner part outside diameter (OD). When ready for assembly, the outer part is expanded using heat, assembled, and allowed to cool in place. When cool, the outer part is locked tightly into place.

Intermetallic compounds. Combinations of two or more metals that differ from the base metal in composition and structure.

Interpass. The period of time on a multipass weld after a weld pass is completed and a new weld pass is started.

Interstitials. Impurities in filler metals.

J

Joggle-type joints. Joints used in cylinder and head assemblies where backing bars or tooling cannot be used.

Joint. The manner in which materials fit together.

K

Keyhole method. Technique in which a keyhole is produced in the root of the weld as the base metal fuses with the filler metal. The keyhole is produced with the right combination of travel speed and heat, and indicates full penetration in the joint.

L

Labor. All of the work required to complete the job.

Laminar flow. A streamlined gas flow that is more concentrated than a turbulent flow.

Lanthanated tungsten. Type of electrode (designated EWLa-1, EWLa-1.5, or EWLa-2) that contains 1%, 1.5%, or 2% lanthanum.

Lap joint. Joint made by laying one piece of material on top or underneath another piece of the same type material, creating a small 90-degree angle weld joint that can be welded in all four positions.

Lathes. Machines that rotate the weldment so welds can be made along circular seams.

M

Macrotest. Type of cross section test in which polished sections of the weld are prepared for viewing with the naked eye or a magnifying glass.

Magnetic particle test (MT). A nondestructive inspection test in which a ferromagnetic material is magnetized and iron particles or powder is applied to the magnetized area. If the magnetic field is interrupted by a defect, the iron particles form a pattern on the surface.

Main welding current timer. Timer that is set for the length of time the weld current is desired.

Manipulators. Auxiliary tooling used to position the welding head at various locations and heights.

Martensite. A metallurgical term that defines a type of grain structure obtained by heating and quenching. The martensitic grain structure makes the metal hard.

Martensitic stainless. Low-carbon steel to which 11.5–18% chromium is added.

Material safety data sheet (MSDS). Document that contains manufacturer's specifications and detailed information about a hazardous product and the dangers associated with it.

Mechanical voltage controls. Controls used where the arc length is fixed and only requires occasional adjustment. The torch is moved up or down by adjusting screws or levers.

Micro bulk systems. Liquefied gas cylinders, which are basically vacuum bottles. The shielding gas in a Dewar flask is changed from liquid form to gas form as it passes through heat exchangers within or on the system. Also called *Dewar flasks*.

Microhardness test. Type of cross section test in which very highly polished sections of the weld are tested on special machines to determine the hardness of a very small area.

Microtest. Type of cross section test in which very highly polished sections of the weld are prepared for viewing with high-power microscopes.

Mismatch. Expansion of the base metal during welding, generally resulting in the weld not penetrating completely through the joint.

N

Natural ventilation. Ventilation that relies on the natural movement of air through the workspace to remove any fumes or smoke developed during welding.

Naturally aged. Refers to a material that has been held at room temperature for a period of time for the purpose of increasing its hardness and strength.

Nick-break tests. Destructive tests used to determine the internal quality of a weld with regard to porosity, lack of fusion, and slag. Notches are cut in the sides of a weld coupon in the weld area, coupons are laid across a support on each end, and force is applied with a hammer to try to break the weld sideways for a simple internal inspection.

Nickel. A metallic element with a silvery-white appearance with a hint of gold coloring. Nickel can be highly polished.

Nonconsumable. Does not melt.

Nondestructive examination (NDE). Inspections and tests made on a weld that do not destroy any portion of the completed weld.

Notch toughness. The ability of a metal to resist cracking failure at a notch during stress loading.

Notch-toughness tests. Tests in which the weldment is cut into test coupons, which are notched, cooled to a low temperature, and put under pressure until they fail. Used to define the ability of welds to resist cracking or crack propagation at low temperatures under loads.

O

Occupational Safety and Health Administration (OSHA). An organization that monitors the safe working practices and conditions of companies to ensure worker safety.

Open joint. A term referring to the type of open root joint found in a butt joint (groove weld).

Open-circuit voltage. The voltage produced when the power switch is turned on, but before the arc is struck.

Oscillators. Mechanical or magnetic devices on semiautomatic and automatic torches that direct the arc back and forth laterally across the weld seam as the weld progresses.

Ovality. Amount of out-of-round condition when referring to pipe, tubing, or round objects.

P

Penetrameter. An indicator of the quality of the radiographic image. Either a wire-type or hole type penetrameter is placed on the test and its image is recorded on the radiograph. Also called *image quality indicator (IQI)*.

Penetrant test (PT). Visual surface inspection completed with dye and developer. May also use fluorescent dye and black light for observing the results.

Phosphor bronze. An alloy of copper, tin, and phosphorus.

Pi tapes. Flexible measuring tapes that provide quick and accurate measurements on round and out-of-round forms.

Planishers. A tool used to flatten and smooth a welded seam and prepare the part for final finishing.

Positioners. Machines that rotate the weldment so welds can be made along circular seams.

Postflow timers. Timers that allow the shielding gas to flow for an adjustable period of time after the arc is broken.

Postheating. A controlled cooling rate of the weld to prevent cracking.

Power source. A machine for producing welding current.

Precipitation hardening. A group of steels that are alloyed with chromium and nickel and smaller quantities of other elements such as copper, titanium, columbium, and aluminum.

Preheating. Bringing the metal up to temperature before the welding process begins.

Pressure tests. Air or fluid tests that subject a vessel, tank, piping, or tubing to internal pressure.

Procedure qualification. The process of performing nondestructive testing (NDT) and destructive testing (DT) on completed test welds.

Pulser. Device that switches between a low and a high current level to reduce the overall heat input used to make a weld.

Pure tungsten. Type of electrode that has the lowest resistance to heat of all the tungsten electrodes.

Purging blocks. Small fabricated boxes equipped with a fitting for connecting shielding gas and many small holes in the face of the box through which the shielding gas can flow. They are used primarily for back purging butt welds on stainless steel to avoid "sugaring."

Q

Quality control. The monitoring of the quality of items produced.

Quenching. The process of rapidly cooling a material to obtain certain properties that increase the material's toughness by changing its crystalline structure.

R

Radiographic test (RT). A nondestructive method that shows the presence and nature of discontinuities in the interior of welds by using short wavelength radiations, such as X-rays or gamma rays, to penetrate the material.

Reference line. The long horizontal line in a welding symbol.

Relief step design. Hold-down bar design that can be used on slightly warped parts.

Remote arc start switch. A switch that is attached to the torch or foot control and is used to start the control sequence. Once the arc is established, the switch can be released.

S

Seam trackers. Machines that use sensors to locate the seam on a weldment and a motorized carriage that moves on two axes to keep the torch positioned directly over the seam.

Seamers. Auxiliary tooling used for making long seam internal or external welds in flat or cylindrical stock.

Second degree burns. Burns that are identified by redness, blisters, and breaks in the skin.

Segmented finger bar. Device that holds the components in place and to control the heat flow from the weld joint. Generally used for small or short welds.

Semiautomatic welding. Welding in which a welding operator monitors and adjusts the equipment to create a satisfactory weld. The equipment may automatically adjust itself, or it may be manually adjusted by the operator.

Shop aids. Tools used where a small number of parts are to be made. These tools may use the weldment as part of the tooling, or they may be a complete tool.

Slope controllers. Devices that control the rise and fall of the welding current at the start and end of the weld operation. Also called *slopers, programmers, electroslopes,* or *sequencers.*

Soft skills. The ability of an employee to be to work on time, work with other people, follow directions, perform shop housekeeping duties, and correctly fill out paperwork such as time cards or reports.

Solid hold-down bar. Device that holds the components in place and to control the heat flow from the weld joint. Used for long welds.

Soluble dams. Paper dams that dissolve in liquid. The dams are inserted into the pipe and held in place with a dissolvable adhesive. When welding is done, the dams are flushed away.

Solution heat-treated. Term that describes material that has been heated to a predetermined temperature for a suitable length of time to allow a certain element in the material to enter into a solid solution with the other elements.

Specifications. Documents that detail types of materials, welding processes, preparation of joints, qualification and welding requirements, and testing that will be used to produce the weldment.

Spitting. The electrode slowly dissolves and begins to release droplets of tungsten into the weld. This can result in significant weld defects.

Stabilized. Term that describes a material that has been strain-hardened and then heated to a predetermined low temperature to slightly lower the strength and to increase the ductility.

Strain-hardened. Term that identifies metal that has been strained by stretching, pulling, or forming to produce a grain structure with more desirable mechanical properties.

Stringer beads. Welds made without any side-to-side movement (oscillation) of the torch.

Surfacing. A series of overlaid welds on the surface of a part to form a weld joint or to protect the base material. Also called *cladding.*

Swaging. Changing the shape of a material with mechanical tools, such as hammers and dies.

T

Tapering down time. The period of elapsed time from weld current to final current.

Temper beads. Where the grain size must be controlled throughout the repair, temper beads are welded on top of the weld. These beads reduce the surface grain size and are removed after welding.

Tempering. A heat treatment in which metal is heated to a certain temperature below its melting temperature for a certain length of time to allow trapped carbon to produce a different crystalline structure of the material.

Tempers. Various mechanical properties that have been imparted in a metal by basic treatments.

Tensile tests. The weldment is sliced into coupons, and then each end of the coupon is pulled in opposite directions until the coupon fails (breaks). Used to compare the weldment to the base metal mechanical values and specification requirements.

Test tooling. Tooling that is made to prove tooling design concepts, make test welds, and establish weld joint shrinkage dimensions.

Third degree burns. Burns in which the surface and the tissue below the skin are white in color or charred and black in color. These are the worst burns and always require professional medical attention.

Thoriated tungsten. Type of electrode in which thorium is added to tungsten to help the arc starting process. EWth-1 electrodes have 1% thorium added and Ewth-2 electrodes have 2% thorium added.

Titanium. An expensive metallic element with good corrosion resistance, light weight, and high strength.

T-joint. Joint created when one piece of material is set on its edge on top of another piece of material that is lying flat, creating a 90° angle.

Torches. Common name for electrode holders.

Trailing shield. A device that attaches to the torch or torch nozzle to deliver a flow of shielding gas over the already-welded area.

Tungsten. Rare metallic element with a very high melting point (approximately 6170°F (3410°C).

U

Ultra-high-purity ferritic steels. Alloys that have a very low carbon and nitrogen content. These alloys have good mechanical properties and corrosion resistance.

Ultrasonic testing (UT). A nondestructive method of detecting discontinuities. Very-high-frequency sound waves are transmitted through the part to be tested. The sound waves then return to the sender and are displayed as a graph on a monitoring screen for interpretation.

Ultraviolet (UV) light. Light that has a wavelength shorter than visible light.

Upslope timer. Timer that determines the period of elapsed time from the start (initial) current to the weld current.

V

Variable-voltage power source. Type of power supply in which the welding current will remain the same if the welder maintains a constant arc length. The movement of the torch raises or lowers the welding power supply output current. Also called *constant-current power scources* or *droopers*.

Visible light. The light that we see, which is the least dangerous type of light.

Visual test (VT). Inspection conducted using equipment including rulers, fillet weld gauges, squares, magnifying glasses, and reference weld samples.

W

Water-cooled torches. Torches that are cooled by the passage of water through the torch head. The water then exits the system through the power cable.

Water-soluble films. Special films that are easily applied and form barriers for purging. They are removed by flushing the pipe with water.

Weave beads. Welds made with side-to-side movement of the torch.

Weld backing. A device or material placed at the back side of a joint to support and shield the molten weld metal.

Weld zone. The area being welded.

Welder certification. The verification that a welder is able to use the weld schedule to make an acceptable weld.

Welding operator. The person who controls a semiautomatic welding operation.

Welding procedure qualification record (WPQR). A document containing the settings used in the WPS and the results of the destructive and nondestructive testing.

Welding procedure specification (WPS). A document that details all of the requirements and instructions for welding a specific joint.

Welding procedures. List of all specific variables and parameters that produce acceptable welds for a production job.

Welding schedule. A record made during a weld test of all the actual welding parameters for use on the production weldment.

Wetted. A type of penetration that has a smooth junction and even flow of the penetration onto the base metal.

Whiskers. Microscopic crystalline filaments that develop on the surface of tungsten when it is exposed to contamination by moisture.

Wrought. Term that identifies material made by processes other than casting and is often used in specifications and codes.

X

X-rays. A form of electromagnetic radiation that penetrates most materials.

Z

Zirconiated tungsten. Type of electrode in which thorium is added to tungsten to help the arc starting process. EWth-1 electrodes have 1% thorium added and EWth-2 electrodes have 2% thorium added.

Index

C

D

E